AF344568

REPUBLIC OF SOUTH AFRICA
REPUBLIEK VAN SUID-AFRIKA

DEPARTMENT OF AGRICULTURE AND WATER SUPPLY
DEPARTEMENT VAN LANDBOU EN WATERVOORSIENING

FLORA OF SOUTHERN AFRICA

VOLUME 28

PART 4

ISBN 0 621 08268 6

FLORA OF SOUTHERN AFRICA

which deals with the territories of

SOUTH AFRICA, CISKEI, TRANSKEI, LESOTHO, SWAZILAND, BOPHUTHA-
TSWANA, SOUTH WEST AFRICA/NAMIBIA, BOTSWANA AND VENDA

VOLUME 28

PART 4   LAMIACEAE

by

L. E. Codd

Edited by

O. A. Leistner

Editorial Committee: B. de Winter, D. J. B. Killick and O. A. Leistner

Botanical Research Institute,
Department of Agriculture and Water Supply

1985

CONTENTS

---

* By L. E. Codd, unless otherwise stated.

A NEW TAXON AND NEW COMBINATIONS PUBLISHED IN THIS PART

Leonotis ocymifolia *(Burm. f.) Iwarsson,* comb. nov., p. 4: 32
L. ocymifolia var. **raineriana** *(Visiani) Iwarsson,* comb. nov., p. 4: 35
L. ocymifolia var. **schinzii** *(Gürke) Iwarsson,* comb. et stat. nov., p. 4: 35
Leucas glabrata *(Vahl) Sm.* var. **linearis** Codd, var. nov., p. 4: 44
Plectranthus hadiensis *(Forssk.) Schweinf. ex Sprenger* var. **tomentosus** *(Benth.) Codd,* comb. nov., p. 4: 153
P. hadiensis *(Forssk.) Schweinf. ex Sprenger* var. **woodii** *(Gürke) Codd,* comb. nov., p. 4: 154

Date of publication: ·May 1985

INTRODUCTION

The Flora of Southern Africa is arranged on the lines of the Engler system. Sequence and numbering of genera are as far as possible in agreement with De Dalla Torre & Harms (Genera Siphonogamarum, 1900—1907). Keys to families are provided in R. A. Dyer's Genera of Southern African Plants.

This part was compiled in accordance with a Guide to Contributors to the Flora of Southern Africa (Ross, Leistner & De Winter, 1977), which is available from the Librarian, Botanical Research Institute, Private Bag X101, Pretoria, 0001.

The following condensed abbreviations for literature references are used:

Burtt Davy, Fl. Transv.	Manual of the Flowering Plants and Ferns of the Transvaal and Swaziland, Vol. 1 (1926) and Vol. 2 (1932).
C.F.A.	Conspectus Florae Angolensis
F.C.	Flora Capensis
F.C.B.	Flore du Congo et du Rwanda-Burundi
F.M.	Flora de Moçambique
F.S.W.A.	Prodromus einer Flora von Südwestafrika
F.T.A.	Flora of Tropical Africa
F.T.E.A.	Flora of Tropical East Africa
F.W.T.A.	Flora of West Tropical Africa
F.Z.	Flora Zambeziaca
R. A. Dyer, Gen.	The Genera of Southern African Flowering Plants by R.A. Dyer, Vol. 1 (1975) and Vol. 2 (1976).

Cited voucher specimens are all housed in PRE (National Herbarium, Pretoria).

Vol. 28 of the Flora, of which the present publication is a component, will appear in parts (see p. ix). The number of the part, namely 4, precedes the page number on all pages marked with Arabic numerals. This was done with a view to binding the entire volume, once completed, and to compiling a combined index to all its component parts. When binding the entire volume the pages marked with Roman numerals may be omitted.

PLAN OF FLORA OF SOUTHERN AFRICA

Cryptogam volumes will in future not be numbered but will be known by the name of the group they cover. The number assigned to the volume on Charophyta therefore becomes redundant.

Alien families are marked with an asterisk.

Published volumes and parts are shown in italics.

Please note that local prices as given below do not include GST. Prices given for other countries include postage.

INTRODUCTORY VOLUMES

The genera of Southern African flowering plants
    Vol.   1: *Dicotyledons* (Published 1975). Price: R11,23. Other countries: R14,00
    Vol.   2: *Monocotyledons* (Published 1976). Price: R8,21. Other countries: R10,00
Botanical exploration of Southern Africa (Published 1981). Price R40,00 (Obtainable from booksellers)

CRYPTOGAM VOLUMES

Charophyta (Published as Vol. 9 in 1978). Price: R4,25. Other countries: R5,30

Bryophyta:

    Part 1: Mosses: Fascicle 1: *Sphagnaceae — Grimmiaceae* (Published 1981). Price: R24,34, Other countries R30,40
    Fascicle 2: Gigaspermaceae — Bartramiaceae
    Fascicle 3: Erpodiaceae — Hookeriaceae
    Fascicle 4: Fabroniaceae — Polytrichaceae

Pteridophyta

FLOWERING PLANTS VOLUMES

Vol.   1: *Stangeriaceae, Zamiaceae, Podocarpaceae, Pinaceae*, Cupressaceae, Welwitschiaceae, Typhaceae, Zosteraceae, Potamogetonaceae, Ruppiaceae, Zannichelliaceae, Najadaceae, Aponogetonaceae, Juncaginaceae, Alismataceae, Hydrocharitaceae* (Published 1966). Price: R1,98. Other countries R2,60

Vol.   2: Poaceae

Vol.   3: Cyperaceae, Arecaceae, Araceae, Lemnaceae, Flagellariaceae

Vol.   4: Part 1: Restionaceae
    Part 2: Xyridaceae, Eriocaulaceae, Commelinaceae, Pontederiaceae, Juncaceae (in press)

Vol.   5: Liliaceae, Agavaceae

Vol.   6: Haemodoraceae, Amaryllidaceae, Hypoxidaceae, Tecophilaeaceae, Velloziaceae, Dioscoreaceae

Vol.   7: Iridaceae: Part 1: Nivenioideae, Iridoideae
    Part 2: Ixioideae: Fascicle 1
    Fascicle 2: *Syringodea, Romulea* (Published 1983). Price: R3,90. Other countries: R5,00

Vol.   8: Musaceae, Strelitziaceae, Zingiberaceae, Cannaceae*, Burmanniaceae, Orchidaceae

Vol.   9: Casuarinaceae*, Piperaceae, Salicaceae, Myricaceae, Fagaceae*, Ulmaceae, Moraceae, Cannabaceae*, Urticaceae, Proteaceae

Vol. 10: Part 1: *Loranthaceae, Viscaceae* (Published 1979). Price: R4,34. Other countries: R5,40
Santalaceae, Grubbiaceae, Opiliaceae, Olacaceae, Balanophoraceae, Aristolochiaceae, Rafflesiaceae, Hydnoraceae, Polygonaceae, Chenopodiaceae, Amaranthaceae, Nyctaginaceae

Vol. 11: Phytolaccaceae, Aizoaceae, Mesembryanthemaceae

Vol. 12: Portulacaceae, Basellaceae, Caryophyllaceae, Illecebraceae, Cabombaceae, Nymphaeaceae, Ceratophyllaceae, Ranunculaceae, Menispermaceae, Annonaceae, Trimeniaceae, Lauraceae, Hernandiaceae, Papaveraceae, Fumariaceae

Vol. 13: *Brassicaceae, Capparaceae, Resedaceae, Moringaceae, Droseraceae, Roridulaceae, Podostemaceae, Hydrostachyaceae* (Published 1970). Price: R10,00. Other countries: R12,00

Vol. 14: Crassulaceae (in press)

Vol. 15: Vahliaceae, Montiniaceae, Escalloniaceae, Pittosporaceae, Cunoniaceae, Myrothamnaceae, Bruniaceae, Hamamelidaceae, Rosaceae, Connaraceae

Vol. 16: Fabaceae: Part 1: *Mimosoideae* (Published 1975). Price: R13,59. Other countries: R16,75
Part 2: *Caesalpinioideae* (Published 1977). Price: R16,04. Other countries: R20,00
    Papilionoideae

Vol. 17: Geraniaceae, Oxalidaceae

Vol. 18: Linaceae, Erythroxylaceae, Zygophyllaceae, Balanitaceae, Rutaceae, Simaroubaceae, Burseraceae, Ptaeroxylaceae, Meliaceae, Aitoniaceae, Malpighiaceae

Vol. 19: Polygalaceae, Dichapetalaceae, Euphorbiaceae, Callitrichaceae, Buxaceae, Anacardiaceae, Aquifoliaceae

Vol. 20: Celastraceae, Icacinaceae, Sapindaceae, Melianthaceae, Greyiaceae, Balsaminaceae, Rhamnaceae, Vitaceae

Vol. 21: Part 1:*Tiliaceae*, (Published 1984). Price R4,30. Other countries: R5,00
Malvaceae, Bombacaceae, Sterculiaceae

Vol. 22: *Ochnaceae, Clusiaceae, Elatinaceae, Frankeniaceae, Tamaricaceae, Canellaceae, Violaceae, Flacourtiaceae, Turneraceae, Passifloraceae, Achariaceae, Loasaceae, Begoniaceae, Cactaceae* (Published 1976). Price: R8,68. Other countries: R10,75

Vol. 23: Geissolomaceae, Penaeaceae, Oliniaceae, Thymelaeaceae, Lythraceae, Lecythidaceae

Vol. 24: Rhizophoraceae, Combretaceae, Myrtaceae, Melastomataceae, Onagraceae, Trapaceae, Haloragaceae, Gunneraceae, Araliaceae, Apiaceae, Cornaceae

Vol. 25: Ericaceae

Vol. 26: *Myrsinaceae, Primulaceae, Plumbaginaceae, Sapotaceae, Ebenaceae, Oleaceae, Salvadoraceae, Loganiaceae, Gentianaceae, Apocynaceae* (Published 1963). Price: R4,53. Other countries: R5,75

Vol. 27: Part 1: Periplocaceae, Asclepiadaceae (Microloma−Xysmalobium)
Part 2: Asclepiadaceae (Schizoglossum−Woodia)
Part 3: Asclepiadaceae (Asclepias−Anisotoma)
Part 4: *Asclepiadaceae (Brachystelma−Riocreuxia)* (Published 1980). Price: R4,43. Other countries: R6,00

Asclepiadaceae (remaining genera)

Vol. 28: Part 1: Cuscutaceae, Convolvulaceae
Part 2: Hydrophyllaceae, Boraginaceae
Part 3: Stilbaceae, Verbenaceae
Part 4: *Lamiaceae*, (Published 1985). Price: R22,00, plus G.S.T. Other countries: R28,00
Part 5: Solanaceae, Retziaceae

Vol. 29: Scrophulariaceae

Vol. 30: Bignoniaceae, Pedaliaceae, Martyniaceae, Orobanchaceae, Gesneriaceae, Lentibulariaceae, Acanthaceae, Myoporaceae

Vol. 31: Plantaginaceae, Rubiaceae, Valerianaceae, Dipsacaceae, Cucurbitaceae

Vol. 32: Campanulaceae, Sphenocleaceae, Lobeliaceae, Goodeniaceae

Vol. 33: Asteraceae: Part 1: Lactuceae, Mutisieae, 'Tarchonantheae'
Part 2: Vernonieae, Cardueae
Part 3: Arctotideae
Part 4: Anthemideae
Part 5: Astereae
Part 6: Calenduleae
Part 7: Inuleae: Fascicle 1: Inulinae
Fascicle 2: *Gnaphaliinae (First part)* (Published 1983). Price: 12,93. Other countries: R16,20
Part 8: Heliantheae. Eupatorieae
Part 9: Senecioneae

LAMIACEAE (Labiatae)

by L. E. CODD*

Herbs, mainly perennial, or shrubs; branches usually 4-angled. *Leaves* opposite or whorled, entire, toothed or sometimes lobed, rarely pinnatifid or digitately compound, usually gland-dotted and aromatic. *Flowers* usually irregular, often bilabiate, bisexual or rarely unisexual *(Tetradenia)*, solitary and opposite or aggregated into cymes or verticils arranged in terminal racemes or panicles, sometimes crowded into a spike or corymb, or more rarely in the axils of foliage leaves; bracts present, leaf-like or reduced, often caducous. *Calyx* tubular, campanulate or funnel-shaped, usually persistent and often enlarged in fruit, rarely becoming fleshy *(Hoslundia)*, regularly or irregularly 3−many-toothed, or with 2 entire or toothed lips, rarely truncate or 5-partite, sometimes with the posterior lobe broadly ovate and decurrent on the tube. *Corolla* gamopetalous, (1−)2-lipped, or oblique, or subregular and 4−5-lobed. *Stamens* rarely 2, usually 4, subequal or in pairs of unequal length (didynamous), all or only 1 pair fertile, inserted at the corolla mouth or in the tube; filaments sometimes connate, sometimes with a crest or projection near the base; anthers 1- or 2-thecous. *Ovary* superior, seated on an entire or lobed disc, deeply or shortly 4-lobed, 4-locular, with a single erect ovule in each locule; style often 2-lobed. *Fruit* composed of 4, or by abortion fewer, dry, 1-seeded nutlets; nutlets rugose or smooth, rarely winged *(Tinnea);* seeds erect.

Characters not applicable in our area: leaves occasionally alternate.

Genera about 170; species up to 5 000, cosmopolitan in warm and temperate areas; 37 genera and 232 species are indigenous or naturalized and are keyed out. In addition a number are cultivated, either as ornamental plants or for their aromatic leaves which are used medicinally, in confectionery or cosmetics, or as culinary herbs. Cultivated plants which have not become naturalized are not included in the keys. If the genus is keyed out, the cultivated members are mentioned after the generic description. This leaves a number of cultivated genera which are not keyed out and for which there are, therefore, no separate generic treatments. The more widely grown of these are listed below.

(a) Ornamental plants: *Molucella laevis* L. (Shell-flower, Bells of Ireland), *Monarda didyma* L. (Oswego Tea, Bee-balm), *M. fistulosa* L. var. *mollis* Benth. (Wild Bergamot), *Perovskia atriplicifolia* Benth., *Phlomis fruticosa* L. (Jerusalem Sage) and *Physostegia virginiana* Benth.; the following are grown for their ornamental effect as well as for their aromatic foliage: *Lavandula* spp. (Lavender), *Nepeta cataria* L. (Catnip or Catmint) and *Rosmarinus officinalis* L. (Rosemary).

(b) Culinary herbs: *Hyssopus officinalis* L. (Hyssop), *Melissa officinalis* L. (Lemon Balm), *Origanum majorana* L. (Marjoram) and *Thymus* spp. (Thyme).

1 Corolla 1-lipped or very unequally 2-lipped, with the upper lip small or absent; ovary shortly 4-lobed or lobed to the middle; nutlets reticulate with oblique or lateral aureole occupying ½ to ¾ the length:

  2 Calyx subequally 5-toothed; stamens exserted, ascending; nutlets not winged:

    3 Inflorescence a spike-like raceme arising from a basal rosette of leaves; flowers in 2−many-flowered verticils, bluish .... 1. **Ajuga**

    3 Inflorescence racemose or paniculate with no basal rosette of leaves; flowers small, white, in 1−several-flowered pedunculate cymes.................. 2. **Teucrium**

  2 Calyx 2-lipped, inflated in fruit, lips entire; stamens not or scarcely exserted; nutlets winged .... 3. **Tinnea**

1 Corolla 2-lipped or nearly regularly lobed; ovary deeply 4-lobed; nutlets smooth with small basal or slightly oblique aureole:

---

* Except *Leonotis* which is by M. Iwarsson of the Institute of Systematic Botany, University of Uppsala, Sweden.

4 Stamens ascending or spreading, never all directed downwards upon the lower
   side of the tube or lower lip of the corolla (absent in female flowers of
   Tetradenia) (second half of couplet on p. 4: 3)

  5 Fertile stamens 2, anther-thecae separated by a long connective; calyx
    2-lipped .... 14. **Salvia**

  5 Fertile stamens 4; calyx 2-lipped or 5−many-toothed:

    6 Stamens spreading, filaments straight, with 2 directed upwards and 2
      downwards (absent in female flowers of *Tetradenia*); corolla small, 2−5
      mm long, subequally 4−5-lobed:

      7 Perennial rhizomatous herbs, monoecious .... 16. **Mentha**

      7 Semisucculent or softly woody shrubs, usually flowering after the leaves
        are shed, dioecious .... 17. **Tetradenia**

    6 Stamens all directed to the upper side of the tube or upper lip of the corolla;
      corolla 5 mm long or longer (sometimes shorter in *Scutellaria*, but then
      calyx distinctly 2-lipped with the upper lip deciduous):

      8 Calyx 2-lipped; perennial decumbent herbs, introduced:

        9 Calyx lips rounded, entire, the upper lip deciduous; inflorescence of lax
          2-flowered verticils; corolla 4,5−6 mm long .... 4. **Scutellaria**

        9 Calyx lips toothed, upper broad with 3 short teeth, lower of 2 longer
          narrow teeth; inflorescence shortly spicate of densely placed
          4−6-flowered verticils; corolla 9−10 mm long .... 8. **Prunella**

      8 Calyx regularly or irregularly 5−many-toothed, the mouth sometimes
      oblique but not distinctly 2-lipped (obscurely 2-lipped in some *Leucas*
      spp. but then calyx 6−10-toothed):

      10 Leaves 3 (−5)-foliolate .... 7. **Cedronella**

      10 Leaves simple:

        11 Corolla usually orange or yellow, rarely cream; upper lobe 12−30
          mm long; calyx 8−10-toothed .... 9. **Leonotis**

        11 Corolla not orange or yellow; upper lobe less than 10 mm long:

          12 Stamens included in the corolla tube:

            13 Anthers and style held together by intermingling hairs; calyx
              glabrous within; teeth not hooked at the apex .... 5. **Acrotome**

           13 Anthers and style not held together by intermingling hairs;
              calyx hairy within; teeth hooked at the apex .... 6. **Marrubium**

          12 Stamens reaching the mouth of the corolla tube or exserted
            (stamens included in cleistogamous flowers of *Lamium
            amplexicaule* but then bracts amplexicaul):

           14 Calyx 6−many-toothed (teeth often very unequal in size):

             15 Calyx glabrous within, 6−10-toothed, often oblique at the
               mouth but limb not spreading .... 10. **Leucas**

             15 Calyx hairy within, 10- or more-toothed, limb eventually
               spreading .... 12. **Ballota**

           14 Calyx subequally 5-toothed:

             16 Calyx 5−10-ribbed; stamens attached within the corolla and
              usually well exserted:

17 Bracts large, amplexicaul (in the naturalized species); upper corolla lip longer than the lower; lower corolla lip with lateral lobes absent or reduced to small acute teeth .... 11. **Lamium**

17 Bracts leaf-like or reduced, not amplexicaul; corolla lips subequal or lower lip the longer with distinct ± obtuse lateral lobes:

  18 Bracts leaf-like; calyx often with additional smaller teeth (5−8-toothed); upper lip of corolla subequal to the lower and beset with stiff brush-like hairs .... 10. **Leucas**

  18 Bracts usually reduced, occasionally leaf-like; calyx always 5-toothed; upper lip of corolla glabrous or pubescent but not with stiff brush-like hairs, usually shorter than the horizontal lower lip .... 13. **Stachys**

16 Calyx 13 (−15)-ribbed; stamens attached near the corolla throat with very short upcurved filaments .... 15. **Satureja**

4 (from p. 4: 2) Stamens directed downwards upon the lower side of the corolla tube or lower lip of the corolla:

19 Calyx enlarged and fleshy in fruit; upper pair of stamens reduced to staminodes .... 28. **Hoslundia**

19 Calyx often somewhat enlarged but not fleshy in fruit; all 4 stamens fertile (upper pair reduced to staminodes in *Plectranthus zuluensis*):

20 Calyx falling away by a clean break above the base in fruit .... 19. **Aeollanthus**

20 Calyx persistent in fruit, 3−5-toothed or bilabiate:

  21 Corolla with 4 subequal or slightly unequal lobes, not distinctly 2-lipped; stamens included in corolla tube .... 20. **Endostemon**

  21 Corolla either distinctly 2-lipped or 5-lobed; stamens exserted:

    22 Calyx with 5 equal rigid spine-like teeth; flowers in a dense terminal spike-like inflorescence .... 21. **Pycnostachys**

    22 Calyx not rigidly spinescent:

      23 Calyx equally or subequally 5-toothed, the uppermost tooth sometimes slightly larger than the other 4:

        24 Bracteoles linear, setose; calyx teeth linear-subulate; corolla small, ± equally 5-lobed .... 18. **Hyptis**

        24 Bracteoles absent or not as above; calyx teeth not linear-subulate; corolla bilabiate:

          25 Lower pair of stamens united for most of their length, upper pair attached in the corolla tube, free.............. 29. **Syncolostemon**

          25 Lower and upper pairs of stamens attached at the corolla throat, free or all shortly united at the base:

            26 Bracts differentiated from and smaller than the leaves:

              27 Style bilobed; inflorescence paniculate or subspicate with flowers in verticils of 3−many-flowered cymes or dichasia .... 23. **Plectranthus**

27 Style entire; inflorescence paniculate with the slender branches somewhat zig-zag towards the ends bearing solitary flowers.. 24. **Holostylon**

26 Bracts leaf-like, becoming gradually smaller towards the apex of the inflorescence or towards the ends of the flower-bearing stems:

28 Stems semi-woody, erect; inflorescences dense, paniculate or spicate:

29 Flowers in 6—8-flowered verticils, densely arranged in short lateral and terminal spikes .... 22. **Neohyptis**

29 Flowers in dichasia arranged in a terminal panicle .... 25. **Rabdosiella**

28 Stems softly herbaceous, decumbent or erect; flowers in slender, lax, axillary racemes borne along almost the entire length of the stem .... 26. **Englerastrum**

23 Calyx bilabiate:

30 Upper lip of calyx 3-toothed (teeth sometimes minute), lower lip 2-toothed or entire:

31 Inflorescence corymbose, flower-heads capitate, 8—10 mm long and almost equally broad.............................. 31. **Acrocephalus**

31 Inflorescence spike-like, 50—100 mm long, flowers in dense many-flowered cymose clusters .... 32. **Geniosporum**

30 Upper lip of calyx consisting of a large, oblong to broadly ovate tooth, often decurrent on the tube, distinctly larger than the remaining teeth:

32 Upper pair of filaments glabrous or pubescent but without a crested or hairy knee-bend near the base:

33 Lower 2 calyx teeth fused for the greater part forming an oblong bifurcate lip; lateral teeth much shorter and rounded.. 27. **Solenostemon**

33 Lower 4 calyx teeth subequal, lanceolate-deltoid, the lower 2 often longer than the lateral 2 and shortly fused at the base:

34 Filaments of lower pair of stamens connate for part of their length; upper pair attached in the corolla tube and free .... 30. **Hemizygia**

34 Filaments all free or all connate at the base:

35 Corolla about 2 mm long, obscurely bilabiate; lower lip almost flat... 33. **Basilicum**

35 Corolla 4 mm long or longer, bilabiate; lower lip concave to boat-shaped:

36 Corolla with the upper lip having 2 ear-like lobes; flowers in 1—many-flowered verticils or cymes, never all solitary:

37 Upper calyx tooth not decurrent on the tube; filaments attached at mouth of corolla tube; mostly herbaceous plants .... 23. **Plectranthus**

37 Upper calyx tooth decurrent on the tube; upper pair of filaments attached within the corolla tube; woody plants .... 36. **Orthosiphon**

36 Corolla with 2 narrow pendulous lobes on each side of and free from the upper lip; flowers solitary in the axils of the somewhat leafy bracts .... 37. **Thorncroftia**

32 Upper pair of filaments with a crested or hairy knee-bend near the base:

38 Lower pair of calyx teeth fused or free but not ending in a small pair of upcurved spinescent teeth; lateral teeth small, deltoid-subulate .... 34. **Ocimum**

38 Lower pair of calyx teeth fused into a lip ending in a small pair of upcurved spinescent teeth; lateral teeth suppressed and replaced by a wide shoulder-like and occasionally fimbriate sinus .... 35. **Becium**

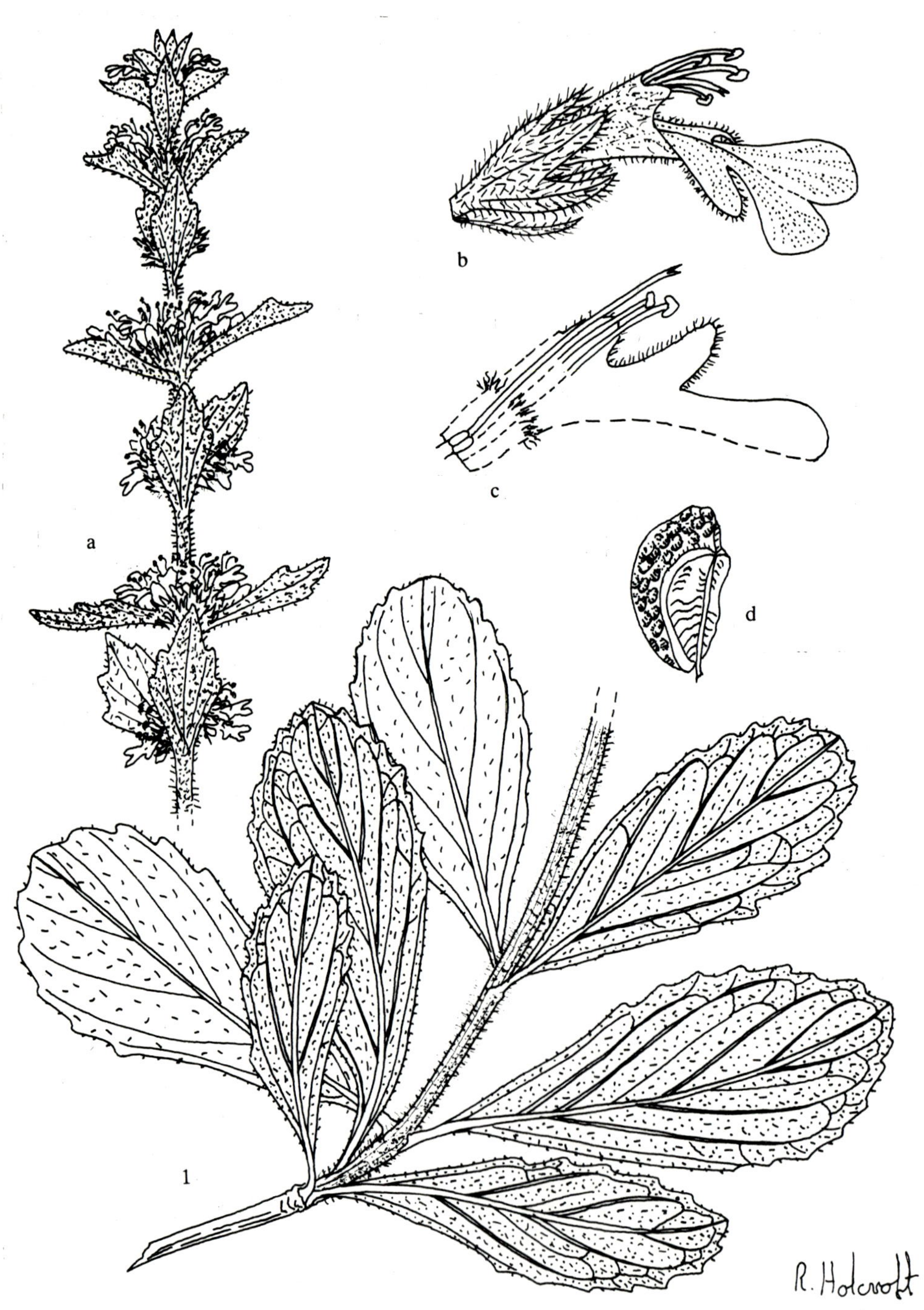

7211

1. AJUGA

Ajuga *L.*, Sp. Pl. 561 (1753); Gen. Pl. edn 5: 246 (1754); Benth. in DC., Prodr. 12: 597 (1848); Briq. in Natürl. PflFam. 4,3a: 209 (1895); Bak. in F.T.A. 5: 501 (1900); Skan in F.C. 5,1: 386 (1910); R. A. Dyer, Gen. 1: 525 (1975). Type species: *A pyramidalis* L.

Perennial herb (S. Africa), often decumbent or stoloniferous. *Leaves* usually coarsely toothed to incised. *Inflorescence* a terminal spike-like raceme; bracts leaf-like, becoming smaller towards the apex; verticils 2−many-flowered; flowers bisexual. *Calyx* short, subequally 5-fid or 5-toothed; tube campanulate. *Corolla* unequally 2-lipped; tube short, constricted near the base with a ring of hairs within the constriction; upper lip short, subentire, emarginate or 2-fid; lower lip large, spreading, 3-lobed, the median lobe the largest, emarginate or 2-fid. *Stamens* 4, curved within the upper lip, the lower pair slightly the longer; filaments linear; anthers 2-thecous, with the thecae divergent, finally confluent. *Ovary* shortly 4-lobed nearly to the middle; style 2-fid, lobes somewhat unequal. *Nutlets* obovoid, reticulate-rugose, attached by a broad oblique areole which extends beyond the middle.

Described species about 100, found chiefly in the extra-tropical regions of the Old World, particularly in the Orient, represented by a few endemic species south of the equator in Australia, Madagascar, the mountains of east tropical Africa and with 1 species occurring naturally in Southern Africa. In addition, *A. reptans* L. has been developed as a ground cover plant and is grown, in several different foliage colour forms, in South African gardens.

Ajuga ophrydis *Burch. ex Benth.*, Lab. 695 (1835); in E. Mey., Comm. 243 (1837); in DC., Prodr. 12: 597 (1848); Skan in F.C. 5,1: 386 (1910); Trauseld, Wild Flow. Natal Drakensberg 156 (1969); Lucas & Pike, Wild. Flow. Witwatersrand 72 (1971); Jacot Guill., Fl. Lesotho 236 (1971); Ross, Fl. Natal 302 (1972). Type: Cape, Bathurst Division, *Burchell* 3700 (K, lecto.).

Perennial low herb; stems several, decumbent-ascending, 60−250 mm tall, arising annually from a short rhizome. *Leaves* mainly forming a basal rosette, sessile or shortly petiolate; blade fairly thick-textured, obovate to obovate-lanceolate, 30−80 × 15−40 mm, subglabrous to sparingly pilose, apex obtuse to rounded, base cuneate, margin coarsely few-toothed to repand or almost entire. *Inflorescence* up to 200 mm long with verticils often starting shortly above the basal leaves, verticils spaced below, crowded towards the apex; pedicels up to 1 mm long. *Calyx* hispid, 6−7 mm long; tube 3−4 mm long; teeth deltoid, 3 mm long. *Corolla* blue to mauve, rarely white; tube 7−8 mm long; upper lip 2 mm long, emarginate; lower lip 7−10 mm long, median lobe broadly obovate, 5−7 mm long and broad, deeply emarginate. Fig. 1.

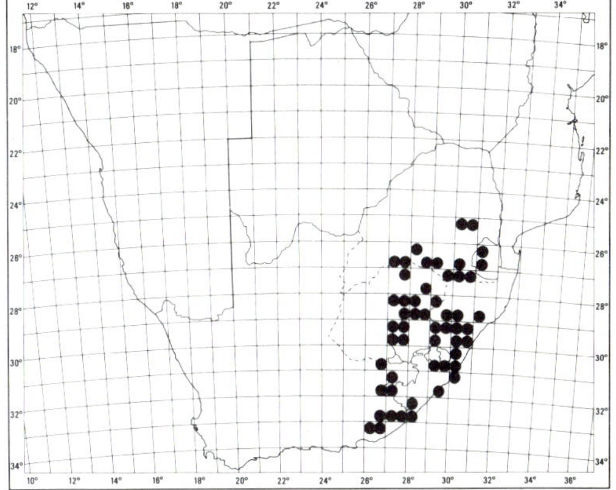

MAP 1. — **Ajuga ophrydis**

Distributed from Grahamstown in the eastern Cape to Transkei, Natal, Lesotho, Orange Free State and southern and eastern Transvaal, mainly in grassland, from near the coast in the Cape to over 2 600 m altitude in Lesotho. Map 1.

Vouchers: *Codd* 8226; *Galpin* 13966; *Killick* 1039; *Schlechter* 3263; *Tyson* 1102.

The specific epithet refers to a resemblance to *Ophrys*, a genus of ground orchids. *A. ophrydis* is related to *A. remota* Benth., a species found in east tropical Africa and India, but the latter has a denser tomentum on the leaves.

FIG. 1. — 1, **Ajuga ophrydis**, lower leaves, × 1; a, inflorescence, × 1; b, flower and bracteole, × 3; c, section through corolla, × 3; d, nutlet, × 7 (*Jenkins* sub TRV 9308).

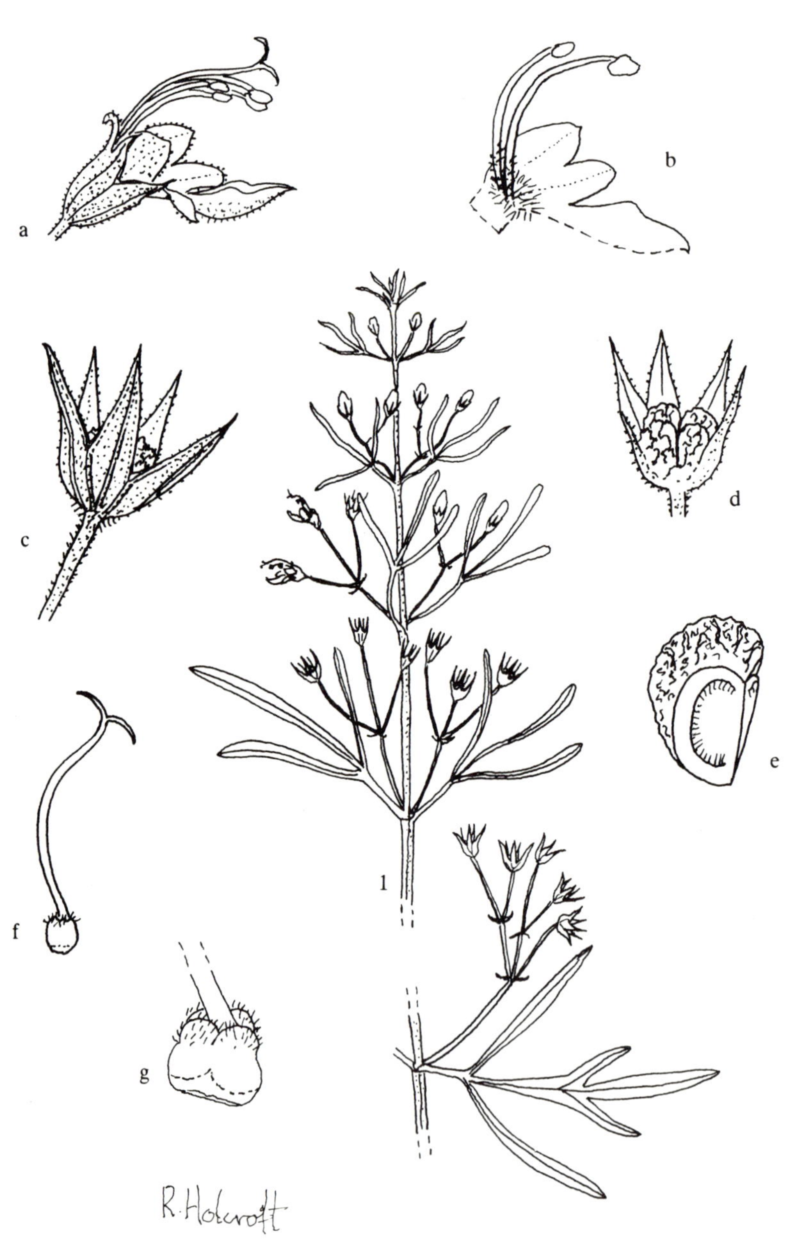

7212   **2. TEUCRIUM**

Teucrium *L.,* Sp. Pl. 562 (1753); Gen. Pl. edn 5: 247 (1754); Benth. in DC., Prodr. 12: 574 (1848); Briq. in Natürl. PflFam. 4,3a: 210 (1895); Bak. in F.T.A. 5: 500 (1900); Skan in F.C. 5,1: 384 (1910); R. A. Dyer, Gen. 1: 525 (1975); Codd in Bothalia 12: 177 (1977). Type species: *T. fruticans* L.

Herbs, undershrubs or shrubs. *Leaves* usually toothed or more or less deeply lobed. *Inflorescence* a terminal raceme or panicle; bracts leaf-like, usually becoming smaller towards the apex of the inflorescence; whorls 2−several-flowered; flowers often borne in pedunculate cymes. *Calyx* about as long as the corolla tube, subequally 5-toothed. *Corolla* small, white in South African spp., 5-lobed, the lowermost the largest, longer than the tube, giving a 1-lipped appearance; tube short, hairy in the throat. *Stamens* 4, ascending-arcuate between the two upper corolla lobes, well exserted; filaments thread-like, villous at the base; anthers 2-thecous. *Ovary* shortly 4-lobed; style slightly exceeding the stamens, terete, 2-fid, lobes subequal. *Nutlets* obovoid, reticulate-rugose, attached by an oblique or lateral areole which extends beyond the middle.

In non-South African species the leaves may be entire, the inflorescence may be a spike-like raceme or terminal head with flowers small or large and showy, white or in shades of yellow, blue or purple, and the upper lobe of the calyx may be larger than the rest. Some are grown as garden plants, but evidently not in Southern Africa.

Species about 200, widely distributed over the temperate and warmer regions of the world, chiefly in the northern hemisphere, represented by a few species in Australia, South America, the mountains of north-east tropical Africa and by 3 species in Southern Africa.

The Southern African species are used medicinally for stomach disorders and haemorrhoids as well as for treating snake-bite and meat suspected of being infected with anthrax. Common names such as Aambeibossie and Maagbossie refer to these properties while Paddaklou and Akkedispoot refer to the lobed leaves.

1 Peduncles usually 1-flowered and much shorter than the internode ....1. *T. africanum*

1 Peduncles 3−7-flowered, cymes often as long as or longer than the internode:

   2 Leaves more or less deeply 3-fid or 3-lobed, if subentire then drying greyish ....2. *T. trifidum*

   2 Leaves entire or few-toothed towards the apex, usually drying dark brown ....3. *T. kraussii*

1. **Teucrium africanum** *Thunb.,* Prodr. 2: 95 (1800); Fl. Cap. edn Schult. 445 (1823); Benth., Lab. 669 (1835); in E. Mey., Comm. 243 (1837); in DC., Prodr. 12: 577 (1848); Skan in F.C. 5,1: 384 (1910); Codd in Bothalia 12: 177 (1977). Type: Cape, without locality, *Thunberg* s.n. (UPS, holo., microfiche 556/13250!).

Ajuga africana (Thunb.) Pers., Syn. 2: 109 (1807).

Greyish, bushy shrublet 0,1−0,25 (−0,3) m tall, branching freely from the base; stems erect to decumbent, simple or sparingly branched, slender, greyish glandular-tomentulose. *Leaves* sessile, grey-green, thinly pubescent, 8−30 mm long, 3-lobed to 3-partite; lobes linear to linear-oblong, 5−25 × 1−3 mm, occasionally the median lobe again 3-fid; margin entire, revolute; basal portion of leaf narrow, up to 3 mm broad, consisting of a narrowly winged midrib. *Inflorescence* leafy, simple, occupying the upper third or half of the stem; flowers solitary or rarely 2 or 3 per peduncle; peduncle 3−8 mm long

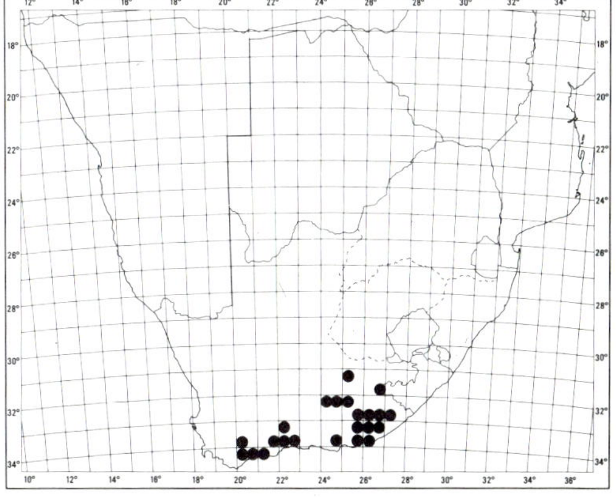

MAP 2. — **Teucrium africanum**

FIG. 2. — 1, **Teucrium trifidum,** flowering stem, × 1; a, flower, × 4; b, section through corolla, × 4; c, d, ripe calyx, × 4; e, nutlet, × 7; f, gynoecium, × 4; g, ovary, × 20 *(Scheepers* 1442).

bearing a pair of minute bracteoles below the middle. *Calyx* greyish, 3—4 mm long; teeth lanceolate-deltoid, about 2 mm long. *Corolla* 5—6 mm long; tube 2 mm long; lower lip obovate, 3—4 mm long, slightly concave, remaining 4 lobes oblong, rounded, 2 mm long. *Anthers* exserted by 6—7 mm.

Found under fairly arid conditions in fynbos, karoo, coastal or thorn scrub from Bredasdorp to near Grahamstown and, inland, to Graaff-Reinet, often among rocks or on overgrazed or disturbed places. Map 2.

Vouchers: *Acocks* 15671; *Galpin* 2012; *Schlechter* 1795.

Commonly known as Aambeibossie in reference to its medicinal use against haemorrhoids, and as Paddaklou or Katjiedriedoring because of the leaf shape.

See note after *T. trifidum* (below).

2. **Teucrium trifidum** *Retz.*, Obs. 1: 21 (1779); Codd in Bothalia 12: 177 (1977). Type: Cape, without locality, right-hand specimen on sheet so named in Hb. Retzius (LD, lecto.!; PRE, photo.!).

T. trifidum Wendl., Bot. Beobacht. 50 (1798), nom. illegit. Type: not indicated.

T. capense Thunb., Prodr. 2: 95 (1800); Fl. Cap. edn Schult. 445 (1823); Benth., Lab. 667 (1835); in E. Mey., Comm. 243 (1837); in DC., Prodr. 12: 577 (1848); Skan in F.C. 5,1: 385 (1910); Wilman, Check List Griq. West 231 (1946); Jacot Guill., Fl. Lesotho 236 (1971); Ross, Fl. Natal 302 (1972). *Ajuga capensis* (Thunb.) Pers., Syn. Pl. 2: 109 (1807). Type: Cape, near "Zeekoerivier" (Humansdorp district), *Thunberg* s.n. (UPS, holo., microfiche 556/13263!).

T. africanum sensu Wilman, l.c. (1946).

An erect soft undershrub 0,3—1,1 m tall, branching freely from the base; stems virgate, branching freely in the upper half or third, woody below, shortly greyish tomentose. *Leaves* drying greyish green to grey-brown, thinly tomentose above, denser to almost canescent and gland-dotted below, usually deeply 3-fid or 3—5-partite, rarely almost entire, 20—60 mm long; lobes linear to lanceolate, 10—35 × 3—8 mm, often again shortly lobed or toothed, margin entire, revolute; basal portion of leaf 3—8 mm broad, consisting of a winged midrib narrowing to a short petiole. *Inflorescence* a leafy panicle occupying the upper third of

the stem; flowers usually in 3—7-flowered pedunculate cymes, rarely solitary; peduncle 5—20 (—25) mm long, pedicels 3—12 mm long; bracteoles usually very small and linear. *Calyx* greyish, 2,5—4 mm long, teeth about 2 mm long. *Corolla* 5—6 mm long; tube 2 mm long; lower lip obovate, 3—4 mm long, slightly concave, remaining 4 lobes oblong, rounded, 2 mm long. *Anthers* exserted by 6—8 mm. Fig. 2.

Common in the central to south-western Transvaal, apparently not extending beyond the Soutpansberg, but extending westwards to the northern Cape Province and just entering Botswana, southwards to northern Natal, central Orange Free State, Transkei and eastern Cape Province, reaching its southernmost limit at about Humansdorp. Usually found in dry woodland where it is often gregarious under thorn trees or in bush groups, particularly on overgrazed or disturbed places. Map 3.

Vouchers: *Bolus* 10842; *Galpin* 1647; *Killick* 1790; *Mogg* 8548; *Scheepers* 1442.

Widely used medicinally for dysentery and haemorrhoids, whence the common names Koorsbossie and Aambeibossie, while the leaf shape has suggested the names Paddaklou and Akkedispootjie. It is also used by native tribes to treat snake-bite and as a measure against anthrax by boiling it with meat which is thought to be infected.

The three species *T. africanum*, *T. trifidum* and *T. kraussii* are almost identical florally but may be separated on vegetative characters. *T. africanum* may be recognized by its smaller stature, rarely exceeding 0,3 m tall and the usually solitary flowers on short peduncles. In the southern Cape Province, where it

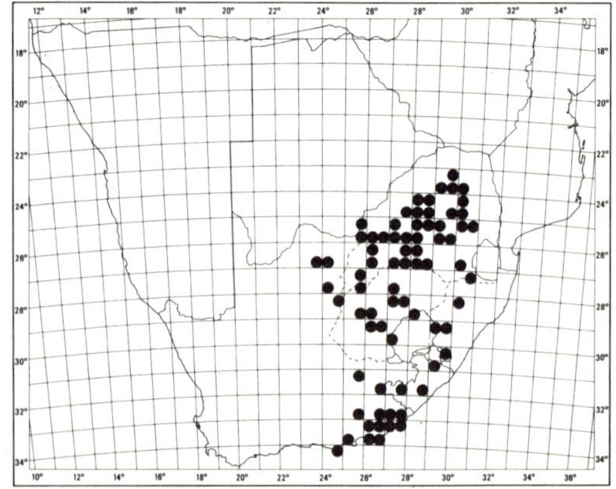

MAP 3. — **Teucrium trifidum**

overlaps with *T. trifidum*, occasional specimens may be intermediate, but these are relatively few and can usually be allocated to one or the other species. Both are characterized by the deeply lobed leaves though occasional specimens of *T. trifidum* may have almost entire leaves which begin to resemble those of *T. kraussii*. In such cases, *T. trifidum* can usually be recognized by the somewhat smaller, greyish green leaves, as against the longer and broader leaves of *T. kraussii*, which tend to dry dark brown.

3. **Teucrium kraussii** *Codd* in Bothalia 12: 179 (1977). Type: Natal, Umlaas River, *Krauss* 153 (K).

T. riparium Hochst. in Flora 28: 66 (1845); Benth. in DC., Prodr. 12: 576 (1848); Skan in F.C. 5,1: 385 (1910); Ross, Fl. Natal 302 (1972); Compton, Fl. Swaziland 491 (1976); nom. illegit., non *T. riparium* Rafin. (1838). Type: as above.

An erect soft undershrub 0,5−1,1 m tall, branching from the base; stems simple below, branched in the upper half or third, softly woody below, 4-angled, finely to fairly densely tomentose, usually with spreading hairs. *Leaves* subsessile, upper surface subglabrous or sparingly hispidulous, lower surface sparingly to fairly densely hispid and minutely gland-dotted, narrowly lanceolate to oblong-lanceolate or oblanceolate, 25−60 × 6−12 mm, remotely 1−few-toothed towards the apex or entire, apex obtuse to acute, base narrowly cuneate, margin flat or slightly revolute. *Inflorescence* a leafy panicle occupying the upper third of the stem, often diffusely branched; flowers in 2−7-flowered pedunculate cymes; peduncle 6−20 mm long, pedicels 3−8 mm long. *Calyx* finely pubescent, 2,5−3,5 mm long; teeth lanceolate-deltoid, 1−1,5 mm long. *Corolla* 5−6 mm long; tube 2−3 mm long; lower lip oblong-obovate, 2,5−3 mm long, slightly concave, remaining 4 lobes oblong, rounded, 2 mm long. *Stamens* exserted by 5−7 mm.

Distributed from Swaziland through semi-coastal and midland Natal, Transkei and to King William's Town district in the Cape, in open bush and grassland. Map 4.

Vouchers: *Acocks* 13306; *Compton* 29695; *Tyson* 1782.

No common names noted, but the plant is used by native tribes in the eastern Cape Province to counteract anthrax by boiling the plant with meat suspected of being infected, in the same way that *T. trifidum* is used, while an infusion is taken to cure snake-bite and as a tonic.

T. kraussii overlaps with *T. trifidum* in the eastern Cape Province and occasional intermediate specimens may be difficult to place with certainty. The main distinguishing characters are discussed under *T. trifidum* (above).

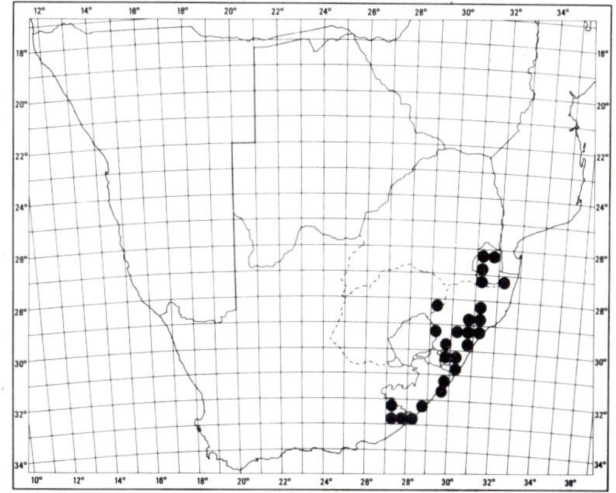

MAP 4. — **Teucrium kraussii**

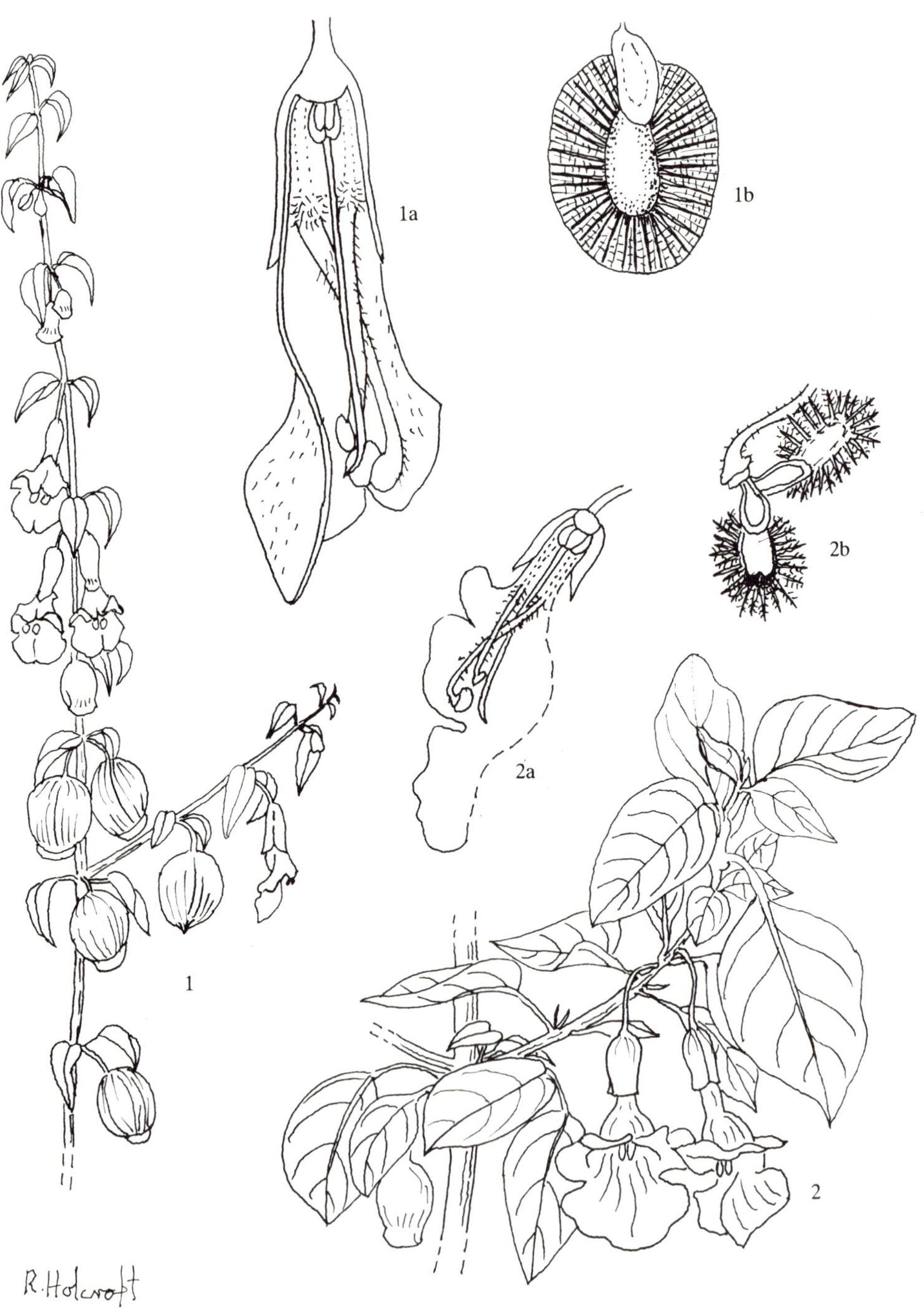
1a
1b
2a
2b
1
2
R. Holcroft

7213 ### 3. TINNEA

Tinnea *Kotschy ex Hook. f.* in Curtis's bot. Mag. t.5637 (April 1867); Kotschy & Peyr., Pl. Tinn. 25, t.11 (July 1867); Benth. & Hook. f., Gen. Pl. 2,2: 1220 (1876); Briq. in Natürl. PflFam. 4,3a: 214 (1895); Bak. in F.T.A. 5: 496 (1900); Skan in F.C. 5,1: 383 (1910); Robyns & Lebrun in Bull. Jard. bot. État Brux. 8: 168 (1930); Launert & Schreiber in F.S.W.A. 123: 31 (1969); Vollesen in Bot. Tidsskr. 70: 13 (1975); R. A. Dyer, Gen. 1: 525 (1975). Type species: *T. aethiopica* Kotschy ex Hook. f.

Perennial herbs or shrubs. *Leaves* opposite or occasionally ternate or subopposite, usually entire or nearly so. *Inflorescence* usually a lax terminal raceme, or flowers borne on short lateral branches; bracts leaf-like, becoming smaller towards the apex of the raceme; flowers produced in 2 (−3)-flowered verticils or in some species often solitary, often scented. *Calyx* 2-lipped, becoming much enlarged, ovoid, inflated and 2-valved with maturity, often densely pubescent; lips entire, broadly rounded. *Corolla* 2-lipped, often liver-coloured or shades of dark reddish purple to mauve; tube short or long, cylindrical at the base, widening near the throat; upper lip short, broad, ascending, emarginate or 2-lobed; lower lip much larger, spreading, 3-lobed, median lobe often emarginate, larger than the lateral rounded lobes. *Stamens* 4, slightly protruding, ascending under the upper lip, pubescent near the base; filaments of the posterior pair thicker, crossing the filaments of the anterior stamens so that the anthers are placed uppermost in the throat, thickened above, the thickened portion yellow and visible in the corolla-throat; anthers 2-thecous, the anthers on the shorter filaments the smaller. *Ovary* divided to half-way or more; style filamentous, shortly bifid, the lobes unequal. *Nutlets* obovoid-clavate, attached by a lateral areole occupying up to ½ the length, furnished on the back with a broad elliptic or orbicular "wing" made up of stiff primary rays interlaced with fine transverse hairs.

An African genus of 19 species, 4 of which occur in Southern Africa. The generic name honours the Tinne family, originally of Holland, three members of whom, Mme Henriette Tinne, her sister and her daughter Alexandrine, organised an ill-fated expedition to the White Nile in 1861−63, during which seeds of *Tinnea aethiopica* were collected and subsequently grown in Europe.

Vollesen, l.c., points out that the publication of the plate of *Tinnea aethiopica* in Curtis's bot. Mag. antedates the description given by Kotschy & Peyritsch, generally accepted in the past as the correct author citation for the genus.

1 Calyx and stems sparingly hispidulous to appressed-tomentose:

  2 Corolla bluish mauve to purple, tube narrow, exceeding 10 mm long; mature calyx membranous; petioles mostly longer than 10 mm .... 1. *T. barbata*

  2 Corolla dark purple-red to chocolate, tube broad, up to 9 mm long; mature calyx coriaceous; petioles usually less than 10 mm long:

    3 Twiggy shrub up to 2,5 m tall; flowers usually 1 (occasionally 2) per verticil, borne on short lateral shoots as well as in short terminal racemes 50−100 mm long .... 2. *T. rhodesiana*

    3 Stems softly woody, sparingly branched, up to 0,6 m long; flowers usually 2 per verticil, borne on slender terminal racemes 80−200 mm long .... 3. *T. galpinii*

1 Calyx and stems densely velvety-lanate .... 4. *T. eriocalyx*

1. **Tinnea barbata** *Vollesen* in Bot. Tidsskr. 70: 25 (1975); Codd in Flower. Pl. Afr. 46: t.1813 (1980). Type: Transvaal, Ida Doyer Nature Reserve near Barberton, *Edwards* 4123 (PRE, holo.!).

T. cf. *rogersii* sensu Compton, Fl. Swaziland 66 (1966).

Shrub 2,8−4 m tall, freely branched; branchlets sericeous. *Leaves* petiolate, soft; blade ovate-lanceolate to broadly ovate, 20−45 × 12−25 mm, sparingly to densely pubescent, freely gland-dotted on both surfaces, apex acute to subobtuse, base obtuse, margin entire or occasionally with a

FIG. 3. — 1, **Tinnea rhodesiana**, flowering branch, × 1; 1a, section through flower, × 4; 1b, winged nutlet, × 4 (after Flower, Pl. Afr. 46: t.1814, 1980). 2, **T. barbata**, flowering branch, × 1; 2a, section through flower, × 2; 2b, winged nutlets, × 3 (after Flower. Pl. Afr. t.1813, 1980).

few weak teeth; petiole 7—17 mm long. *Inflorescence* terminal or on short side shoots, of few to several spaced verticils; verticils 1—2-flowered; pedicels 8—10 mm long with a pair of minute bracteoles below the middle. *Calyx* enlarging to 15 mm long, membranous, lips 3—4 mm long. *Corolla* mauve to violet, 22—27 mm long, sparsely pubescent without; tube 12—15 mm long, 2 mm broad at the base; upper lip 2—3 mm long, about 5 mm broad; lower lip 10—12 mm long, 12—14 mm broad. *Nutlets*, excluding the wing, 6 mm long, glabrous; wing broadly elliptical, about 8 × 7 mm. Fig. 3:2.

In riverine scrub and forest margins at about 1 400 m altitude in the mountains of the Barberton district and adjoining northern Swaziland. Map 5.

Vouchers: *Buitendag* 753; *Compton* 28736.

A distinctive species allied to *T. rhodesiana* (below) but has larger, softer leaves, membranous calyx, longer mauve to purple corolla and glabrous nutlets.

It appears to have first been collected by an officer of the Department of Forestry near Louws Creek in the Barberton district in 1956.

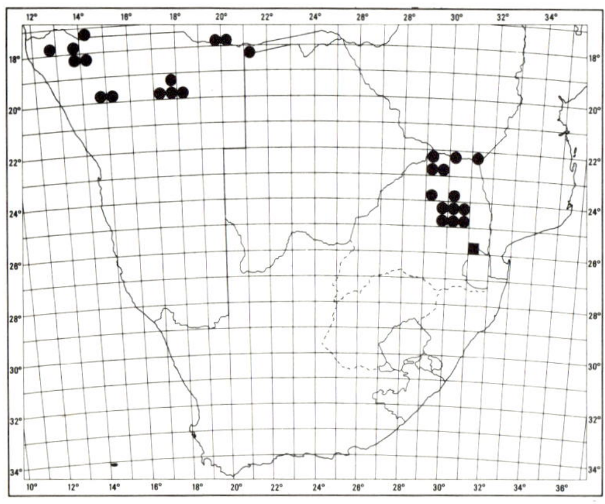

MAP 5. — ■ **Tinnea barbata**
● **T. rhodesiana**

2. **Tinnea rhodesiana** *S. Moore* in J. Bot., Lond. 43: 51 (1905); Robyns & Lebrun in Bull. Jard. bot. État Brux. 8: 180 (1930); Vollesen in Bot. Tidsskr. 70: 23 (1975); Codd in Flower. Pl. Afr. 46: t.1814 (1980). Type: Zimbabwe, Matoppos, *Eyles* 159 (BM, holo.).

T. juttae Dinter, Fl. Forst- u. landw. Fragm. 118 (1909), as *Timea*; Robyns & Lebrun, l.c. 176 (1930); Letty, Wild Flow. Transv. 288 (1962); Launert & Schreiber in F.S.W.A. 123: 32 (1969). Type: S.W.A./Namibia, Grossbarmen, *Dinter* 507.

T. galpinii sensu Skan in F.C. 5,1: 383 (1910), partly, as to *Rehmann* 5288; 5289.

T. rehmannii Schinz in Vjschr. naturf. Ges. Zürich 57: 561 (1913); Robyns & Lebrun, l.c. 175 (1930). Type: Transvaal, Klippan, *Rehmann* 5288 (Z, holo.).

T. dinteri Gürke ex Dinter in Feddes Reprium 24: 13 (1927); Ullbrich, Biologie Früchte & Samen 194 (1928), nom. nud.

Twiggy soft shrub 0,6—2,5 m tall; branches pale brown, minutely tomentulose, often glabrescent with age. *Leaves* petiolate; blade subcoriaceous, ovate to ovate-lanceolate, 8—20 (—30) × 3—8 (—12) mm, upper surface dark greenish brown, subglabrous to tomentulose, lower surface paler, gland-dotted; secondary nerves not visible above, 2—3 pairs faintly visible below, apex subacute to obtuse, base obtuse to truncate, margin entire; petiole 3—10 mm long. *Inflorescence* lax, 50—100 mm long, of few to several verticils borne terminally and on twiggy side-shoots; verticils 1—2-flowered; pedicels about 7 mm long with a pair of minute bracteoles about the middle. *Calyx* becoming ovoid, inflated, coriaceous, straw-coloured, 12—18 × 7—12 mm, lips 2—3 mm long. *Corolla* violet-scented, chocolate to purplish brown, 14—18 mm long; tube 6—9 mm long; upper lip 2—3 mm long, about 5 mm broad; lower lip broadly 3-lobed, 5—8 mm long, 8—10 mm broad. *Nutlets*, excluding the wing, 5—7 mm long, minutely tomentulose; wing broadly elliptical, about 8 × 6 mm. Fig. 3:1.

Found usually on stony hillsides in dry open woodland in north-eastern and northern Transvaal and nothern S.W.A./Namibia; occurs also in Zimbabwe and Angola. Map 5.

Vouchers: *De Winter & Leistner* 5905; *Galpin* 9188; *Merxmüller & Giess* 30333.

T. rhodesiana is related to *T. aethiopica* Kotschy ex Hook. f., a variable species distributed from Tanzania to Ethiopia, but differs mainly in the pubescent nutlets and terete branches. From *T. galpinii* (below) it differs in the taller, erect and more twiggy growth form and the tendency of the flowers to be borne singly on short side-shoots. The above description is based mainly on Southern African material.

3. **Tinnea galpinii** *Briq.* in Bull. Herb. Boissier sér. 2,3: 1094 (1903); Skan in F.C. 5,1: 383 (1910), partly, excl. *Rehmann* 5288, 5289; Robyns & Lebrun in Bull. Jard. bot.

État Brux. 8: 187 (1930); Phillips in Flower. Pl. S. Afr. 13: t.517 (1933); Letty, Wild Flow. Transv. 285, t.142, 4 (1962); Ross, Fl. Natal 302 (1972) Vollesen in Bot. Tidsskr. 70: 31 (1975); Compton, Fl. Swaz. 492 (1976). Type: Transvaal, Upper Moodies near Barberton, *Galpin* 1212 (PRE!).

Stems few to several arising from a perennial woody rootstock, sparingly branched, suberect to decumbent, 0,15–0,6 m long, softly woody below, densely and shortly tomentose. *Leaves* subsessile to shortly petiolate; blade subcoriaceous, elliptic-lanceolate to ovate, 15–25 × 6–10 mm, sparingly tomentulose and gland-dotted on both surfaces, particularly below, secondary nerves obscure above, 2–3 pairs visible below, apex subacute to obtuse, sometimes minutely apiculate, base obtuse, margin entire; petiole up to 5 mm long. *Inflorescence* lax, terminal, unbranched, 80–200 mm long, of many spaced verticils; verticils usually 2-flowered; pedicels 4–12 mm long with a pair of minute bracteoles about the middle. *Calyx* densely pilose, gland-dotted, often purple-tinged, becoming ovoid, inflated, membranous, 10–14 × 8–10 mm; lips rounded, 2–3 mm long. *Corolla* violet-scented, maroon to chocolate, 12–18 mm long, finely tomentose; tube 6–9 mm long; upper lip 2–3 mm long, about 5 mm broad; lower lip 3-lobed, 5–8 mm long, 8–10 mm broad. *Nutlets*, excluding the wing, 5–6 mm long, subglabrous; wing broadly elliptical, about 8 × 6 mm.

Found among rocks in grassland on the mountains of eastern Transvaal, extending along the Lebombo Mts to Swaziland and northern Natal. Map 6.

Vouchers: *Codd* 7973; *Compton* 26377; *Schlechter* 3994.

Differs from *T. rhodesiana* (above) in the smaller stature with softer, sparingly branched, suberect to spreading stems with denser tomentum, particularly on the calyx, and flowers borne in slender terminal racemes.

4. Tinnea eriocalyx *Welw.* in Trans. Linn. Soc. Lond. 27: 59 (1869); Engl., Hochgebirgsfl. Trop. Afr. 371 (1892); Hiern, Cat. Afr. Pl. Welw. 1,4: 880 (1900); Bak. in F.T.A. 5: 499 (1900); Robyns & Lebrun in Bull. Jard. bot. État Brux. 8: 196 (1930); Launert & Schreiber in F.S.W.A. 123: 32 (1969); Vollesen in Bot. Tidsskr. 70:

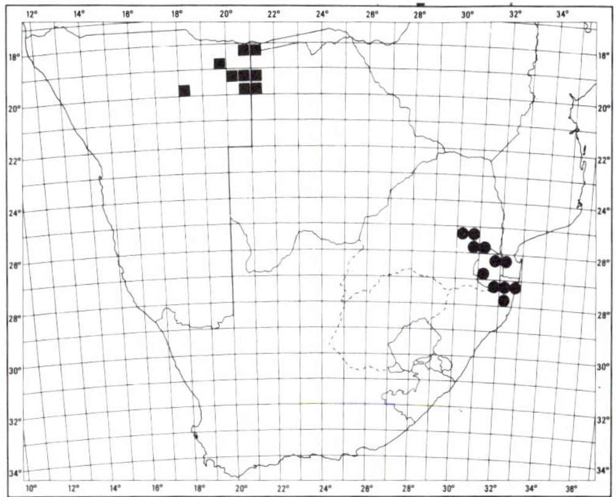

MAP 6. —   ● **Tinnea galpinii**
            ■ **T. eriocalyx**

42 (1975). Type: Angola, Huilla distr., Lopollo, *Welwitsch* 1635 (BM).

Soft shrublet 0,6–1 m tall, branching near the base from a perennial woody rootstock; stems erect, sparingly branched, densely lanate. *Leaves* often ternate or subopposite, subsessile; blade subcoriaceous, lanceolate-elliptic to broadly ovate, 20–45 × 8–20 mm, finely pubescent to densely velvety, secondary nerves indistinct, apex subacute to obtuse, often minutely apiculate, base obtuse, margin entire; petiole up to 5 mm long. *Inflorescence* a lax terminal raceme 100–350 mm long; verticils usually 2–3-flowered; pedicels 3–8 mm long with a pair of linear-lanceolate bracteoles near the base. *Calyx* subglobose, densely yellowish lanate-velutinous, ovoid, enlarging to 20 mm long in fruit; lips 2–3 mm long. *Corolla* mauve to almost purple, 15–25 mm long; tube 8–14 mm long; upper lip 3–5 mm long, about 5 mm broad; lower lip broadly 3-lobed, 4–6 mm long, 8–10 mm broad. *Nutlets*, excluding the wing, 7–8 mm long, sparsely pubescent; wing about 12 × 9 mm.

Found in dry open woodland on sandy soil in northern S.W.A./Namibia; also recorded from Botswana, Angola and Zaire. Map 6.

Vouchers: *De Winter & Marais* 4826; *Wild & Drummond* 6915.

Readily distinguishable from other species in Southern Africa by its densely lanate-velvety stems and calyx. Like *T. galpinii* (above), the flowers are produced in slender terminal racemes.

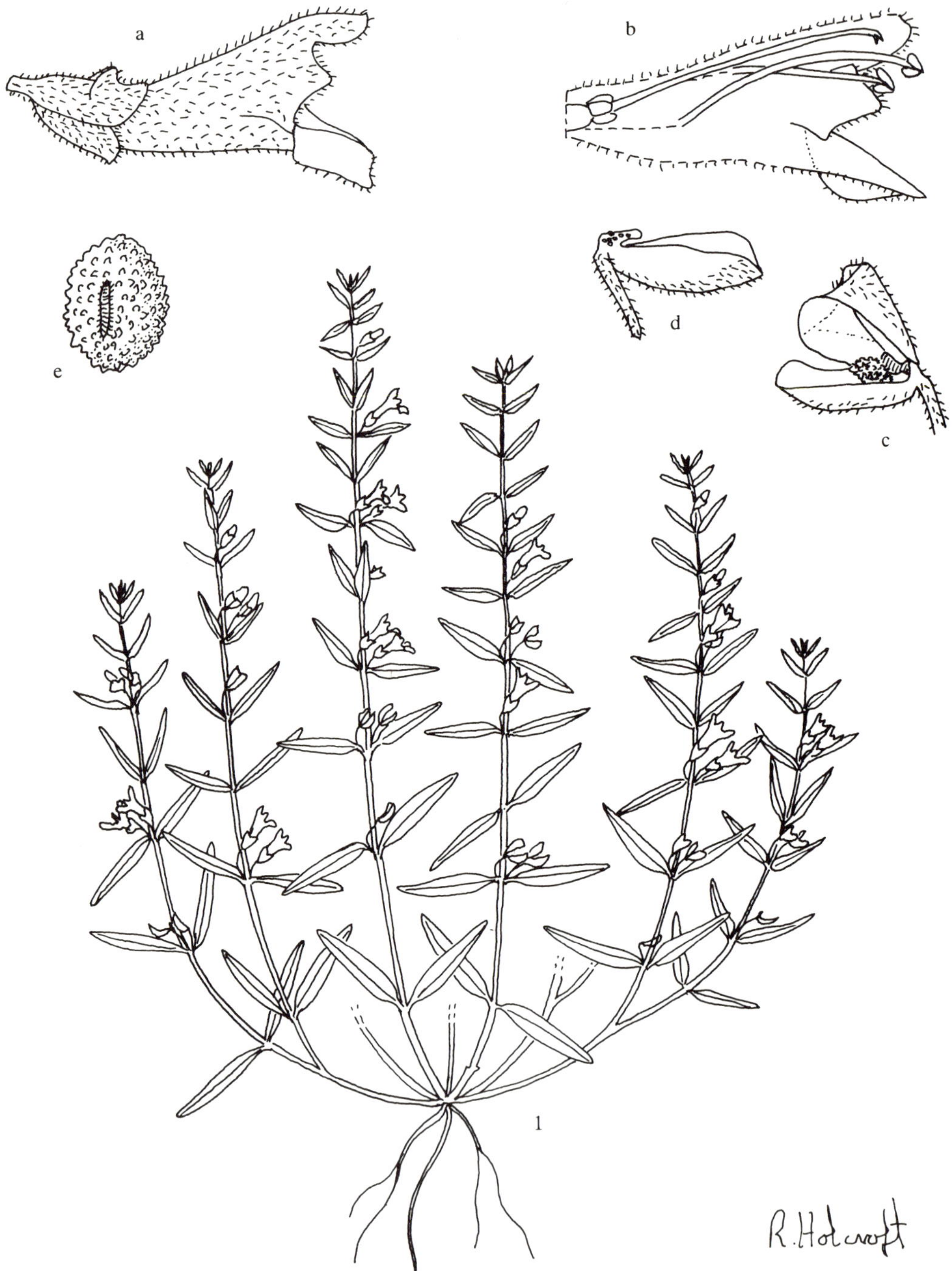

FIG. 4. — 1, **Scutellaria racemosa**, habit, × 1; a, flower, × 7; b, section through flower, × 7; c, d, calyx, × 7; e, nutlet, × 20 (*Theron* 1903).

7234

<h2 style="text-align:center">4. SCUTELLARIA</h2>

Scutellaria *L.*, Sp. Pl. 598 (1753); Gen. Pl. edn 5: 260 (1754); Benth. in DC., Prodr. 12: 412 (1848); Benth. & Hook. f., Gen. Pl. 2,2: 1201 (1876); Briq. in Natürl. PflFam. 4, 3a: 225 (1896); Bak. in F.T.A. 5: 461 (1900); Epling, Univ. Calif. Publs Bot. 20: 16 (1942); Richardson in Fl. Europ. 3: 135 (1972); R. A. Dyer, Gen. 1: 526 (1975). Type species: *S. peregrina* L.

Perennial decumbent herb. *Leaves* usually small and soft, entire or toothed. *Inflorescence* racemose, lax or dense; bracts leaf-like, often smaller towards the apex of the inflorescence; verticils 2-flowered. *Calyx* pouch-shaped, 2-lipped; lips equal in length, entire, broad, rounded, the upper bearing a transverse protruding outgrowth and finally deciduous, the lower persistent. *Corolla* 2-lipped; tube ascending-arcuate or somewhat sigmoid, upper lip erect, concave or galeate, usually obliquely joined to the lateral lobes of the lower lip; lower lip spreading with an oblong central lobe and 2 smaller lateral lobes. *Stamens* 4, didynamous, ascending under the upper lip and included by it, the lower pair the longer; anthers of the upper pair 2-celled, of the lower with one cell imperfect or obsolete. *Ovary* on a variably developed gynobase; style slender, included in the upper lip, subequal to the stamens, apex subulate with one branch wanting or short and appressed to the longer. *Nutlets* ovoid to subglobose, rarely obovoid, borne on a raised gynophore, tuberculate or variously sculptured.

In non-Southern African species, annual or perennial herbs or subshrubs, occasionally erect.

Species about 150 widely distributed in the north temperate zones of the Old and New Worlds, extending to South America, to Malaysia and Australia, and to tropical Africa as far south as Zimbabwe; one species of South American origin naturalized in Southern Africa.

Scutellaria racemosa *Pers.*, Syn. Pl. 2: 136 (1807); Epling, Univ. Calif. Publs Bot. 20: 18 (1942); Hilliard & Burtt in Notes R. bot. Gdn Edinb. 30: 127 (1970). Type: Uruguay, Montevideo, *Commerson* (P).

Weak-stemmed perennial herb, freely branched, rhizomatous; stems 0,15−0,3 m long, slender, quadrangular, glabrous to sparingly pilose. *Leaves* petiolate; blade lanceolate to lanceolate-hastate, 10−30 × 3−6 mm, glabrous or nearly so, gland-dotted beneath, apex acute, base obtuse to rounded, margin entire, often inrolled; petiole 1−3 mm long. *Inflorescence* slender, lax, 5−12 mm long; pedicels 2−3 mm long. *Calyx* puberulous to hispidulous, at flowering 1,5−2 mm long, enlarging to 3 mm long in fruit. *Corolla* small, 4,5−6 mm long, variously coloured, violet to red or white with purple spots, the lower lip often paler than the upper; tube ascending, 3−4 mm long, widening to 2 mm wide at the throat; lips subequal, 1,5−2 mm long. *Stamens* reaching the apex of the upper lip, the lower pair attached about 2 mm above the base of the tube. *Nutlets* 1 mm long, pale brown, minutely tuberculate. Fig. 4.

Naturalized on river banks and moist places in central Transvaal and coastal parts of the Cape Province. Originally from South America.

Vouchers: *M. de Winter* s.n.; *Smart* sub TRV 26683.

The first known gathering of the species in South Africa was at Plettenberg Bay in 1921, followed by a collection near Middelburg, Transvaal, in 1933.

7236   **5. ACROTOME**

Acrotome *Benth.* in Endl., Gen. Pl. 1: 627 (1838); in DC., Prodr. 12: 435 (1848); Benth. &
Hook. f., Gen. Pl. 2,2: 1206 (1876); Briq. in Natürl. PflFam. 4,3a: 229 (1896); Bak. in
F.T.A. 5: 471 (1900); Skan in F.C. 5,1: 335 (1910); G. Tayl. in J. Bot., Lond. 73: 1 (1935);
Launert & Schreiber in F.S.W.A. 123: 5 (1969); R. A. Dyer, Gen. 1: 526 (1975). Type
species: *A. pallescens* Benth.

Annual or perennial herbs or subshrubs. *Leaves* entire or toothed. *Inflorescence*
terminal, of spaced few- or densely many-flowered, often glomerate verticils; bracts
leaf-like, becoming smaller towards the apex of the inflorescence; bracteoles linear, often
bristle-like and spine-tipped, arcuate and often conspicuous at the base of the verticil. *Calyx*
as long as or shorter than the corolla tube, tubular-campanulate, 10- or 11-nerved, slightly
oblique or symmetrical at the mouth, equally or unequally 5−11-toothed. *Corolla* 2-lipped;
tube slightly exceeding the calyx, tubular, often with a ring of hairs or glands within about
the middle, pubescent without; upper lip ascending or slightly arched, almost flat, without a
fringe of hairs; lower lip spreading, 3-lobed, with the middle lobe the largest. *Stamens* 4,
didynamous, inserted at about the same level near the middle of the corolla tube, included
within the tube, held together by intertwining hairs; filaments of posterior pair more or less
straight, those of the anterior pair recurved; anthers obovoid or oblong, 1-celled by
confluence, bearing a glandular crest. *Ovary* 4-lobed, truncate and glandular on the
truncate surface; style hairy, included with the anthers, entire, oblique at the apex. *Nutlets*
obovoid, triquetrous, truncate at the apex.

A genus of 8 species occurring in Africa south of the equator; 6 species recorded from Southern Africa, one or
two of which tend to become weeds of waste places. There is a superficial resemblance to the genus *Leucas* (no.
10), but *Acrotome* is distinguished by the peculiar arrangement of the stamens, which are included in the corolla
tube and held together by intermingling hairs, and by the upper lip of the corolla being shorter than the lower
(which is usually, but not always, longer than the lower in *Leucas*).

1 Plants annual, unbranched or branched shortly above the base; verticils usually 15−many-flowered in
  dense globose clusters (occasionally fewer in *A. fleckii*); bracteoles usually well developed:

  2 Verticils 1 or 2 (rarely 3) per flowering branch; calyx regular, 5-toothed:

    3 Leaves ovate to ovate-lanceolate or oblong-lanceolate; pubescence on stems appressed-retrorse
    .... 1. *A. inflata*

    3 Leaves linear-lanceolate to linear; pubescence on stems appressed-antrorse .... 2. *A. angustifolia*

  2 Verticils (3−) 4−10 per flowering branch; calyx oblique, 8−10-toothed.................................. 3. *A. fleckii*

1 Plants perennial, branching at the base from a woody rootstock; verticils 2−10-flowered; bracteoles
  usually poorly developed:

  4 Plant puberulous or shortly pubescent:

    5 Calyx mouth oblique, 5-toothed .... 4. *A. pallescens*

    5 Calyx mouth not oblique, 8−11-toothed.. 5. *A. thorncroftii*

  4 Plant densely hispid-villous .... 6. *A. hispida*

1. **Acrotome inflata** *Benth.* in DC.,
Prodr. 12: 436 (1848); Oliv. in Hooker's
Icon. Pl. 15: t.1467 (1884); N.E. Br. in Kew
Bull. 1909: 132 (1909); Skan in F.C. 5,1: 335
(1910); G. Tayl. in J. Bot., Lond. 73: 8
(1935); Wilman, Check List Griq. West 229
(1946); Launert & Schreiber in F.S.W.A.
123: 7 (1969); Jacot Guill., Fl. Lesotho 236
(1971). Type: Cape "Zuurebergen", *Burke*
s.n. (K, holo.).

Leucas eenii Hiern, Cat. Afr. Pl. Welw. 1,4: 878
(1900). *Lasiocorys eenii* (Hiern) Bak. in F.T.A. 5: 469
(1900). Syntypes: Angola, Mossamedes, *Welwitsch*
5486 (BM); S.W.A./Namibia, *Een* s.n. (BM).

Acrotome amboensis Briq. in Bull. Herb. Boissier
sér. 2,3: 1095 (1903). Syntypes: S.W.A./Namibia,
several cited.

Annual erect herb 0,15−0,7 (−1) m
tall, usually freely branched shortly above
the base; stems densely appressed retrorse
villous. *Leaves* subsessile to shortly petiol-

ate; blade ovate to ovate-lanceolate or oblong-lanceolate, (20−) 30−100 (−120) × (5−) 10−25 mm, shortly and densely pilose, apex acute, base cuneate, margin remotely and sparingly crenate-serrate, mainly above the middle. *Inflorescence* of 1 or 2 (rarely 3) spaced verticils; verticils densely many-flowered, globose, 15−35 mm in diam.; bracteoles numerous, filiform, arcuate-erect, villous, spine-tipped, up to 10 mm long; flowers sessile. *Calyx* widest about the middle, 7 mm long at flowering, enlarging to 12−15 mm, hispid, symmetrical at the mouth, subequally 5-toothed; teeth deltoid-subulate, spine-tipped, eventually 3 mm long. *Corolla* small, white or pale mauve; upper lip oblong, 3−4 mm long; lower lip 4−5 mm long. *Nutlets* 2,5−3 mm long. Fig. 5:1.

Widespread in the semi-arid parts of S.W.A./Namibia, Botswana, northern and western Transvaal, northern Cape Province and Orange Free State, entering Lesotho and reaching as far south as Middelburg and Queenstown in the eastern Cape; locally common in open woodland, especially under trees and disturbed places in grassland, and found as a weed of roadsides and cultivation. Also recorded in Angola, Zambia and Zimbabwe. Map 7.

Vouchers: *Codd* 4146; *Dieterlen* 89; *Galpin* 1505; *Rodin* 9299; *Smith* 3902.

Closely related to the next species, *A. angustifolia*, and the differences are discussed under that species.

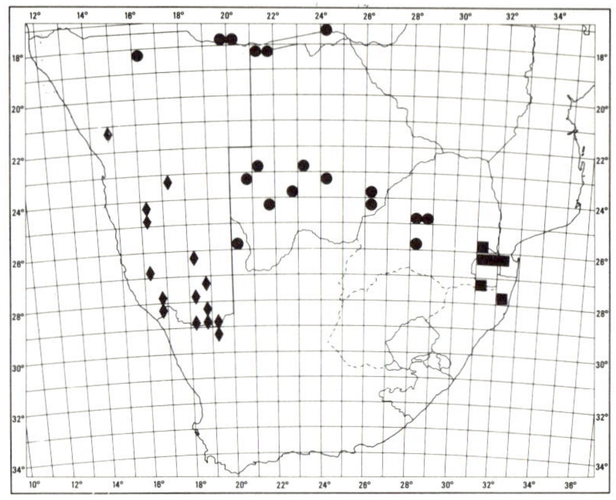

MAP 8. — ● **Acrotome angustifolia**
◆ **A. pallescens**
■ **A. thorncroftii**

There is a superficial resemblance between *A. inflata* and *Leucas martinicensis* (p. 4: 40) but in the latter species the calyx is bent at the apex and obliquely 10-toothed, the upper corolla lip is as long as the lower with the stamens shortly exserted from the tube, and the nutlets are not truncate at the apex. Both species are recorded as weeds of cultivation and disturbed places.

Although listed by Compton, Fl. Swaziland 66 (1976), no material of this species from Swaziland has been seen.

2. **Acrotome angustifolia** *G. Tayl.* in J. Bot., Lond. 73: 9 (1935); Launert & Schreiber in F.S.W.A. 123: 6 (1969). Type: Transvaal, Mosdene, *Galpin* M 602 (BM, holo.; K; PRE!).

A. lancifolia Brem. & Oberm. in Ann. Transv. Mus. 16: 431 (1935). Type: Botswana, Kaotwe, *Van Son* sub TRV 28919 (PRE, holo.!; BM).

Similar in habit and appearance to *A. inflata* (above) but pubescence on the stems antrorse not retrorse and leaves narrower, linear-lanceolate or narrowly lanceolate-oblong, 40−90 × 5−14 mm.

Found in open woodland on deep sandy soil in northern S.W.A./Namibia, Botswana, north-western Transvaal and just entering the northern Cape Province. Also in Zimbabwe and Zambia. It does not

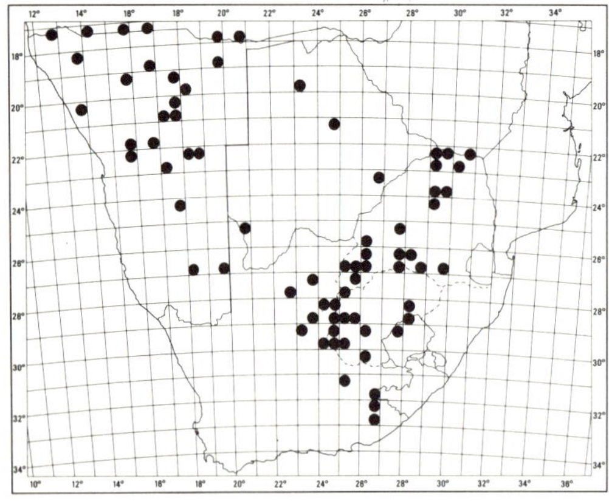

MAP 7. — **Acrotome inflata**

FIG. 5. — 1, **Acrotome inflata,** flower clusters, × 1; 1a, calyx, × 3; 1b, nutlet, × 7 (*Mason & Boshoff* 2533). 2, **A. hispida,** lower part of plant, × 1; 2a, mature calyx, × 3 (*Codd* 8433); 2b, calyx and bracteole, × 3; 2c, corolla, × 3; 2d, section of corolla tube showing position of anthers and style, × 7 (*Mrs Jenkins* s.n., Pretoria District). 3, **A. pallescens,** calyx, × 3 (*Oliver & Muller* 6429). 4, **A. fleckii,** calyx, × 3 (*De Winter & Hardy* 3210). 5, **A. thorncroftii,** calyx, × 3 (*Compton* 31936).

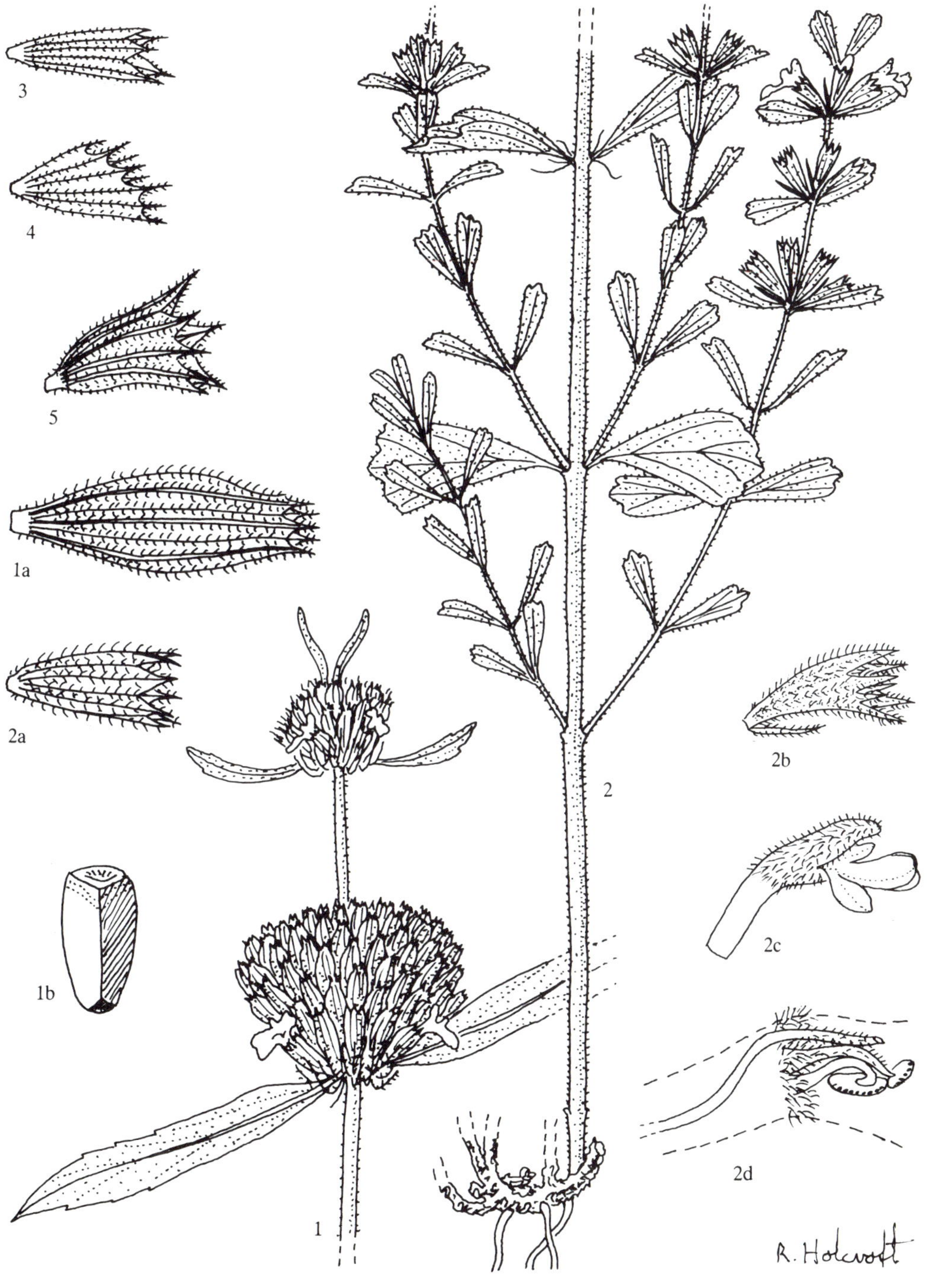
3
4
5
1a
2a
1b
1
2
2b
2c
2d
R. Holcroft

show the same tendency to become a weed as *A. inflata* and is restricted to the particular environment indicated. Mainly for these reasons, *A. angustifolia* is retained as a separate species. Map 8.

Vouchers: *Rodin* 9142; *Story* 5555; *Wild & Drummond* 6990.

According to Rodin a tea is made from the leaves and flowers in the Okavango area and given to children with upset stomachs.

3. Acrotome fleckii *(Gürke) Launert* in Mitt. bot. StSamml., Münch. 2: 360 (1957); Launert & Schreiber in F.S.W.A. 123:6 (1969). Lectotype: S.W.A./Namibia, Damaraland, Tiras, *Schinz* 43 (Z).

Leucas fleckii Gürke in Bot. Jb. 22: 140 (1895).

Acrotome belckii Gürke in Bull. Herb. Boissier 6: 549 (1898); Bak. in F.T.A. 5: 471 (1900). Type: S.W.A./Namibia, Kaokoveld, near Otjitambi, *Belck* 40.

Annual erect herb 0,1—0,6 (—0,8) m tall, unbranched or branched shortly above the base; stems shortly retrorse and somewhat crisped tomentose. *Leaves* subsessile or shortly petiolate; blade linear to linear-lanceolate, 25—70 × 5—10 mm, sparingly to fairly densely pilose, apex subacute, base attenuate, margin with a few small teeth in the upper half. *Inflorescence* of (3—) 4—10 spaced verticils, often starting near the base of the plant; verticils dense, few- to many-flowered, 10—22 mm in diam.; bracteoles numerous, filiform, arcuate-erect, villous, 5—8 mm long, spine-tipped. *Calyx* hispid, widest at the oblique mouth, 5 mm long at flowering, enlarging to 8—10 mm, subequally 8—10-toothed; teeth deltoid-subulate to filiform, spine-tipped, eventually 3—4 mm long. *Corolla* small, white; upper lip oblong, 2,5—3 mm long; lower lip 4—5 mm long. *Nutlets* 2,5 mm long. Fig. 5:4.

Distributed in the western half of S.W.A./Namibia from Damaraland to Bethanien district in the south, in sandy soil usually among rocks or under trees. Map 9.

Vouchers: *De Winter* 2685; *Schlieben* 10324.

Differs from *A. inflata* (no. 1) and *A. angustifolia* (no. 2) in its usually smaller stature and more numerous but smaller verticils which often start near the base of the plant. It also has a more westerly distribution.

4. Acrotome pallescens *Benth.* in DC., Prodr. 12: 436 (1848); Briq. in Bull. Herb. Boissier sér. 2,3: 1096 (1903); Skan in F.C. 5,1: 335 (1910); G. Tayl. in J. Bot., Lond. 73: 11 (1935); Launert & Schreiber in F.S.W.A. 123: 7 (1969). Type: Cape,

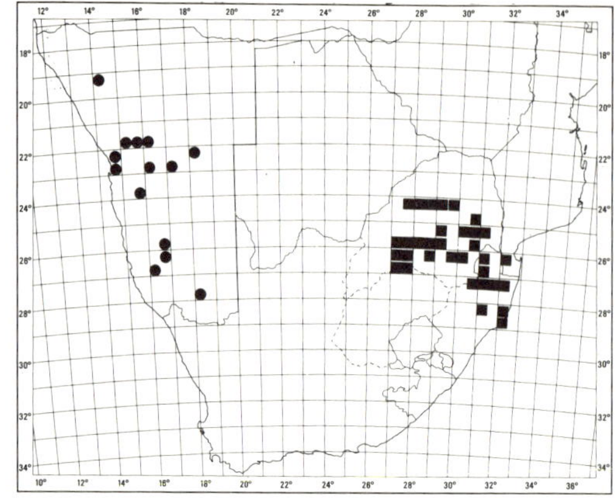

MAP 9. — ● **Acrotome fleckii**
■ **A. hispida**

probably Namaqualand, *Drège* s.n. (K, holo.).

Stachys steingroeveri Briq. in Bot. Jb. 19: 193. Type: S.W.A./Namibia, *Steingröver* 11.

Shrublet 0,45—0,5 m tall, usually freely branched at the base from a perennial woody rootstock; stems slender, erect to spreading, softly woody below, minutely glandular-puberulous. *Leaves* very shortly petiolate; blade ovate to ovate-lanceolate or linear-oblong, 10—25 × 4—10 mm, subglabrous to puberulous or minutely hispidulous, apex subacute, base cuneate, margin entire or with a few small teeth near the apex. *Inflorescence* of 2—6 spaced verticils; verticils up to 12-flowered, usually fewer; bracteoles few, linear-filiform, minutely hispidulous, somewhat spine-tipped, 3—5 mm long; pedicels 1—2 mm long. *Calyx* puberulous to hispidulous, widest near the slightly oblique mouth, 6—7 mm long at flowering, enlarging slightly at maturity, 5-toothed; teeth deltoid-subulate, spine-tipped, eventually 3 mm long. *Corolla* small, white; upper lip oblong, 2—3 mm long; lower lip 4—5 mm long. *Nutlets* dark brown, 1,5 mm long. Fig. 5:3.

Grows in sandy soil among rocks in southern S.W.A./Namibia and northern Namaqualand. Map 8.

Vouchers: *Acocks* 18025; *Leistner* 2480; *Strey* 2309.

5. Acrotome thorncroftii *Skan* in F.C. 5,1: 335 (1910); G. Tayl. in J. Bot., Lond.

73: 10 (1935); Ross, Fl. Natal 303 (1972). Type: Transvaal, Barberton, *Thorncroft* sub TRV 3124 (K, holo.; PRE!).

Perennial herb 0,1−0,2 m tall, branching from the base; stems arising annually from a woody rootstock, ascending to spreading, slender, sparingly branched, fairly densely hispidulous. *Leaves* very shortly petiolate; blade oblong-elliptic to linear-oblanceolate, 14−20 × 2−4 mm, hispidulous, apex obtuse, base cuncate, margin entire or with a pair of minute teeth near the apex. *Inflorescence* simple or occasionally with a pair of branches near the base, consisting of 2−7 spaced or fairly crowded verticils; verticils 4−10-flowered; bracteoles few, lanceolate to linear-filiform, 3−6 mm long, hispidulous, somewhat spine-tipped; flowers sessile. *Calyx* hispidulous, 5−7 mm long at flowering, enlarging slightly at maturity, mouth symmetrical, 8−11-toothed; teeth deltoid-subulate, spine-tipped, 1−1,5 mm long. *Corolla* small, white; upper lip 2,5 mm long; lower lip 5−6 mm long. Fig. 5:5.

Known from only a few gatherings in grassland in the south-eastern Transvaal, Swaziland and northern KwaZulu. Map 8.

Vouchers: *Compton* 28847; *Ward* 3598.

See note after *A. hispida* (below).

6. **Acrotome hispida** *Benth.* in DC., Prodr. 12: 436 (1848); Skan in F.C. 5,1: 336 (1910); G. Tayl. in J. Bot., Lond. 73: 12 (1935); Ross, Fl. Natal 303 (1972); Compton, Fl. Swaziland 492 (1976). Type: Transvaal, "Schoenstrome" (Mooi River), *Burke* s.n. (K, holo.).

A. hispida var. *elongata* Benth., l.c. (1848). Type: Transvaal, Vaal River, *Burke* s.n. (K, holo.).

— var. *obliqua* Benth., l.c. (1848). Type: Transvaal, Aapies River, *Burke* s.n. (K, holo.).

Perennial herb 0,1−0,25 m tall, branching from the base; stems few to many from a woody rootstock, ascending to spreading, sparingly branched, sometimes fairly woody at the base, densely hispid. *Leaves* very shortly petiolate; blade obovate to elliptic, 8−25 × 5−14 mm, densely hispid, apex obtuse to rounded, base cuneate, margin usually few-toothed at the apex or entire. *Inflorescence* simple, of 2−8 spaced or sometimes fairly crowded verticils, often starting low down on the stems; verticils 6−10-flowered; bracteoles few, linear to linear-filiform, 3−6 mm long, hispid, somewhat spine-tipped; flowers sessile. *Calyx* densely hispid, 5−7 mm long, enlarging slightly at maturity, mouth symmetrical, 7−10-toothed; teeth deltoid-subulate, spine-tipped, 1−2 mm long. *Corolla* small, white; upper lip 2−3 mm long; lower lip 4−6 mm long. *Nutlets* nearly 2 mm long, 1 mm broad. Fig. 5:2.

Found in central and southern Transvaal, extending through Swaziland to northern KwaZulu, in grassy places, often among rocks. Map 9.

Vouchers: *Acocks* 21014; *Galpin* 14631; *Schlechter* 3722.

In calyx characters *A. hispida* is similar to *A. thorncroftii* (no. 5), but can be recognized by the denser and more hispid pubescence and the broader leaves, though there appears to be some introgression between the two. Thus in the Nelspruit area specimens of *A. hispida* tend to have narrower leaves, while in northern KwaZulu some specimens have somewhat shorter pubescence, approaching the condition found in *A. thorncroftii*; perhaps varietal status for the latter may be more appropriate.

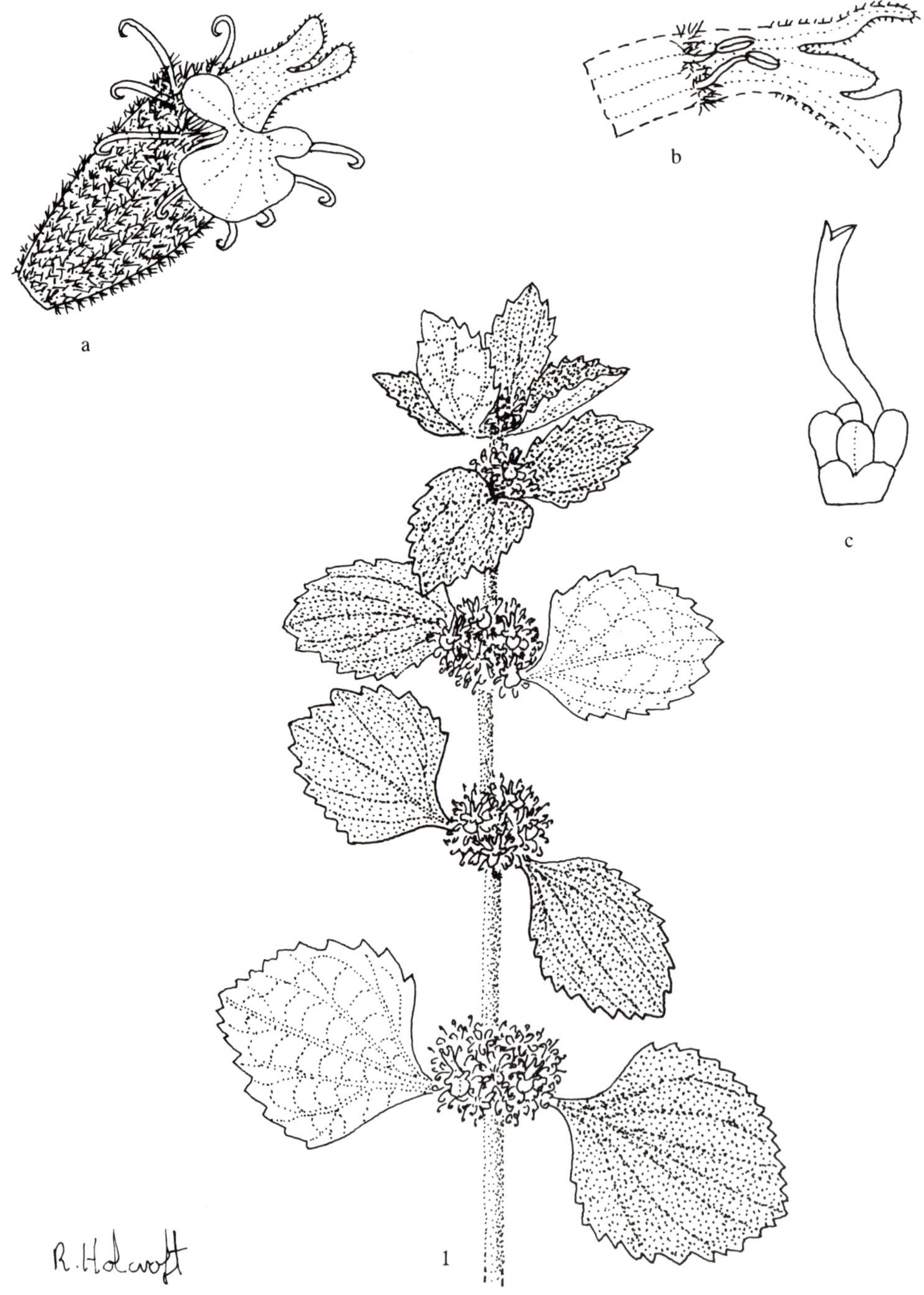

FIG. 6. — 1, **Marrubium vulgare,** flowering stem, × 1; a, calyx and corolla, × 10; b, section through upper part of corolla, × 10; c, gynoecium, × 10 (*Bruce* 359).

7238   **6. MARRUBIUM**

Marrubium *L.*, Sp. Pl. 582 (1753); Gen. Pl. edn 5: 254 (1754); Benth. in DC., Prodr. 12: 447 (1848); Benth. & Hook. f., Gen. Pl. 2,2: 1206 (1876); Briq. in Natürl. PflFam. 4,3a: 230 (1896); Cullen in Fl. Europ. 3: 137 (1972); Standley & Williams in Fieldiana Bot. 24,9: 264 (1973); R. A. Dyer, Gen. 1: 526 (1975). Type species: *M. vulgare* L.

Perennial herbs, usually tomentose or lanate. *Leaves* petiolate, rugose, toothed. *Inflorescence* of spaced verticils; verticils dense, often glomerate, many-flowered; bracts leaf-like, becoming smaller towards the apex; bracteoles usually present, linear-subulate; flowers sessile. *Calyx* tubular, widening towards the mouth, 5−10-nerved, densely hairy inside the mouth; teeth 10, subequal, spine-tipped, recurved or hooked. *Corolla* small, white to purplish; tubc included in the calyx, glabrous or with a ring of hairs within; upper lip erect, almost flat or concave, almost entire to deeply 2-fid; lower lip slightly longer than the upper, spreading, 3-fid, the middle lobe broader and emarginate. *Stamens* 4, didynamous, included in the corolla tube, the anterior pair the longer; anthers 2-celled, the cells divaricate, confluent. *Nutlets* ovoid, smooth, truncate at the apex.

In non-Southern African species the calyx may be 5−10-toothed.

About 30 species, natives of Europe, northern Africa and Asia; 1 species now a widespread weed and naturalized in Southern Africa.

Marrubium vulgare *L.*, Sp. Pl. 583 (1753); Benth. in DC., Prodr. 12: 453 (1848); Salter in Fl. Cape Penins. 698 (1950); Butcher, New Illustr. Brit. Fl. 2: 352 (1961); Standley & Williams in Fieldiana Bot. 24,9: 264 (1973). Type: from Europe.

Aromatic herb 0,3−0,6 m tall; stems few to several from a perennial rhizome, erect, lanate, somewhat woody at the base. *Leaves* petiolate; blade broadly ovate to subrotund, 20−45 × 18−45 mm, sparingly to densely tomentose above, densely lanate below, apex rounded to obtuse, base cuneate, margin crenate-dentate; petiole 5−14 mm long. *Inflorescence* simple, of 4−12 spaced verticils; bracteoles linear-filiform, 5−7 mm long, villous. *Calyx* 5−8 mm long at flowering, scarcely enlarging; tube about 4−5 mm long, stellate-hispid; teeth 10, subulate, spine-tipped, 1,5−3 mm long, spreading, bent or hooked at the apex. *Corolla* 7−8 mm long; upper lip 2,5 mm long; lower lip 2,5−3 mm long. *Nutlets* smooth, 2 mm long. Fig. 6.

A native of Europe and Asia, now a widespread weed, occasionally cultivated; naturalized in parts of the Orange Free State, Lesotho and the Cape Province.

Vouchers: *Brink* 226; *Bruce* 359.

Commonly known as Horehound or Hoarhound, the plant has long been used medicinally for treating colds. The hooked calyx teeth adhere to wool and the plant tends to be distributed in this way.

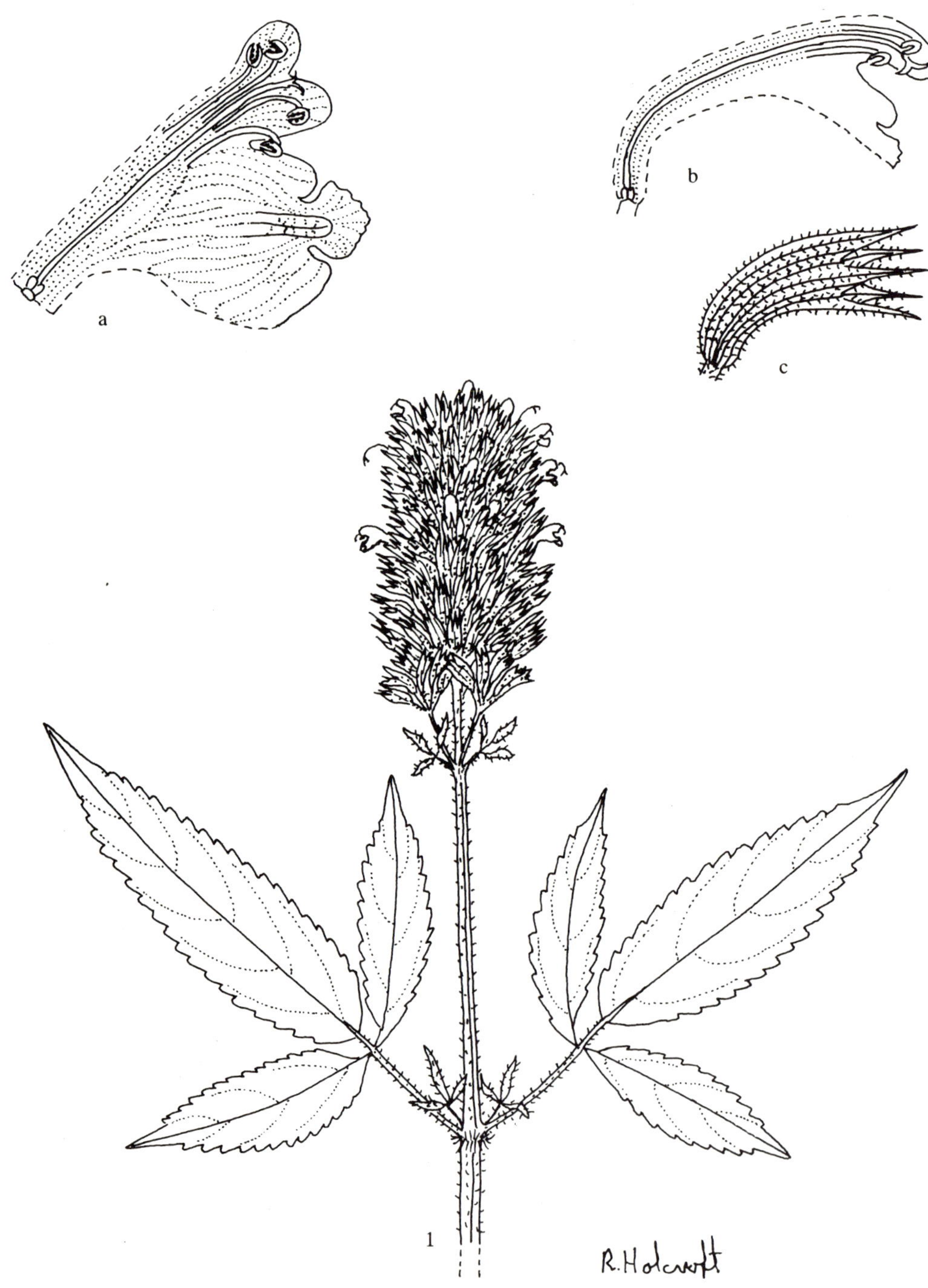

FIG. 7. — 1, **Cedronella canariensis,** flowering stem, × 1; a, corolla, opened longitudinally, × 3; b, section through corolla, × 3; c, calyx, × 3 (*Rourke* 1494).

7245

7. CEDRONELLA

Cedronella *Moench,* Meth. 411 (1794); Benth. in DC., Prodr. 12: 405 (1848); Benth. & Hook. f., Gen. Pl. 2,2: 1200 (1875); Briq. in Natürl. PflFam. 4,3a: 235 (1896); Skan in F.C. 5,1: 333 (1910); Salter in Fl. Cape Penins. 698 (1950); R. A. Dyer, Gen. 1: 526 (1975). Type species: *C. canariensis* (L.) Webb & Berth.

Perennial herbs, woody at the base. *Leaves* petiolate, thin-textured, digitately 3 (−5)-foliolate, toothed. *Inflorescence* dense, of many crowded verticils, the lowest verticil often a distance below the rest; verticils densely many-flowered, made up of opposite much-branched cymes, the lower cymes pedunculate; bracts narrow, simple or 3-foliolate; bracteoles linear-filiform. *Calyx* tubular-campanulate, 13−15-nerved, equally 5-toothed; teeth narrowly lanceolate-deltoid, subulate, erect. *Corolla* tube exserted, somewhat widened at the throat, glabrous within; upper lip erect, somewhat hooded, 2-fid or emarginate, about equal in length to the lower; lower lip spreading, 3-fid with the median lobe the largest. *Stamens* 4, didynamous, ascending under the upper lip or slightly exserted, the upper pair longer than the lower; anthers 2-celled; cells parallel, distinct. *Style* shortly 2-fid; lobes equal. *Nutlets* ovoid, smooth.

The present tendency is to limit the genus to 1 species; originally from Madeira and the Canary Islands, now a widespread weed, naturalized in the south-western Cape Province.

Cedronella canariensis *(L.) Webb & Berth.,* Phyt. Canar. 3:87 (1847); Salter in Fl. Cape Penins. 698 (1950); Bramwell in Fl. Europ. 3: 157 (1972). Type: from the Canary Islands.

Dracocephalum canariense L., Sp. Pl. edn 2,2: 829 (1763). *C. triphylla* Moench, Meth. 412 (1794); Benth. in DC., Prodr. 12: 406 (1848); Skan in F.C. 5,1: 334 (1910); Bailey, Cycl. Hort. 1: 698 (1963), nom. illegit.

Erect perennial to 2,5 m tall; stems slender, 4-angled, glabrous except for a ring of hairs at the nodes. *Leaflets* lanceolate, median leaflet the largest, 40−70 × 14−24 mm; lateral leaflets 25−45 × 8−14 mm, base occasionally with a lobe or conspicuous tooth. *Inflorescence* 30−80 mm long; verticils 6−12-flowered. *Calyx* 11−13 mm long, pubescent, gland-dotted; tube 8−9 mm long; teeth 3−4 mm long. *Corolla* purplish, 17−18 mm long; tube about 14 mm long, sparingly pubescent without; upper lip 3 mm long; lower lip 3 mm long. Fig. 7.

Naturalized in the south-western Cape Province, mainly along streams in forest clearings; originally from Madeira and the Canary Islands.

Vouchers: *Bolus* 4624; *Rourke* 1494.

Commonly known as Balm of Gilead; the plants have a pleasant cedar-like smell from which the generic name *Cedronella* ("little cedar") is derived.

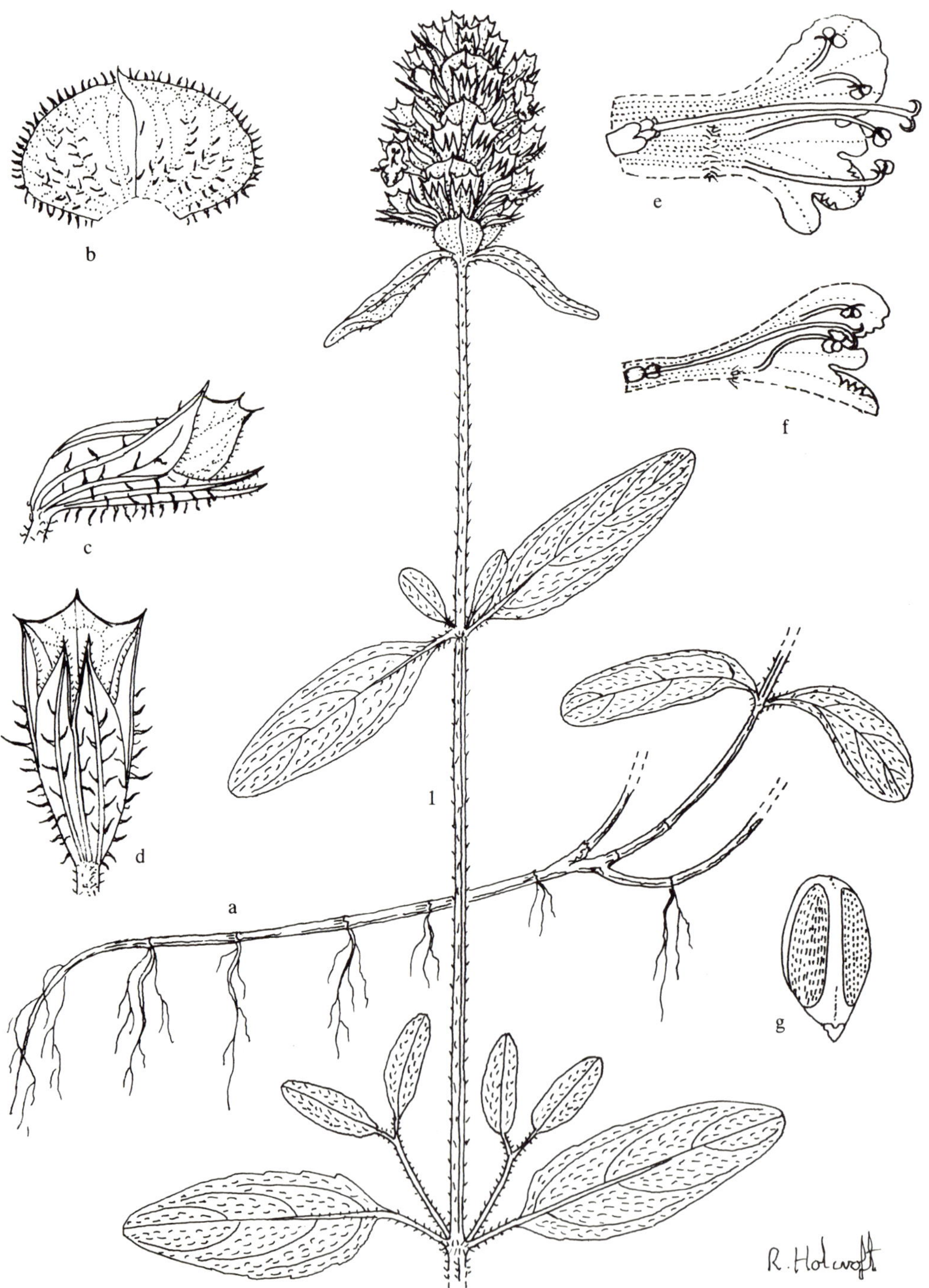

FIG. 8. — 1, **Prunella vulgaris,** flowering stem, × 1; a, creeping stem, × 1; b, bract, × 3; c, d, calyx, × 4; e, flower opened longitudinally, × 4; f, section through corolla, × 4; g, nutlet, × 8 (*Hugo* 7254).

7254 **8. PRUNELLA**

Prunella *L.*, Sp. Pl. 600 (1753); Gen. Pl. edn 5: 261 (1754), as *Brunella;* Willd., Sp. Pl. 3: 176 (1800); Benth. in DC., Prodr. 12: 409 (1848), as *Brunella;* Benth. & Hook, f., Gen. Pl. 2,2: 1203 (1876), as *Brunella;* Briq. in Natürl. PflFam. 4,3a: 241 (1896), as *Brunella;* Smith in Fl. Europ. 3: 162 (1972); Standley & Williams in Fieldiana Bot. 24,9: 271 (1973); R. A. Dyer, Gen. 1: 526 (1975). Type species: *P. vulgaris* L.

Soft, perennial herbs, usually decumbent or prostrate. *Leaves* petiolate, entire. *Inflorescence* dense, spike-like or subcapitate; verticils 4—6-flowered, closely placed; bracts differing from the leaves, usually large, ovate to orbicular; flowers very shortly pedicellate. *Calyx* tubular-campanulate, subcompressed, 10-nerved, reticulate-veined, bilabiate; upper lip broad, shortly 3-toothed; lower lip narrower with 2 long subulate teeth. *Corolla* tube shortly exserted from the calyx; upper lip erect, somewhat hooded, entire; lower lip slightly shorter, deflexed, 3-lobed. *Stamens* 4, didynamous, ascending beneath the upper lip, the lower pair the longer; filaments with a subterminal tooth or claw; anthers 2-celled. *Style* bifid, lobes subulate. *Nutlets* ovoid or oblong, keeled, smooth.

Leaves in non-South African species may be dentate to pinnatifid.

About 5 species, in temperate regions of both hemispheres; 1 species widely naturalized in wet or moist places and recently recorded in South Africa. The usual pre-Linnaean spelling was *Brunella* but Linnaeus consistently spelt it *Prunella* in his Sp. Pl. (1753) and subsequent editions, though in his Gen. Pl. edn 5 (1754) he spelt it *Brunella.*

Prunella vulgaris *L.*, Sp. Pl. 600 (1753); Benth. in DC., Prodr. 12: 410 (1848); Butcher, New Illustr. Brit. Fl. 2: 329 (1961); Salisbury, Weeds and Aliens 219 (1961); Smith in Fl. Europ. 3: 162 (1972); Standley & Williams in Fieldiana Bot. 271, t.57 (1973). Type: from Europe.

Straggling herb, rooting at the nodes; stems up to 0,3 m long, glabrous or sparingly pubescent. *Leaves* petiolate; blade ovate to broadly ovate, 20—30 × 12—20 mm, sparingly pilose, apex rounded, base obtuse, margin entire; petiole 10—20 mm long. *Inflorescence* dense, 20—40 mm long, subtended by a pair of leaf-like bracts; floral bracts ovate-orbicular, about 5 mm long and 6 mm broad, abruptly acute at the apex, often purple-tinged. *Calyx* 6—7 mm long at flowering, enlarging only slightly at maturity. *Corolla* dark blue to purple, rarely white, 9—10 mm long; tube about as long as the calyx; upper lip 2,5 mm long, sparingly pubescent, lower lip about 2 mm long. Fig. 8.

Originally from Europe and now widespread; recorded from Natal midlands in wet vleis or moist forest margins.

Vouchers: *Edwards* 3083; *Moll & Mauve* 2443.

Commonly known as Self-heal or Heal-all, though this claim is no doubt overrated.

7264

9. LEONOTIS

by M. Iwarsson

Leonotis *(Pers.) R. Br.*, Prodr. Fl. Nov. Holl. 504 (1810); in Ait. f., Hort. Kew. edn 2, 3: 409 (1811); Benth., Lab. 618 (1834); Spach, Hist. Nat. 9: 210 (1840); Benth. in DC., Prodr. 12: 534 (1848); Benth. & Hook. f., Gen. Pl. 2: 1214 (1876); Briq. in Natürl. PflFam. 4, 3a: 246 (1896); Bak. in F.T.A. 5: 490 (1900); Skan in F.C. 5,1: 374 (1910); Morton in F.W.T.A. 2: 470 (1963); Launert & Schreiber in F.S.W.A. 123: 15 (1969). Lectotype species: *L. leonitis* R. Br., now included in *L. ocymifolia* (Burm. f.) Iwarsson (selected by Britton, Fl. Bermuda 324, 1918).

Phlomis L. sect. *Leonotis* Pers., Syn. Pl. 2: 127 (1807), the rank of this taxon was not indicated.

Hemisodon Raf., Fl. Tellur. 3: 88 (1837). Type species: *H. leonurus* (L.) Raf.

The lectotypes are selected here, unless otherwise indicated.

Annual or perennial robust herbs, or shrubs up to 5 m tall. *Stem* rounded at base, 4-angled and 4-grooved at apex, the upper 10−25 nodes green, without lenticels, nodes thicker and more hairy than internodes, often with prominent leaf scars. *Inflorescence* composed of 3−11 verticils per shoot, dense, spherical, axillary, many-flowered; bracts leaf-like; bracteoles linear, spinescent. *Calyx* tubular, 10-nerved, 8−10-toothed; teeth usually rigid, spinescent or rarely almost obsolete; the dorsal calyx tooth sometimes dominating, supported by three calyx veins. *Corolla* tubular, 2-lipped, white, covered by orange-coloured (rarely white) hairs; tube with 1−3 transverse fringes of hairs inside, 2−8 mm above the abscission zone; upper corolla-lip entire, almost as long as the tube, a fringe of longer hairs covers anthers and stigma; lower corolla-lip 3-lobed, soon withering and becoming patent or reflexed, lobes subequal or the middle one larger and faintly retuse. *Stamens* 4, inserted at the mouth of the corolla, didynamous, the lower pair longer; thecae 2, divaricate, subconfluent. *Disc* ventrally enlarged. *Style* not bifid, only ventral branch developed and the dorsal stigma surface sessile. *Nutlets* glabrous, oblong, 3-angled in transverse section, distally truncate and glandular.

About 10 species in Africa south of the Sahara, all bird-pollinated; 3 species (one with 3 varieties) are recognized in Southern Africa, one of which, *L. nepetifolia*, is a pantropical weed.

The generic name is derived from the Greek words "leon" and "otis", i.e. "lion's ear". This was originally a specific epithet given to one of the species by Linnaeus, certainly alluding to the morphology of the hair-fringed upper corolla lip.

1 Calyx shorter than 15 mm with 10 subequal teeth; leaves linear, length/width ratio 5:1−10:1; lower lip of corolla with (three) separate lobes, reflexed .... 1. *L. leonurus*

1 Calyx longer than 15 mm, 2-lipped, dorsal tooth supported by 3 veins, usually more than twice as long as the other teeth; leaves usually broader; lower lip of corolla with the lobes united at the base, patent:

  2 Shrub, with many branches from a thick woody base; nodes without conspicuously long hairs; corolla-tube with 1 ring of hairs inside .... 2. *L. ocymifolia*

  2 Annual or short-lived perennial herb (to 3 m high), not branched at the base; nodes usually with a tuft of long hairs; corolla-tube usually with 3 rings of hairs inside .... 3. *L. nepetifolia*

1. Leonotis leonurus *(L.) R. Br.* in Ait. f. Hort. Kew. edn 2,3: 410 (1811); Benth., Lab. 620 (1834); in E. Mey., Comm. 1: 243 (1837); Krauss in Flora 28: 66 (1845); Skan in F.C. 5,1: 375 (1910). Type: Cape of Good Hope, Herb. Linnaeus 740.19 (LINN, lecto.!).

Phlomis leonurus L., Sp. Pl. 586 (1753); Mant. 2: 412 (1767); Bergius, Descr. Pl. Cap. 151 (1767); Thunb., Prodr. 2: 95 (1800); Sims in Curtis's bot. Mag. 478 (1800); Pers., Syn. Pl. 2: 127 (1807). *Leonurus afri-canus* Mill., Dict. (1768). *Hemisodon leonurus* (L.) Raf., Fl. Tellur. 3: 88 (1837).

Leonurus grandiflorus Moench, Meth. 400 (1794). No type material known.

Leonotis leonurus (L.) R. Br. var. *albiflora* Benth. in E. Mey., Comm. 1: 243 (1837); in DC., Prodr. 12: 537 (1848); Skan in F.C. 5,1: 376 (1910). Type: Cape, Hexrivier, *Drège* s.n. (K, lecto.!; K!; S!; W!). The specimens at K have the number 4829.

Shrub 2−5 m tall, branching from a thick woody base; internodes 10−50 mm long, in the inflorescence region up to 85 mm long; cortex pale brown, densely pubescent with antrorse hairs, striate by elongated lenticels. *Leaves* petiolate; blade 50−100 × 10−20 mm, linear, acute at apex and base, serrate (15−30 teeth/leaf) in the distal half; upper surface green and rough, more densely pubescent beneath; petiole up to 10 mm long, densely pubescent with short hairs. *Inflorescence* of 3−11 compact, subspherical (flattened from below) verticils, 25−40 mm in diameter; verticil branches less than 4 mm long; bracts 40−80 × 4−9 mm; bracteoles 6−20 × 0,5−1 mm, apiculate; pedicels shorter than 4,5 mm. *Calyx* 12−16 mm long, 4 mm in diameter; calyx-teeth 10, 0,9−3 mm long, subequal, spreading. *Corolla* 40−49 mm long, covered with orange-coloured (rarely white) hairs, tube bent forward, 26−30 mm long, with 1 to 3 diffuse fringes of hairs on the inside; lobes of lower lip 4,2−7,2 mm long, the lateral ones distinctly retuse. *Lateral stamens* with orange hairs at base; fresh pollen pale yellow. *Nutlets* 4,8−6 × 1,6−1,9 mm, brown, distal surface conspicuously oblique, glandular and without distinct delimitation against the two ventral sides. Fig. 9:2.

Recorded from Transvaal, Natal and Cape. In Transvaal the species occurs at altitudes between 900 and 2 000 m, while in the southern provinces it descends to sea level. It is locally common at forest margins, often on river banks, on rocky hillsides or in tall grassland. Map 10.

Vouchers: *Galpin* 10792; *MacOwan & Bolus* 591; *Rodin* 1003; *Thorncroft* sub TRV 19188.

This species is one of the Cape plants brought to Europe at an early stage, often described and depicted, for example in Bartolin, Acta Med. & Philos. Hafn. 2: 57 (1673), Breynius, Exot. Pl. Cent. t. 86 (1678) and Hermann, Hort. Lugd.-Bat. Cat. 115 (1687). It is still grown as an ornamental in various parts of the world.

An albinistic form of the species was described as *L. leonurus* var. *albiflora* Benth., based on *Drège* s.n. (K, S, W), and is also represented by *Marloth* 7424 (BOL), *Wood* 164 (BM, K, SAM), *Ecklon & Zeyher* s.n. (S) and *Zeyher* s.n. (SAM).

2. **Leonotis ocymifolia** *(Burm. f.)* *Iwarsson,* comb. nov.

Phlomis ocymifolia Burm. f., Prodr. Fl. Cap. 16 (1768). Type: Cape of Good Hope, Herb. N. L. Burman s.n. (G, holo.!).

Shrub 1−5 m tall, branching from a thick woody base; internodes 20−80 mm long, in the inflorescence 45−325 mm long, sometimes with a few leafy nodes in between the verticils; nodes prominent; leaf scars prominent, sometimes with a marginal rim. *Leaves* petiolate; blade 9−170 × 4−85 mm, broadly ovate to ovate or obovate, apex acute to rounded, base cordate, truncate or angustate, margin crenate, upper surface green, loosely pubescent to velvety, rarely almost smooth, lower surface silvery velvety to pubescent or rarely almost smooth, except on nerves; when indumentum is sparse, the surface is covered by sessile, colourless glands; petiole 4−110 mm long. *Inflorescence* of 2 to 5 spherical to subspherical (horizontally flattened below) verticils; verticils (excluding corollas) 28−78 mm in diameter with 10−18 verticil branches 5−20 mm long, dichasially branched at base; pedicels 0,5−7 mm long; bracts leaf-like, sometimes early deciduous, 8−85 × 2−25 mm; petiole 1−25 mm long; bracteoles 6−22 × 0,3−2,5 mm, linear, green with acuminate white apex. *Calyx* 14−30 mm long, 4−5,5 mm in diameter, usually curved forwards, slightly enlarging in fruit, bilabiate or without produced lips, 8(−11)-toothed or sometimes all teeth

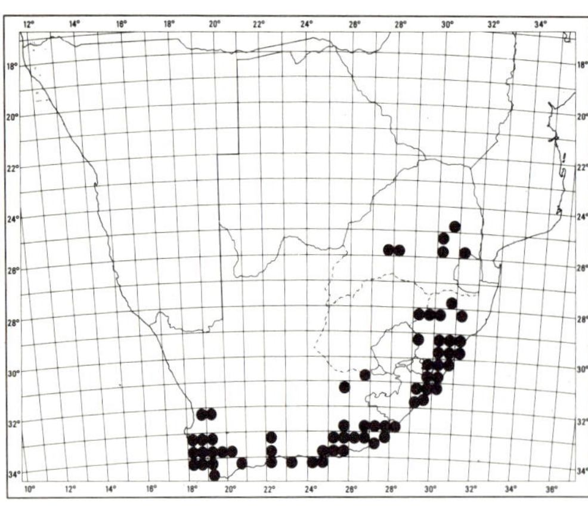

MAP 10. — **Leonotis leonurus**

FIG. 9. — 1, 1a, **Leonotis ocymifolia** var. **schinzii,** parts of flowering stem, × 1; 1b, calyx and corolla, × 1,5; 1c, section through corolla, × 1,5; 1d, mature calyx, × 1,5 (from living plant in Meyerspark). 2, **L. leonurus,** flowering stem, × 0,5; 2a, mature calyx, × 1,5 (living plant, PRE garden). 3, **L. nepetifolia,** mature calyx, × 1,5 (living plant, BRI garden).

1b
1d
1a
1c
2a
3
1
2
R. Holcroft

obsolete, shortly pubescent to velutinous; calyx teeth rigid, deltoid, with apiculate white apex, the dorsal one 2—14 mm long, the 3 or 5 lower teeth bend downwards, more or less united to a lower lip. *Corolla* 24—45 mm long, covered by orange-rufous hairs (albinistic forms are rare in Southern Africa); tube 10—25 mm long, with one distinct ring of hairs inside, lower lip 6—10 mm long, the median lobe retuse, 2,5—4,5 mm long. *Fresh pollen* orange-coloured. *Nutlets* 2,4—4,3 × 1,2—2,1 mm, blackish brown, glossy.

Widespread in Eastern and Southern Africa. An extremely variable species in which three varieties are here recognized.

1 Mature leaves typically longer than 50 mm, velvety to almost glabrous; stem sparsely branched with long internodes .... (c) var. *raineriana*

1 Mature leaves typically shorter than 50 mm, shortly pubescent to velvety; main stem apically with many leafy short-shoots, internodes generally short:

  2 Leaves orbicular to broadly ovate, truncate to cuneate-attenuate at base; petiole more than half as long as the leaf blade .... (a) var. *ocymifolia*

  2 Leaves ovate to narrowly ovate, attenuate at base; petiole less than half as long as the leaf blade .... (b) var. *schinzii*

(a) var. **ocymifolia.**

Leonotis ocymifolia (Burm. f.) Iwarsson. *Phlomis ocymifolia* Burm. f., Prodr. Fl. Cap. 16 (1768). Type: Cape of Good Hope, Herb. N. L. Burman s.n. (G, holo.!).

P. leonotis L., Mant. 1: 83 (1767); Thunb., Prodr. 96 (1800); Willd., Sp. Pl. 3: 128 (1800), as *"P. leonitis"*; Pers., Syn. Pl. 2: 127 (1807), as *"P. leonitis"*. *L. leonitis* R. Br. in Ait. f., Hort. Kew. 3: 410 (1811); Skan in F.C. 5,1: 377 (1910). *L. ovata* Spreng., Syst. Veg. 2: 744 (1825). *L. capensis* Raf., Fl. Tellur. 3: 88 (1836). Type: Cape of Good Hope, Herb. Linnaeus 740: 21 (LINN, lecto.!).

L. parvifolia Benth., Lab. 619 (1834). *L. dubia* E. Mey., Comm. 1: 242 (1837); Skan in F.C. 5,1: 380 (1910); Compton, Fl. Swaziland 493 (1976). Type: Cape of Good Hope, *Masson* in Herb. Banks (BM, lecto.!).

L. mollis Benth. in E. Mey., Comm. 1: 242 (1837); Skan in F.C. 5,1: 378 (1910). Type: Cape, Nieuweveldsbergen near Beaufort West, *Drège* 7953a (K, left specimen lecto.!).

L. hirtiflora Benth. in DC., Prodr. 12: 536 (1848). *L. leonitis* R. Br. var. *hirtiflora* (Benth.) Skan in F.C. 5,1: 378 (1910). Type: Cape, Cape Town, "Ludwigsburg" (cultivated in the garden?), *Zeyher* 206 (K. lecto.!; BM!).

Slender shrub 1—3 m tall; stems with many short-shoots (up to 6 shoots/node) apically; median shoot sometimes surviving to next season and developing a new inflorescence; lower internodes 30—50 mm long, on side-branches less than 10 mm, below the inflorescence 45—320 mm, between verticils 35—105 mm. *Leaves* petiolate; blade orbicular to broadly ovate, 9—45 × 6—30 mm, upper surface green, shortly pubescent, undersurface greyish green often more densely pubescent to almost velvety, apex usually rounded, base angustate, cuneate-angustate, truncate or cordate, margin crenate with 7—27 teeth/leaf; leaf with 2—3 side-veins on each side of the midrib, the basal ones reach the distal half of the leaf; petiole 5—25(—45) mm long. *Verticils* 28—62 mm in diameter with 10—12 verticil branches and 5—7 flowers/branch; bracts 8—30 × 4—22 mm, petiole 3—12 mm long; pedicels 0,5—2,5 mm. *Calyx* shortly pubescent, sometimes with longer spreading ("*L. hirtiflora* Benth.") hairs distally; dorsal calyx tooth 3,5—8,5 mm long, the others 0,7—3,5 mm. *Corolla* 26—37 mm long, tube 11—20 mm long with one distinct ring of hairs inside.

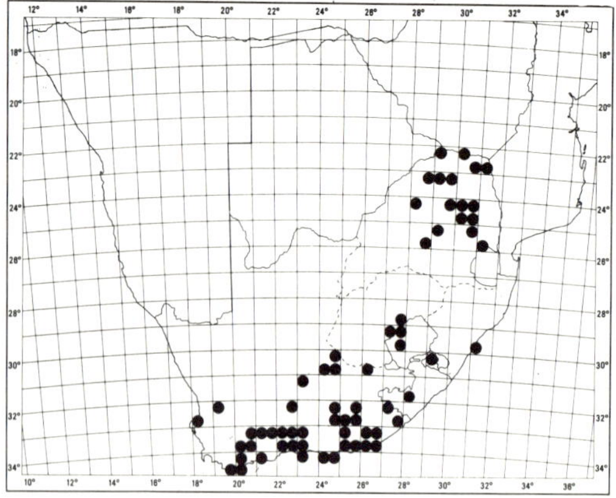

MAP 11. — **Leonotis ocymifolia** var. **ocymifolia**

Scattered in south-eastern and eastern Africa northwards to Kenya. Found on rocky outcrops and in well-drained soil on rocky hillsides at altitudes from 1 000 m to 2 000 m in Transvaal (in Tanzania to 3 000 m), descending to sea level in the Cape Province. Map. 11

Vouchers: *Acocks* 17996; *Bolus* 31131; *Bos* 957; *Codd & De Winter* 5547; *Werdermann & Oberdieck* 1007.

The circumscription of this taxon agress with that of "*L. leonitis*" in Fl. Cap. Although the leaf shape varies much, this variety exhibits a rather narrow range of variation in other characters. In Transvaal intermediates to *L. ocymifolia* var. *schinzii* are known, e.g. *Bredenkamp* 337 (PRE), *Codd* 935 (PRE) and *Leistner* 156 (B, K, PRE).

An albinistic form is represented by *Compton* 20329 (NBG).

(b) var. **schinzii** *(Gürke) Iwarsson,* comb. et stat. nov.

Leonotis schinzii Gürke in Bot. Jb. 22: 143 (1895). Type: S.W.A./ Namibia, Nomeib ("Homeib"), *Schinz* 40 (B†; Z, lecto.!).

L. randii S. Moore in J. Linn. Soc., Bot. 40: 465 (1900). Type: Zimbabwe, Bulawayo, XII, 1897, *Rand* 165 (BM, holo.!).

L. microphylla Skan in F.C. 5,1: 377 (1910). Type: Transvaal, Johannesburg, Jeppestown Ridges, 1800 m, XI, 1898, *Gilfillan* in Herb. Galpin 6169 (K, lecto.!; BOL!).

Shrub 1−2 m tall; stems slender, c. 5 mm in diameter, pubescent, sometimes also with longer spreading hairs, much branched with leafy short-shoots apically; internodes 10−50 mm long, on side-branches 10−25 mm; nodes thicker than internodes, notably at the point of branching. *Leaves* petiolate; blade 12−50 × 4−20 mm, narrowly ovate to ovate, apex acute, base attenuate, margin crenate with 3−15 teeth/leaf, upper surface green, usually sparsely pubescent, lower surface more densely pubescent, 2−5 side-veins on each side of the midrib; petiole 4−23 mm long. *Verticils* 34−78 mm in diameter; bracts 8−30 × 3−5 mm, sometimes deciduous; pedicels 1−5 mm long. *Calyx* 18−28 mm long, shortly pubescent on veins with surface glossy, notably in fruit; dorsal calyx tooth 4−9 mm long, the others 0,5−4 mm, usually faintly curved downwards. *Corolla* 32−44 mm long, tube 15−25 mm long with 1 complete ring of hairs and fragments of another distal ring inside at the base. Fig. 9:1.

Known from three almost disjunct areas around the Kalahari: S.W.A./Namibia, south-western Zimbabwe−Botswana, and Transvaal. Occurs on rocky slopes and hills, and along rivers, often in sandy soil, at an altitude of 1 000−2 000 m. Map 12.

Vouchers: *De Winter* 2612; *Rogers* 6234; *Schlieben* 7044; *Van Vuuren* 998.

In S.W.A./Namibia intermediates with *L. ocymifolia* var. *raineriana* are frequent. Additional characters

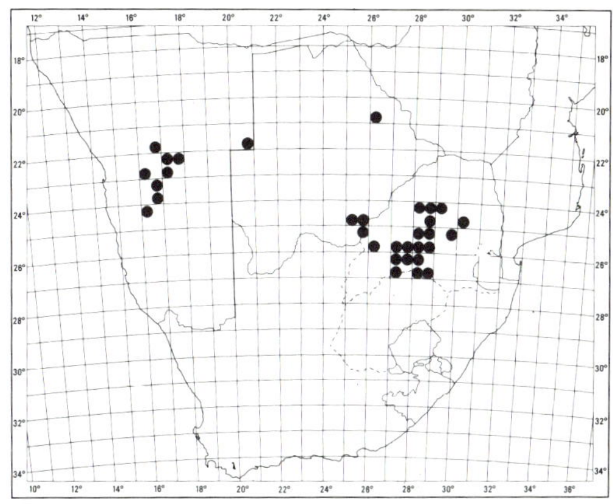

MAP 12. — **Leonotis ocymifolia** var. **schinzii**

for separating the taxa in this region are: *L. ocymifolia* var. *schinzii* has pubescent (not velvety) leaves and stems, minimal leaf length/width ratio 3:1, and maximal leaf width 20 mm, while var. *raineriana* generally has wider leaves.

(c) var. **raineriana** *(Visiani) Iwarsson,* comb. nov.

Leonotis raineriana Visiani, Orto Bot. Padova 1842: 142 (1842). *L. velutina* Fenzl ex Benth. (nom. superfl.) var. *raineriana* (Visiani) Benth. in DC., Prodr. 12: 535 (1848). Type: ex hort., seeds from *Kotschy* 519, Sudan, near Camamil and Kassan, Tumat (no cultivated material seen, not in PAD; K, lecto.!; BM!; FI!; FI-W!; K!; M!; P!; W!).

L. intermedia Lindl., Bot. Reg. t. 850 (1824); Skan in F.C. 5,1: 381 (1910). Type: ex hort., seeds from *Forbes* s.n., Cape, Algoa Bay (erroneously Delagoa Bay in Bot. Reg.) (CGE, holo.!; BM!; BR!; G-DC!; K!).

L. dysophylla Benth. in E. Mey., Comm. 1: 242 (1837); Skan in F.C. 5,1: 380 (1910); Prain in Curtis's bot. Mag. t.8404 (1911); Launert & Schreiber in F.S.W.A. 123: 15 (1969); Compton, Fl. Swaziland 493 (1976). Type: between "Omsamwubo and Omcomas", *Drège* 4832a (K, lecto.!).

L. laxifolia MacOwan in Kew. Bull. 1893: 13 (1893); Skan in F.C. 5,1: 381 (1910); Ross, Fl. Natal 303 (1972); Compton, Fl. Swaziland 494 (1976). Type: Cape, Malowe, *Tyson* 2766, in Herb. A.A. 1300 (GRA, holo.!; BM!; BOL!; K!; SAM!; UPS!; W!; Z!).

L. malacophylla Gürke in Bot. Jb. 22: 142 (1895). Type: Natal, Clydesdale, *Tyson* 2729, in Herb. A.A. 1508 (K, lecto.!; SAM!; UPS!; W!; Z!).

L. bachmannii Gürke in Bot. Jb. 22: 143 (1895); Skan in F.C. 5,1: 382 (1910). Type: Transvaal, Barberton, *Galpin* 922 (Z, lecto.!).

L. latifolia Gürke in Bot. Jb. 22: 143 (1895); Skan in F.C. 5,1: 379 (1910); Ross, Fl. Natal 303 (1972);

Compton, Fl. Swaziland 493 (1976). Type: Natal, Biggarsberge, *Rehmann* 7057 (Z, lecto.!).

L. laxifolia MacOwan f. *pilosa* Gürke in Bot. Jb. 22: 144 (1895); Skan in F.C. 5,1: 382 (1910). Type: Natal, Karkloof, *Rehmann* 7374 (Z, lecto.!).

L. dinteri Briq. in Bull. Herb. Boissier sér. 2,3: 1090 (1903). Type: S.W.A./Namibia, Hereroland, near Okahandja, Tabakstuin, *Dinter* 249 (Z!).

L. urticifolia Briq. in Bull. Herb. Boissier sér. 2, 3: 1091 (1903). Type: Natal, *Cooper* 1182 (as *Cooper* 1152 in Briquet, 1903) (BM!; K!; W!; Z!).

L. hereroensis Briq. in Bull. Herb. Boissier sér. 2,3: 1092 (1903). Type: S.W.A./Namibia, Hereroland, *Nels* s.n. (Z!).

L. brevipes Skan in F.C. 5,1: 378 (1910). Type: Transvaal, Zoutpansberg, Medingen, *Burtt-Davy* 2657 (K, holo.!).

L. galpinii Skan in F.C. 5,1: 379 (1910). Type: Cape, near Queenstown, *Galpin* 1825 (K, holo.!; Z!, partly, as indicated by me).

L. westae Skan in F.C. 5,1: 382 (1910). Type: Cape, Port Elizabeth, *E. West* 75 (K, holo.!).

L. mollis Benth. var *albiflora* Skan in F.C. 5,1: 378 (1910). Type: Cape, Boschberg, *MacOwan* s.n. (K, lecto.!).

L. intermedia Lindl. var. *natalensis* Skan in F.C. 5,1: 381 (1910). Type: Natal, near Durban, *Peddie* s.n. in Hb. Harvey (K, lecto.!). (N.B. Peddie was never in Natal; the specimen was probably collected by Williamson, one of the men in Col. Peddie's regiment, and who was included in a small detachment of the regiment which occupied Port Natal during 1838—39).

Shrub 1—5 m tall; stem not much branched, leafy, non-flowering internodes sometimes present between verticils, internodes 20—80(—110) mm long, velutinous, pilose or patently pubescent, sometimes almost smooth. *Leaves* petiolate; blade (40—)50—170 × 20—85 mm, broadly ovate to obovate, apex rounded or acute to acuminate, margin crenate with 21—65 teeth/leaf, upper surface usually green with sessile glands, loosely pubescent to densely velvety, rarely almost smooth, lower surface white silvery to yellowish velvety, pubescent or rarely almost smooth, if so then generally densely covered with sessile glands, 5—8 side-veins on each side of midrib; petiole 30—110 mm long. *Verticils* 28—67 mm in diameter, loose to compact, with 10—18 verticil branches, 6—12 mm long and with 5—19 flowers/branch; pedicels 0,5—7 mm long. *Calyx* 14—30 mm long, 4—5,5 mm in diameter, curved forwards, shortly pubescent to velvety, bilabiate or without lips, 8(—11)-toothed sometimes all teeth obso-

lete; dorsal calyx tooth 2—14 mm long usually supported by 3 veins, the 3 or 5 lower teeth usually more or less united into a lip 0,5—3 mm long, teeth curved downwards, 0,5—5,5 mm long. *Corolla* 24—45 mm long, orange-rufous (rarely cream-buff), tube 10—25 mm long with one distinct ring of hairs inside.

This variety is widespread in eastern and Southern Africa. Map 13.

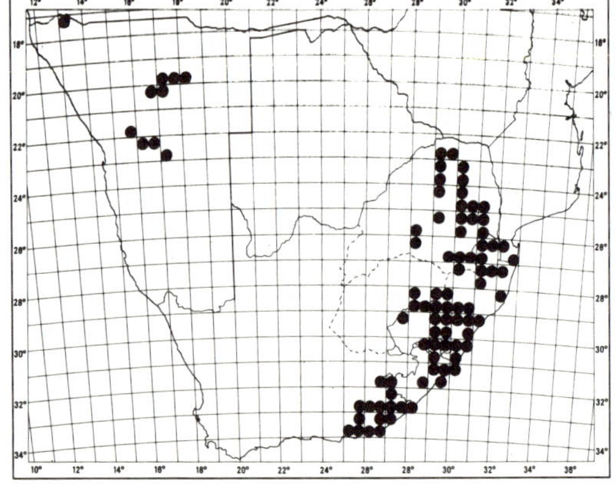

MAP 13. — **Leonotis ocymifolia** var. **raineriana**

Vouchers: *Compton* 29994; *Dinter* 5449; *Galpin* 10658; *Meeuse* 10208; *Scheepers* 1563; *Schlechter* 2847.

Var. *ocymifolia* and var. *raineriana* are morphologically distinct and behave as species in East Africa and southwards to Zimbabwe. In Southern Africa this pattern is confused and intermediates in all the separating characters occur. In Natal and eastern Cape deviating populations have been discerned under the name *L. laxifolia*. The latter, when typical, have large (c. 100 mm wide) thin leaves with long petioles and conspicuously lax verticils. In the material some deviating specimens should be noted. *Devenish* 634 might represent a new taxon, closely related to var. *raineriana*. The leaves are similar to those of var. *ocymifolia*: petiole c. 25 mm long, leaf length/width 35/25 mm; the inflorescence is unusually lax with verticil branches c. 8 mm long and pedicels 4—5 mm long, and the calyces are long (23—25 mm) and green. Another extreme form recorded from two localities in the Cape is represented by *Lewis* 67480, *Bayliss* 7414 and *Barker* 7880. Features in common for these collections are: extremely long petioles (1—2 times leaf length) general leaf shape as in var. *ocymifolia*, short (c. 50 mm long) internodes between the verticils, and bracts 5—20 mm long and 1—4 mm wide.

An albinistic form is represented by *Jacot Guillarmod* 895 (PRE), *MacOwan* s.n. (K) and *Stewart* 182 (SAM).

3. **Leonotis nepetifolia** *(L.) R. Br.* in Ait. f., Hort. Kew. edn 2, 3: 409: (1811); Ker-Gawler in Bot. Reg. t. 281 (1818); Benth., Lab. 618 (1834); in DC., Prodr. 12: 535 (1848); Bak. in F.T.A. 5: 491 (1900); Launert & Schreiber in F.S.W.A. 123: 16 (1969). Type: plate p. 117 in P. Hermann, Horti Academi Lugduno-Batavi Catalogus (1687), figured from a plant raised from seeds originally from Surinam (lecto!; the material in Herb. Linnaeus 740:17 and in S are post-1753 collections).

Phlomis nepetifolia L., Sp. Pl. 586 (1753); Willd., Sp. Pl. 3: 128 (1800); Pers., Syn. Pl. 2: 127 (1807).

L. kwebensis N.E. Br. in Kew Bull. 1909: 132 (1909). Type: Botswana, Kwebe Hills, *Lugard* 222 (K, holo.!).

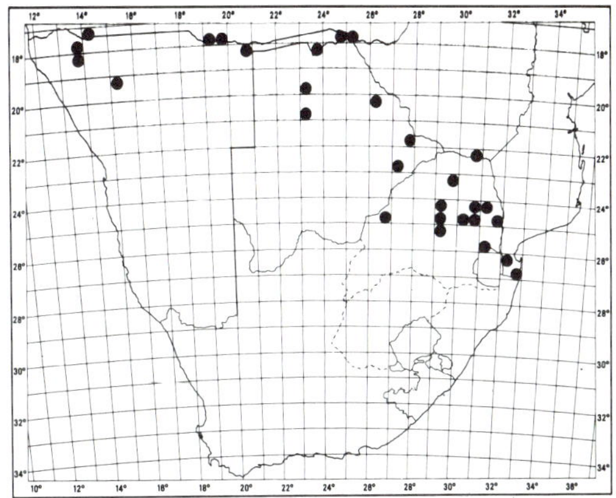

MAP 14. — **Leonotis nepetifolia**

Annual or short-lived perennial slender herb, 1−3 m tall, with easily uprooted c. 100 mm long taproot; stem branched at upper nodes only, green, shortly pubescent, deeply 4-furrowed apically; upper nodes with a tuft of 1−4 mm long hairs; internodes 20−150 mm long, in the inflorescence 70−280 mm. *Leaves* petiolate; blade broadly ovate, 50−200 × 40−150 mm, apex acuminate, base cuneate-attenuate to cordate-attenuate, margin deeply crenate with 25−51 teeth/leaf, green, pubescent and with colourless glands; petiole 30−80 mm long. *Inflorescence* of 2−5 spherical to cylindrical verticils 38−65 mm in diameter, with 20−28 verticil branches 2−16 mm long; pedicels 0,5−2 mm long; bracts 25−90 × 4−13 mm, linear; petiole 12−60 mm long; bracteoles stout, linear, 7−15 × 0,5−1,5 mm, acuminate. *Calyx* 17−25 mm long, 4−5,6 mm in diameter, bilabiate, bent forwards, shortly pubescent, basal half stiff, yellowish white, distal half flexible, green; calyx teeth straight, stiff, apiculate, the dorsal one 4−7 mm long, the others 1−3 mm long. *Corolla* 19−38 mm long, tube 9−20 mm long with three distinct rings of hairs inside the tube. *Fresh pollen* pale yellow. *Nutlets* 2,9−4,3 × 1,1−1,9 mm, surface not glossy, marmorated in grey and brown. Fig. 9:3.

Pantropical weed. In Africa it is recorded from Sierra Leone to Kenya and southwards to S.W.A./Namibia, Botswana, Transvaal and northern Natal. Often found at roadsides and in abandoned cultivations at altitudes of up to 2 000 m. In Southern Africa usually found at altitudes of 1 000−1 500 m. Map 14.

Vouchers: *Codd* 5238; *De Winter & Leistner* 5283; *Rodin* 3992.

In the Flora area the species is represented only by var. *nepetifolia* with orange-coloured corolla. Var. *africana* (P. Beauv.) J. K. Morton with yellow corolla is known from West Africa to Ethiopia. Two specimens with a somewhat intermediate corolla colour have been noted from S.W.A./Namibia: *Merxmüller & Giess* 1987; 30373 (M).

7268   **10. LEUCAS**

Leucas *Burm. ex R. Br.*, Prodr. Fl. Nov. Holl. 504 (1810); Benth. in DC., Prodr. 12: 523 (1848); Benth. & Hook. f., Gen. Pl. 2,2: 1213 (1876); Gürke in Bot. Jb. 22: 129 (1895); Briq. in Natürl. PflFam. 4,3a: 250 (1896); Bak. in F.T.A. 5: 472 (1900); Skan in F.C. 5,1: 369 (1910); Launert & Schreiber in F.S.W.A. 123: 16 (1969); R.A. Dyer, Gen. 1: 527 (1975); Sebald in Stuttgarter Beitr. Naturk. A, 308: 1−42 (1978); A,341: 1−200 (1980). Type species: *L. flaccida* R. Br.

Lasiocorys Benth., Lab. 600 (1834); in DC., Prodr. 12: 534 (1848); Benth. & Hook. f., Gen. Pl. 2,2: 1213 (1876); Bak. in F.T.A. 5: 469 (1900); Skan in F.C. 5,1: 372 (1910); Phill., Gen. edn 2: 645 (1951). Type: *L. capensis* Benth., fide Phillips, l.c.

Annual or perennial herbs or subshrubs. *Leaves* entire or toothed, thin or thick-textured. *Inflorescence* usually simple, of few to many spaced or fairly crowded verticils; verticils 2−many-flowered, often in glomerate clusters; bracts leaf-like, often smaller towards the apex of the inflorescence; bracteoles linear, ascending, small or conspicuous. *Calyx* shorter or longer than the corolla tube, tubular or tubular-campanulate, rarely inflated, 10-nerved, straight or curved, sometimes oblique at the mouth; teeth 5−10, equal or unequal, often spine-tipped. *Corolla* bilabiate, white (in Southern Africa); tube tubular, widening above, sometimes shortly constricted at the base, annular-pilose or annular-papillose within, rarely glabrous; upper lip ascending or spreading, concave or flattish, entire or rarely emarginate or 2-lobed, longer than, subequal to or shorter than the lower lip, usually with stiff brush-like hairs; lower lip spreading or deflexed, 3-lobed, the median lobe the largest. *Stamens* 4, didynamous, the lower pair longer than the upper, ascending under the upper lip, included or shortly exserted; anthers approximate in pairs, 2-celled with the cells divaricate and finally confluent. *Style* terete, ascending under the upper lip, unequally lobed at the apex. *Nutlets* ovoid-triquetrous, smooth, obtuse or somewhat flattened at the apex.

Described species over 160, found mainly in tropical Africa and Asia, extending to Australia, and 1 species a world-wide weed; 8 species in Southern Africa.

Bentham separated *Lasiocorys* from *Leucas* mainly on the basis of the 5-toothed calyx, while other workers claim that the corolla tube is shortly narrowed at the base with a shortly stipitate ovary. All these characters break down and there is no reliable way of separating the two genera.

As in *Acrotome* (no. 5), plants of different habit are included in *Leucas:* erect annuals with subglobose, many-flowered verticils, and annual or perennial plants with few-flowered verticils. The two genera may be distinguished by the stamens: in *Acrotome* they are included in the corolla tube and held together by intermingling hairs, whereas in *Leucas* they reach the mouth of the tube or, more often, ascend in the upper lip; also, in *Leucas* the upper lip of the corolla is often longer than the lower and is usually supplied with stiff brush-like hairs, which is not the case in *Acrotome.*

1 Verticils many-flowered, glomerate, distantly spaced, usually about 20 mm or more in diameter (sometimes less than 10-flowered in *L. ebracteata,* but then verticils 25−50 mm apart and leaves grey-green); bracteoles linear, 5−12 mm long:

  2 Leaves linear, 40−60 × 3−6 mm .... 1. *L. lavandulifolia*

  2 Leaves ovate to ovate-lanceolate, usually wider than 10 mm:

    3 Calyx tube abruptly curved near the apex .... 2. *L. martinicensis*

    3 Calyx tube straight:

      4 Calyx teeth 7−8, spreading; corolla about 6 mm long .... 3. *L. ebracteata* var. *kaokoveldensis*

      4 Calyx teeth 6, not spreading; corolla about 15 mm long.. 4. *L. sexdentata*

1 Verticils 2−12-flowered, usually closely placed (up to 30 mm apart in *L. glabrata*); bracteoles apparently absent or minute, 1−2 mm long;

FIG. 10. — 1, **Leucas martinicensis,** upper part of flowering stem, × 1; a, base of plant, × 1; b, section through part of corolla, × 3; c, flowering calyx, × 3; d, mature calyx, × 3; e, nutlet × 10 (after Henderson & Anderson, Mem. bot. Surv. S. Afr. 37: 263, t.130, 1966).

5 Calyx clothed in long white villous hairs .... 6. *L. pechuelii*

5 Calyx glabrous to hispid:

  6 Calyx teeth 10:

    7 Corolla 12—15 mm long:

      8 Leaves ovate to ovate-lanceolate .... 5(a). *L. glabrata* var. *glabrata*

      8 Leaves linear .... 5(b). *L. glabrata* var. *linearis*

    7 Corolla 6—7 mm long.. 7. *L. neuflizeana*

  6 Calyx teeth 5 (occasionally with 1 or 2 small intermediate teeth) .... 8. *L. capensis*

1. **Leucas lavandulifolia** *Sm.* in Rees, Cyclop. 20,2 (1812); Ross, Fl. Natal 303 (1972); Sebald in Stuttgarter Beitr. Naturk. A,341: 188 (1980). Type: India (LINN 739.8, holo.).

Leonurus indicus L., Sp. Pl. edn 2: 817 (1763); Burm. f., Fl. Ind. 127 (1768). *Leucas indica* (L.) Vatke in Öst. bot. Z. 25: 95 (1875), nom. illegit., non *L. indica* (L.) Sm. (1812). Type: as above.

Phlomis linifolia Roth, Nov. Pl. Sp. 260 (1821). *Leucas linifolia* (Roth) Spreng., Syst. Veg. 2: 743 (1825); Benth. in DC., Prodr. 12: 533 (1848); Hook. f., Fl. Brit. Ind. 4: 690 (1885). Type: India, *Heyne* (B).

Annual, erect, branched herb 0,6—1 m tall; stems somewhat woody below, tomentulose. *Leaves* shortly petiolate; blade linear, 40—60 × 3—6 mm, minutely tomentulose, apex acuminate, base attenuate, margin entire or with a few distant minute teeth. *Inflorescence* of 1—few spaced verticils; verticils many-flowered, glomerate, 15—25 mm in diam.; bracteoles numerous,

5—10 mm long. *Calyx* tubular-obconical, minutely tomentulose, 7—8 mm long, mouth very oblique, 8-toothed, produced on the upper side into a conspicuous deltoid-ovate spine-tipped tooth, 2 mm long; remaining 7 teeth minute. *Corolla* 15—16 mm long; tube 5—6 mm long, annular-papillose just below the middle; upper lip horizontal, 5 mm long, hooded, with a dense fringe of stiff white brush-like hairs; lower lip 10 mm long. Fig. 11:5.

A native of India, recently naturalized in and around Durban and as far north as Empangeni; first recorded in 1960. Map 15.

Vouchers: *Strey* 4862; *Ward* 4793.

Readily distinguished from other species by the long, linear leaves and the obconical, oblique calyx, produced in the upper part to a single large tooth, with 6 or 7 small lateral teeth.

2. **Leucas martinicensis** *(Jacq.) R. Br.* in Ait. f., Hort. Kew. edn 2,3: 409 (1811); Benth. in E. Mey., Comm. 242 (1838); in DC., Prodr. 12: 533 (1848); Hook. f., Fl. Brit. Ind. 4: 688 (1885); Bak. in F.T.A. 5: 479 (1900); Skan in F.C. 5,1: 371 (1910); Henderson & Anderson, Common Weeds S. Afr. 262, t.130 (1966); Launert & Schreiber in F.S.W.A. 123: 18 (1969); Ross, Fl. Natal 303 (1972); Compton, Fl. Swaziland 494 (1976); Sebald in Stuttgarter Beitr. Naturk. A,341: 179 (1980). Type: from West Indies.

Clinopodium martinicense Jacq., Enum. Pl. Carib. 25 (1760). *Phlomis martinicensis* (Jacq.) Swartz, Prodr. Veg. Ind. Occ. 88 (1788).

P. caribaea Jacq., Ic. Pl. Rar. 1: 11, t.110 (1785?). Type: from West Indies.

Annual, erect herb 0,15—1,2 m tall; stems simple or sparingly branched, finely tomentulose. *Leaves* petiolate; blade ovate to ovate-lanceolate, 25—80 × 12—45 mm, tomentulose, apex long-acute, base cuneate to obtuse, margin coarsely crenate-serrate;

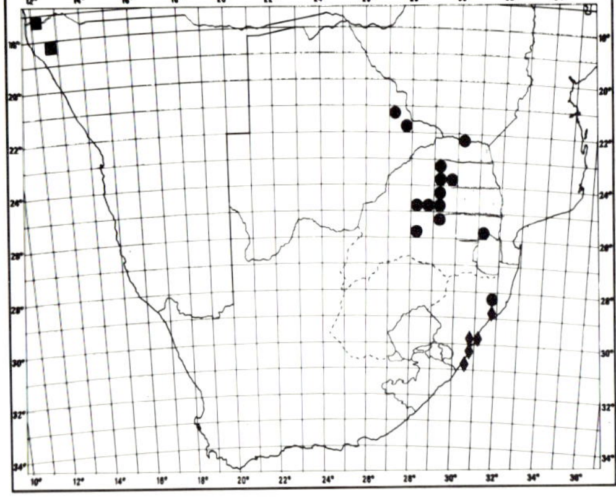

MAP 15. — ♦ **Leucas lavandulifolia**
    ■ **L. ebracteata** var. **kaokoveldensis**
    ● **L. neuflizeana**

petiole 5−20 mm long. *Inflorescence* of few to several spaced verticils; verticils many-flowered, glomerate, 20−25 mm in diam.; bracteoles numerous, 6−11 mm long. *Calyx* tubular, hispid, abruptly curved near the apex, somewhat inflated near the base, about 7−8 mm long at flowering stage, enlarging to 15 mm long in fruit, oblique at the mouth, 10-toothed, the upper tooth the longest, lanceolate-subulate, 2,5−3 mm long, the remaining 9 teeth subequal, deltoid-subulate, 1 mm long. *Corolla* 6 mm long; tube 4 mm long, exannulate or imperfectly annular-papillose; upper lip 2 mm long, lacking a fringe of stiff hairs; lower lip 2−2,5 mm long. Fig. 10.

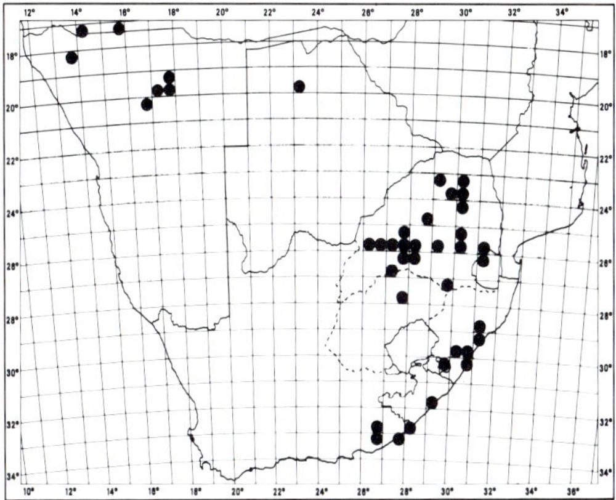

MAP 16. — **Leucas martinicensis**

A weed of cultivated land and disturbed places in the warmer parts of all four provinces, S.W.A./Namibia, Botswana and Swaziland; absent from Lesotho and the western Cape Province; indigenous in South America, the West Indies and, possibly, Africa, now world-wide. Map 16.

Vouchers: *Compton* 25109; *Dinter* 7468; *Schlieben & Strey* 8275; *Tyson* 1156.

Sometimes confused with *Acrotome inflata* (p. 4: 19) but may be recognized by the shape of the calyx tube, which is curved near the apex, and by the long, subulate upper calyx tooth.

3. **Leucas ebracteata** *Peyr.* in Sber. Akad. Wiss. Wien 38: 577 (1860). Type: Angola, Benguella, *Wawra* 292 (W, holo.).

var. **kaokoveldensis** *Sebald* in Stuttgarter Beitr. Naturk. A,341: 141 (1980). Type: S.W.A./Namibia, Kaokoveld Reserve, *Kers* 1748 (M, holo.).

L. *ebracteata* sensu Launert & Schreiber in F.S.W.A. 123: 18 (1969).

Annual, erect herb 0,3−0,8 m tall, unbranched or sparingly branched near the base; stems pilose. *Leaves* petiolate; blade grey-green, ovate-elliptical to broadly ovate, 25−50 × 13−30 mm, tomentose, apex obtuse to rounded, base obtuse, margin somewhat obscurely crenate-serrate. *Inflorescence* of several spaced verticils, often occupying almost the whole length of the stem; verticils (8−) many-flowered, usually glomerate, about 20 mm in diam.; bracteoles few, 5−6 mm long. *Calyx* tubular, pubescent, 6 mm long at flowering, wider and oblique at the mouth, 8-toothed, produced below and more or less 2-lipped; teeth somewhat spreading, the upper lip of 5 short deltoid-subulate teeth about 1−1,5 mm long; lower lip about 3,5 mm long of 3 teeth, narrowly-deltoid, spine-tipped, 2 mm long. *Corolla* 6−9 mm long, annular-papillose about the middle; upper lip spreading, concave, 2−2,5 mm long with a short fringe of hairs; lower lip 2,5−3 mm long. Fig. 11:6

An annual weed of waste places and water courses in south-western Angola and northern S.W.A./Namibia. Map 15.

Voucher: *De Winter & Leistner* 5781.

In the typical variety the plants are more robust with longer calyx and corolla, forming glomerate verticils about 30 mm in diameter, while the bracteoles are minute, 1−3 mm long.

4. **Leucas sexdentata** *Skan* in F.C. 5,1: 371 (1910); Sebald in Stuttgarter Beitr. Naturk. A,341: 141 (1980). Type: Transvaal, probably Marico District, *Holub* s.n. (K, holo.; PRE, fragment!).

Annual, erect herb, 0,15−0,6 m tall, usually with a few spreading branches near the base; stems whitish, pilose. *Leaves* petiolate; blade ovate, 14−40 × 10−25 mm, tomentose, apex rounded, base obtuse, margin somewhat coarsely crenate except in the lower third; petiole 5−15 mm long. *Inflorescence* of 1−4 spaced verticils; verticils many-flowered, glomerate, 20−40 mm

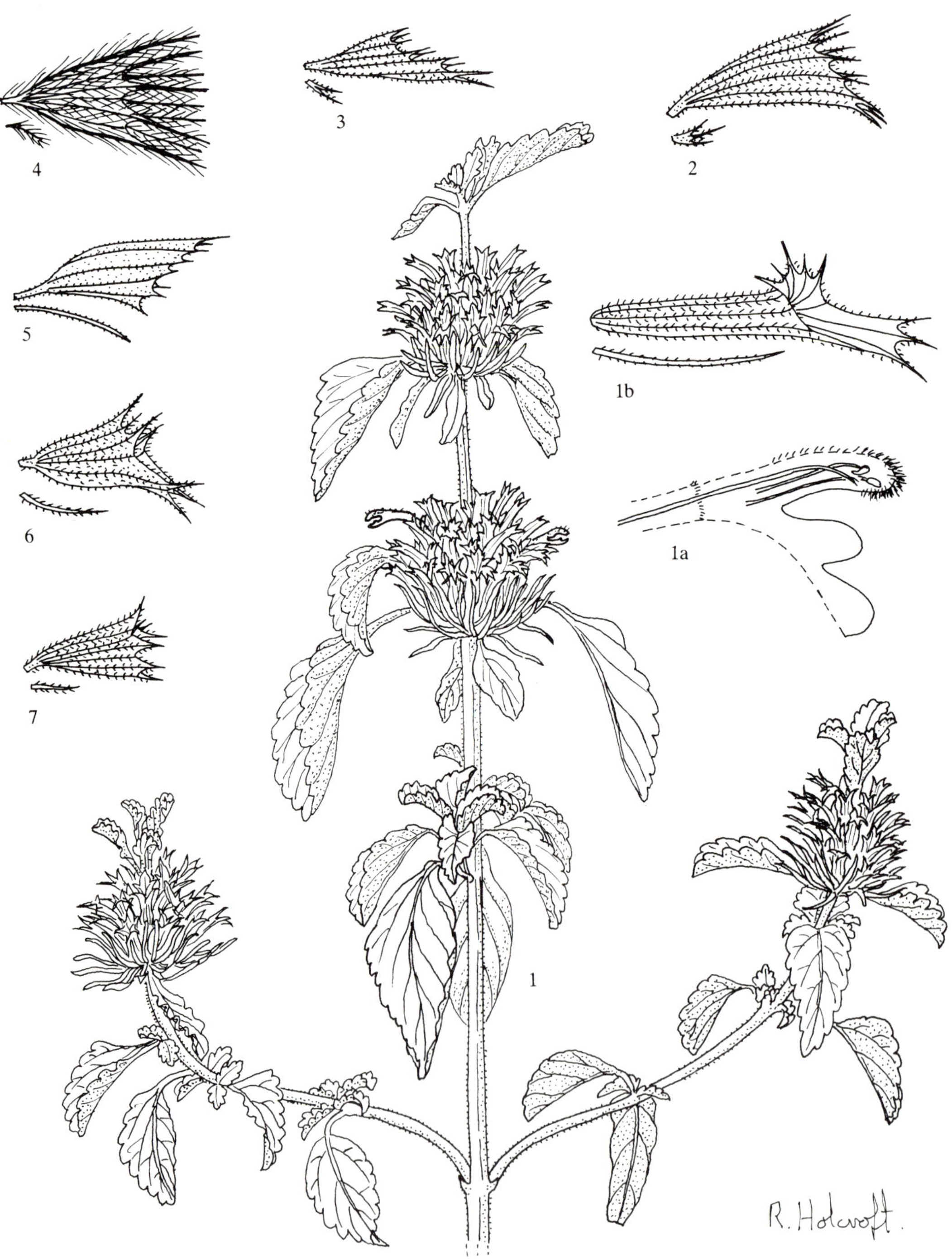
4
3
2
5
1b
6
1a
7
1
R. Holcroft.

in diam.; bracteoles numerous, 8—14 mm long. *Calyx* tubular, densely pubescent, 11—14 mm long at flowering, enlarging slightly at maturity, oblique and somewhat bilabiate at the mouth, 6-toothed; upper lip 3-toothed, 2—3 mm long, median tooth the largest, ovate-lanceolate, lateral teeth narrowly deltoid, setaceous; lower lip 4—5 mm long, 3-toothed, teeth ovate-deltoid, spine-tipped, the median tooth narrower and shorter than the lateral pair. *Corolla* 18—20 mm long; tube 10—11 mm long, annular-pilose about the middle; upper lip horizontal, 7—8 mm long, with a dense fringe of bristle-like hairs on the apical part; lower lip 10—11 mm long. Fig. 11:1

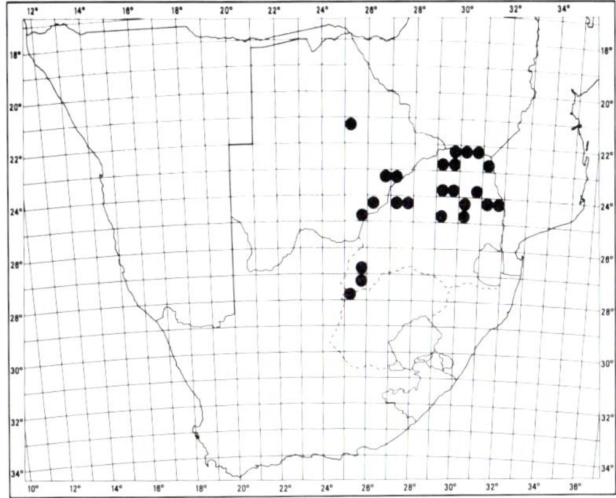

Ḿap 17. — **Leucas sexdentata**

In semi-arid grassland and open woodland, usually on sandy soil, often locally common in disturbed places and under trees, in eastern, northern, central and western Transvaal, extending into the adjoining parts of Botswana and Zimbabwe. Map 17.

Vouchers: *Codd* 4045; *Galpin* 12167.

Among the species with large, glomerate, many-flowered verticils, *L. sexdentata* is characterized by the 6-toothed calyx and the relatively large corolla nearly 20 mm long.

5. **Leucas glabrata** *(Vahl) Sm.* in Rees, Cyclop. 20,2 (1812). Type: Arabia, *Forsskål* (C, holo.).

Perennial (rarely annual) herb or soft shrublet 0,25—0,8 m tall, rarely scandent up to 1,5 m, branched from the base; stems spreading to suberect, sparingly branched, glabrous to pilose. *Leaves* shortly petiolate; blade ovate to ovate-lanceolate, or linear (see vars.), glabrous to pilose, apex obtuse to subacute, base obtuse to truncate or tapering, margin coarsely few-toothed, rarely entire; petiole up to 15 mm long. *Inflorescence* of several verticils, spaced below, somewhat crowded towards the apex; verticils 2—10 (—12)-flowered; bracteoles small, setaceous, 1—2 mm long, usually 2 or 3 arising from a common base and often persisting after the flowers are shed. *Calyx* subglabrous to hispid, 7—9 mm long, somewhat oblique at the mouth, 10-toothed; teeth more or less subequal, lanceolate-deltoid, 1,5—2 mm long, the lower 3 forming a protruding lower lip. *Corolla* 12—15 mm long; tube 6—7 mm long, annular-papillose about the middle; upper lip horizontal, 7—8 mm long, with a fringe of bristle-hairs on the apical part; lower lip 5—7 mm long. *Anthers* orange or red.

Widespread from the Arabian Peninsula through tropical East Africa to the warmer parts of S.W.A./Namibia and Botswana, the northern and eastern Transvaal, Swaziland, the valley bushveld of Natal, and the eastern Cape Province, usually in woodland and among rocks in grassy places.

Two varieties are recognized in Southern Africa (see key to species). In addition, Sebald in Stuttgarter Beitr. Naturk. A, 341: 101 (1980) maintains an annual form of restricted distribution in Somalia and Kenya as var. *chiatelliana* (Chiov.) Sebald.

(a) var. **glabrata.**

Sebald in Stuttgarter Beitr. Naturk. A, 341: 95 (1980).

Phlomis glabrata Vahl, Symb. Bot. 1: 42 (1790).

Leucas glabrata (Vahl) Sm. in Rees, Cyclop. 20,2 (1812); Benth. in DC., Prodr. 12: 524 (1848); Bak. in F.T.A. 5: 482 (1900); Skan in F.C. 5,1: 370 (1910); Launert & Schreiber in F.S.W.A. 123: 18 (1969); Ross, Fl. Natal 303 (1972); Compton, Fl. Swaziland 494 (1976); Sebald, l.c. 93 (1980).

FIG. 11. — 1, **Leucas sexdentata**, upper part of plant, × 1; 1a, section through part of corolla, × 3; 1b, mature calyx and bracteole, × 3. 2 – 7, mature calyces and bracteoles of the following species: 2, **L. glabrata** var. **glabrata**; 3, **L. neuflizeana**; 4, **L. pechuelii**; 5, **L. lavandulifolia**; 6, **L. ebracteata** var. **kaokoveldensis**; 7, **L. capensis**; all × 3.

L. natalensis Sond. in Linnaea 23: 85 (1850). Type: Port Natal, *Gueinzius* 363 (S, holo.).

L. junodii Briq. in Annu. Conserv. Jard. bot. Genève 2: 109 (1898). Type: Mozambique, Rikatla, *Junod* 92 (G, holo.).

L. dinteri Briq. in Bull. Herb. Boissier sér. 2,3: 1088 (1903). Type: S.W.A./Namibia, Quassiputs, *Dinter* 200 (Z).

Stems 0,25—0,8 m long, spreading to suberect, rarely scandent up to 1,5 m. *Leaves* petiolate; blade ovate to ovate-lanceolate, (15—)20—65(—80) × 10—25 (—30) mm, usually pilose, apex obtuse to subacute, base obtuse to truncate, margin coarsely few-toothed, rarely almost entire; petiole up to 15 mm long. *Inflorescence* of several verticils, spaced below, somewhat crowded towards the apex; verticils 2—10(—12)-flowered. Fig. 11:2.

Distribution as for the species. Map 18.

Vouchers: *Acocks* 13480; *Codd & De Winter* 5114; *Compton* 27022; *Merxmüller & Giess* 30047.

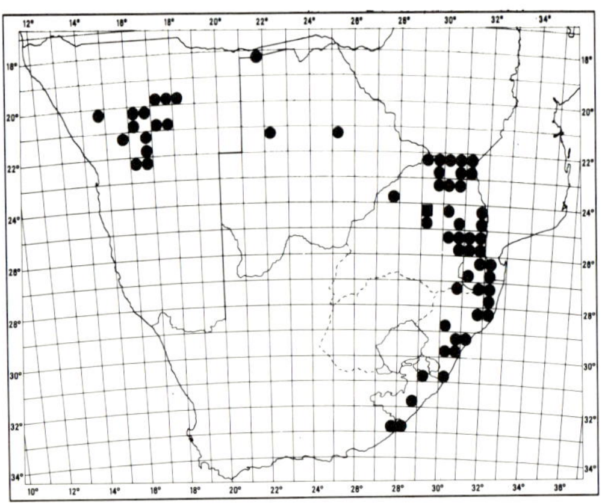

MAP 18. — ● *Leucas glabrata* var. **glabrata**
■ **L. glabrata** var. **linearis**

There is a good deal of variation in pubescence and certain extreme specimens with markedly hirsute stems are found in the Transvaal, e.g. *De Winter* 2219 and *Pott* 5692. However, the floral characters are uniform and there are intermediates in degree of pubescence. It is grazed by game and domestic livestock and is suspected of causing a taint in milk.

See Sebald, l.c. for full synonymy.

(b) var. **linearis** *Codd*, var. nov., a typo foliis linearibus integris differt.

Type: Transvaal, 10 km from Potgietersrus on road to Pietersburg, *Germishuizen* 1360 (PRE, holo.).

Short-lived perennial herb; stems slender, ascending to erect, 4-angled, 0,3—0,5 m long, subglabrous. *Leaves* shortly petiolate to subsessile; blade linear, 20—30 × 1,5—2 mm, glabrous, tapering at the base, margin entire, petiole up to 2 mm long. *Inflorescence* of 3—5 verticils, spaced below; verticils 3—8-flowered; bracteoles, calyx and corolla as in var. *glabrata*.

Known only from the type gathering, in open savanna on rocky soil. Map 18.

Although the leaves of var. *glabrata* are variable in shape and size, there is no indication that they approach the linear leaves of var. *linearis*.

6. **Leucas pechuelii** *(Kuntze) Gürke* in Bot. Jb. 22: 135 (1895); Bak. in F.T.A. 5: 477 (1900); Launert & Schreiber in F.S.W.A. 123: 19 (1969). Type: S.W.A./Namibia, Hereroland, *Pechuel-Loesche* (B†).

Lasiocorys pechuelii Kuntze in Jb. K. bot. Gart. Mus. Berl. 4: 271 (1886).

Leucas altissima Engl. in Bot. Jb. 10: 268 (1888); Hiern, Cat. Afr. Pl. Welw. 1,4: 878 (1900); Bak. in F.T.A. 5: 478 (1900). Type: S.W.A./Namibia, near Otjimbingwe, *Marloth* 1410 (PRE!).

Perennial, erect, branched shrub, 0,25—1 m tall. *Leaves* sessile or shortly petiolate; blade grey-green, ovate to ovate-oblong or obovate-oblong, 15—40 × 10—15 mm, pilose, apex rounded, base cuneate, margin subentire or with a few small teeth in the upper third. *Inflorescence* 100—200 mm long, of several to many fairly crowded verticils; verticils 3—10-flowered; bracteoles about 3 mm long, densely villous. *Calyx* densely villous, 7—9 mm long, symmetrical at the mouth, 10-toothed, the alternate ones shorter; teeth lanceolate-deltoid, spine-tipped, the longer 3 mm long, the shorter 1,5—2 mm long. *Corolla* 11—12 mm long; tube 5—6 mm long, annular-papillose about the middle, shortly constricted near the base; upper lip horizontal, 6 mm long with a dense fringe of stiff brush-like white hairs; lower lip 6 mm long. *Anthers* red. Fig. 11: 4.

Found in dry watercourses, stony hillsides and sandy places in S.W.A./Namibia and Angola, sometimes locally common. Map 19.

Vouchers: *De Winter & Leistner* 5188; *Leach & Bayliss* 12930.

Easily recognized among South African species by the densely villous calyx.

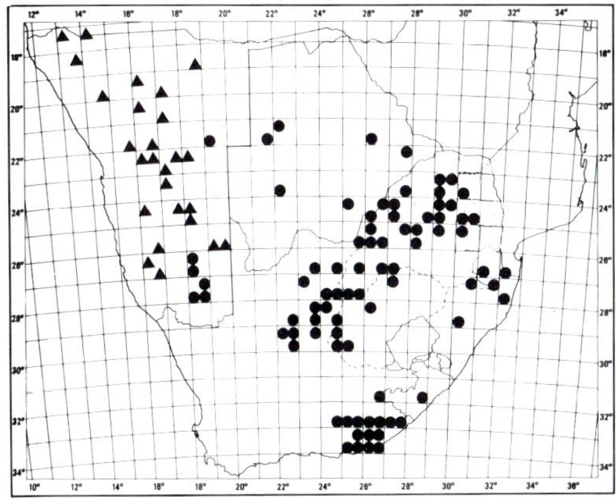

MAP 19. — ▲ Leucas pechuelii
● L. capensis

7. Leucas neuflizeana *Courbon* in Annls Sci. nat. sér. 4,18: 145 (1862); Balf. f., Bot. Socotra 242 (1888); Bak. in F.T.A. 5: 480 (1900); Skan in F.C. 5,1: 377 (1910); Sebald in Stuttgarter Beitr. Naturk. A,341: 81 (1980). Type: Dessi Island in the Red Sea, *Courbon* 389 (P, holo.).

Annual or weakly perennial, erect or decumbent, pubescent herb, 0,15—0,4 m tall. *Leaves* sessile or subsessile; blade oblanceolate to oblanceolate-oblong, 10—50 × 5—8 mm, pubescent, apex obtuse, base cuneate, margin subentire or sparingly toothed in the upper third. *Inflorescence* simple or with several short branches near the base, of many verticils, crowded above and spaced below, occupying almost the entire length of the plant; verticils 4—12-flowered; bracteoles few, setaceous, 1—2 mm long. *Calyx* hispid, 6—7 mm long, very oblique at the mouth, 10-toothed, the lower part much produced into a 3-toothed lip; teeth narrowly deltoid-subulate, spine-tipped, 0,5—1 mm long. *Corolla* 6 mm long; tube 3,5 mm long, annular-papillose about the middle; upper lip 3 mm long with a dense short brush-like fringe; lower lip 3 mm long. Fig. 11: 3.

Found in Botswana, the northern, central and eastern Transvaal, Swaziland and northern Natal, in dry woodland; extends through tropical East Africa to the Red Sea and also in Socotra. Map 15.

Vouchers: *Galpin* 14848; *Schlechter* 4171.

No specimens have been seen from Swaziland though the species should occur there. The specimens cited by Compton, Checklist Fl. Swaziland 66 (1966) and Fl. Swaziland 495 (1976) prove to be either *Acrotome thorncroftii* or *A. hispida*.

Sebald, l.c., separated a variety, var. *princei* Sebald, with a restricted distribution in Zambia.

8. Leucas capensis *(Benth.) Engl.* in Bot. Jb. 10: 268 (1888); Gürke in Bot. Jb. 22: 129 (1895); Launert & Schreiber in F.S.W.A. 123: 18 (1969); Sebald in Stuttgarter Beitr. Naturk. A,308: 12 (1978). Type: Cape, *Burchell* 1820 (K, lecto.).

Phlomis capensis Thunb., Prodr. 95 (1800); Fl. Cap. edn Schult. 446 (1823). *Leucas capensis* (Thunb.) Engl. ex Juel, Pl. Thunb. 406 (1918), nom. illegit. Type: Cape, *Thunberg* s.n. (UPS, holo.).

Lasiocorys capensis Benth., Lab. 600 (1834); in E. Mey., Comm. 241 (1838); in DC., Prodr. 12: 534 (1848); Skan in F.C. 5,1: 373 (1910); Ross, Fl. Natal 303 (1972); Compton, Fl. Swaziland 495 (1976). Type: as for *Leucas capensis*.

Perennial shrublet 0,25—1,5 m tall, sparingly or freely branched, often rather twiggy; stems whitish-buff, finally terete, canescent-tomentulose, eventually glabrescent. *Leaves* shortly petiolate; blade linear-spathulate or oblanceolate to elliptic or lanceolate, 8—20(—40) × 2—5(—10) mm, apex rounded to obtuse, occasionally apiculate, base cuneate, margin entire or rarely few-toothed near the apex. *Inflorescence* elongate, terminal, or on short, often fascicled, lateral shoots, of few to many verticils; verticils 2—6-flowered; bracteoles subulate, 0,5—2,5 mm long; flowers sessile or shortly pedicellate. *Calyx* canescent, 6—7 mm long, almost symmetrical at the mouth, 5(—8)-toothed; teeth ovate-deltoid, 1,5—2,5 mm long, shortly acuminate-spinescent, the additional intermediate teeth, when present, smaller. *Corolla* 12—14 mm long; tube 5—7 mm long, annular-papillose about the middle; upper lip 6—7 mm long with a dense short brush-like fringe in the upper half; lower lip 6—7 mm long. *Anthers* orange-red. Fig. 11: 7.

Apparently restricted to Southern Africa, occurring in S.W.A./Namibia, Botswana, central and western Transvaal, Orange Free State, Swaziland, Natal, Transkei and the northern and eastern Cape Province; found in the drier types of grassland and low woodland, often locally common on surface limestone and among rocks. Map 19.

Vouchers: *Acocks* 15609; *Codd* 3424; *Medley Wood* 10789; *Schlechter* 4213.

An aromatic plant which is grazed by game and stock and is suspected of tainting milk and dairy products. In Sekukuniland an infusion of the plant is used to treat headaches and sore eyes.

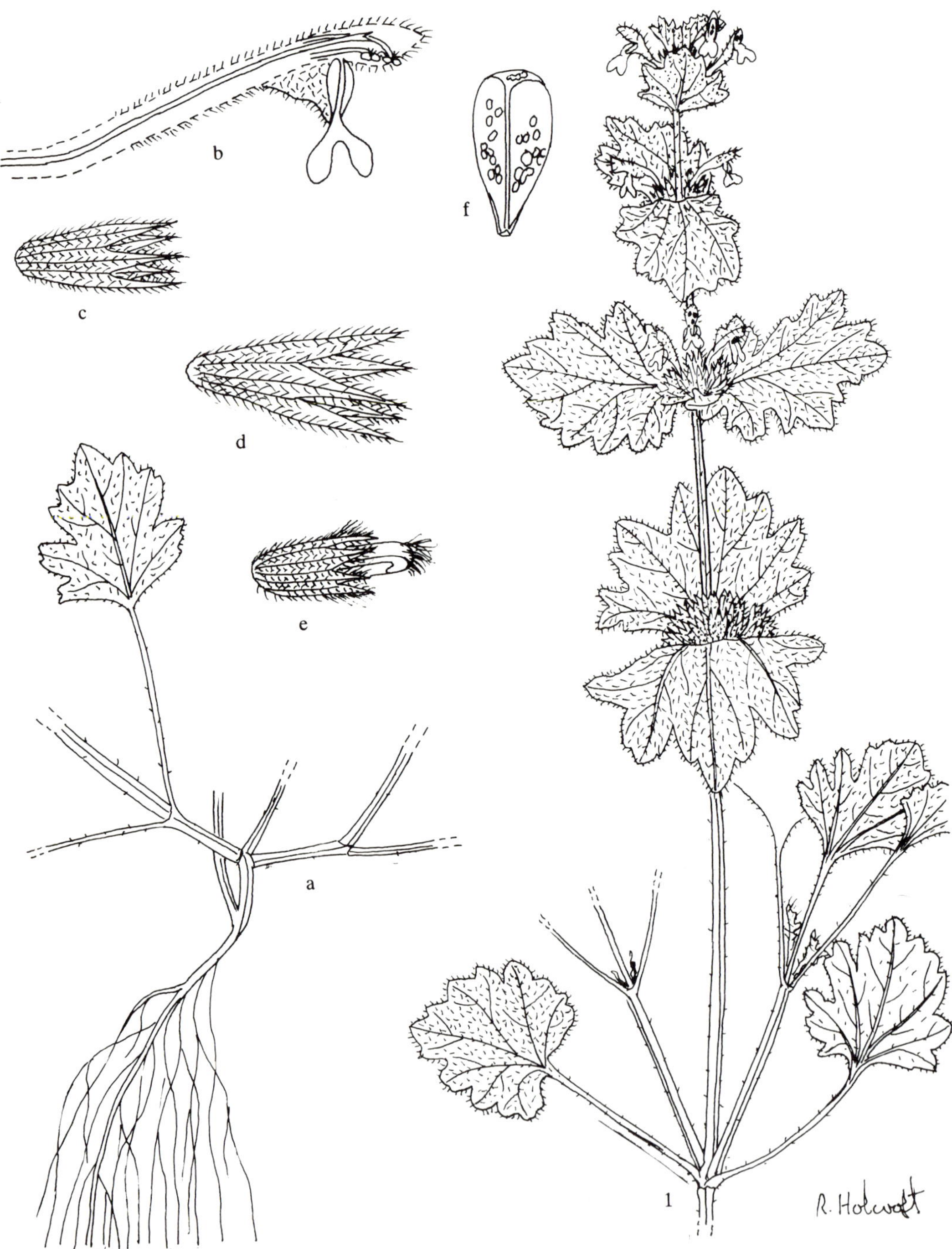

FIG. 12. — 1, **Lamium amplexicaule,** flowering stem, × 1; a, base of plant, × 1; b, section through normal corolla, × 3; c, young calyx, × 6; d, mature calyx, × 6; e, cleistogamic flower, × 6; f, nutlet, × 10 (*Mauve* 5243).

7271

11. LAMIUM

Lamium *L.*, Sp. Pl. 579 (1753); Gen. Pl. edn 5: 252 (1754); Benth. in DC., Prodr. 12: 503 (1848); Benth. & Hook. f., Gen. Pl. 2,2: 1210 (1876); Briq. in Natürl. PflFam. 4,3a: 254 (1896); Bak. in F.T.A. 5: 468 (1900); Ball in Fl. Europ. 3: 147 (1972); R. A. Dyer, Gen. 1: 528 (1975). Type species: *L. purpureum* L.

Annual or perennial soft herbs. *Leaves* toothed or incised, more or less cordate. *Inflorescence* usually simple, of one to several spaced or fairly crowded verticils; verticils few- to many-flowered; bracts broad-based, often clasping; bracteoles not evident. *Calyx* tubular-campanulate, usually 5-nerved with 5 equal or subequal teeth, the uppermost often the longest. *Corolla* white, pink or purple, bilabiate; tube dilated towards the mouth, usually longer than the calyx; upper lip ascending, concave, ovate or oblong, usually entire; lower lip spreading or deflexed, obovate, emarginate, with or without small lateral lobes. *Stamens* 4, didynamous, the anterior pair longer, arcuate in the upper lip; anthers 2-celled, divaricate, often hirsute on the back. *Style* 2-lobed. *Nutlets* oblong, triquetrous, somewhat truncate at the apex, smooth or tuberculate.

About 40 species, mainly in the North Temperate zone of the Old World, a few of which have become widespread weeds, 1 of these being found in Southern Africa.

Lamium amplexicaule *L.*, Sp. Pl. 579 (1753); Benth. in DC., Prodr. 12: 508 (1848); Britton & Brown, Ill. Fl. N. United States 3: 94 (1898); Bak. in F.T.A. 5: 469 (1900); Salisbury, Weeds and Aliens 293 (1961); Cornell & Johnston, Man. Vasc. Pl. Texas 1363 (1970); Ball in Fl. Europ. 3: 147 (1972). Type: from Europe (LINN).

Annual soft herb, freely branched from the base; stems ascending or decumbent, 0,1–0,3 (–0,4) m long. *Leaves* petiolate; blade subrotund to reniform, 10–25 mm long and equally broad, sparingly to fairly densely pubescent, apex rounded, base cordate to truncate, margin coarsely and often deeply crenate or lobed; petiole 15–40 mm long. *Inflorescence* of 2–6 verticillasters, spaced below, crowded above; verticillasters 3–10-flowered; bracts clasping, broader than long, resembling the leaves. *Calyx* villous, 6 mm long, 5-toothed; teeth lanceolate-subulate, 2,5 mm long, not spinescent. *Corolla* purple to whitish, 15–16 mm long in normal flowers (cleistogamic flowers much shorter); tube slender, 9–11 mm long, exannulate; upper lip 4–5 mm long; lower lip 3 mm long. *Nutlets* smooth, often mottled, 2 mm long. Fig. 12.

A native probably of southern Europe and south-west Asia, now a widespread weed and fairly widely distributed in gardens and waste places in Southern Africa.

Vouchers: *Acocks* 8985; *Repton* 2038.

The earliest specimen seen was collected near Kimberley in 1936 but it was probably already widely distributed by then because it was recorded from several scattered localities in the 1940's.

Commonly known as Henbit or Dead-nettle. As mentioned by Salisbury, l.c., and others, it produces, in addition to normal flowers, many small cleistogamic flowers which do not open, are self-pollinated, and produce quantities of nutlets.

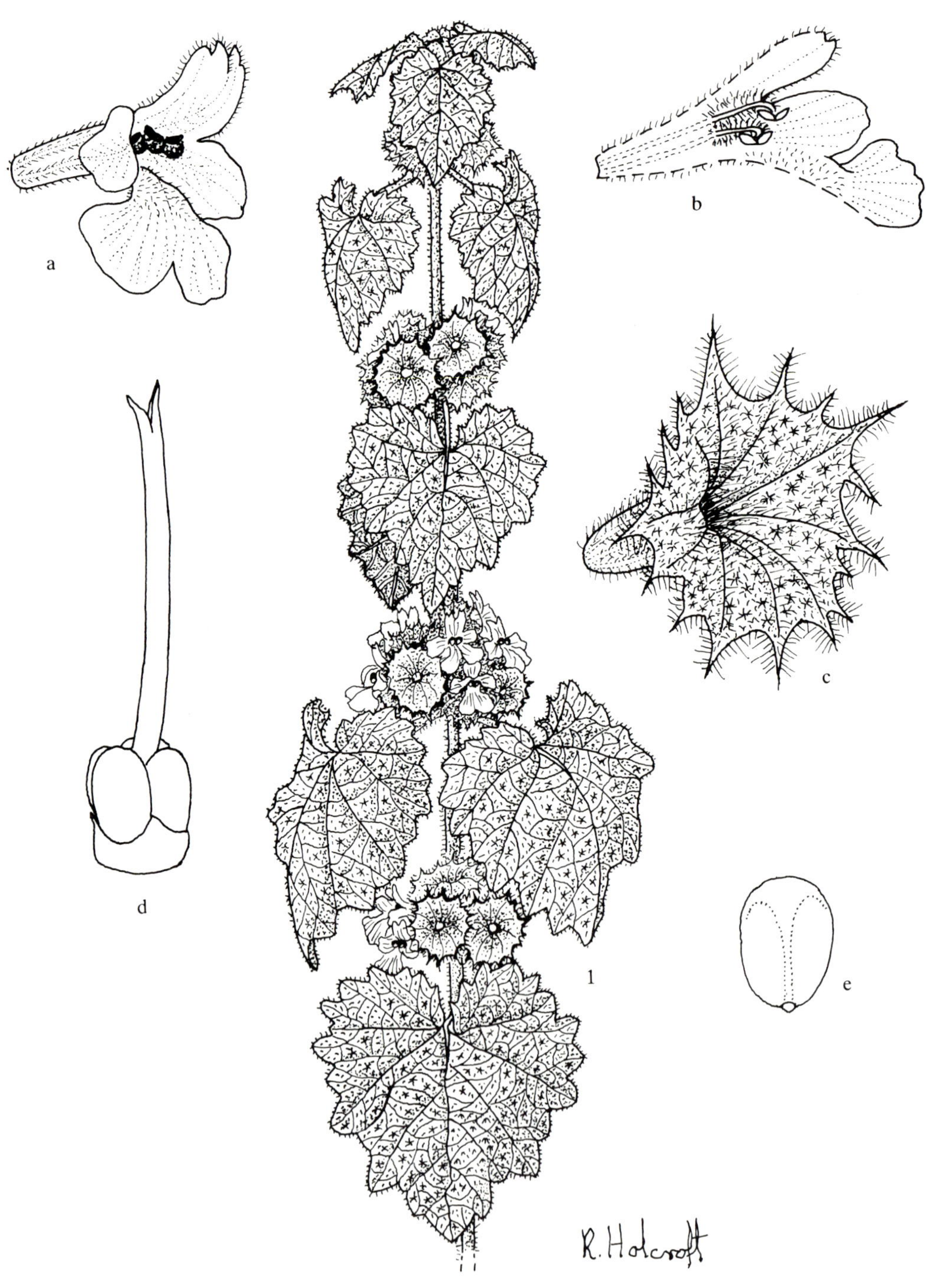
a
b
c
d
e
1
R. Holcroft

7279   12. **BALLOTA**

Ballota *L.*, Sp. Pl. 582 (1753); Gen. Pl. edn 5: 253 (1754); Benth. Lab. 592 (1834); in DC., Prodr. 12: 517 (1848); Benth. & Hook. f., Gen. Pl. 2,2: 1212 (1876); Briq. in Natürl. PflFam. 4,3a: 259 (1896); Bak. in F.T.A. 5: 472 (1900); Skan in F.C. 5,1: 368 (1910); Salter in Fl. Cape Penins. 699 (1950); Patzak in Annln naturh. Mus. Wien 63: 33 (1959); ibid. 64: 42 (1961); Fl. Europ. 3: 149 (1972); R. A. Dyer, Gen. 1: 528 (1975). Type species: *B. nigra* L.

Perennial herbs or small shrubs, usually markedly pubescent. *Leaves* often rugose, toothed. *Inflorescence* usually simple, of several to many verticils; verticils few- to many-flowered; bracts similar to the leaves; bracteoles linear to spathulate, ascending, often somewhat spine-tipped or subulate. *Calyx* funnel-shaped, 10-nerved, 10−20-toothed, villous, glandular; teeth subequal or unequal, spreading, ovate-deltoid, shortly acuminate or narrowed into an awn. *Corolla* bilabiate; tube shorter than or equalling the calyx, with a ring of hairs in the throat; upper lip shorter than the lower lip, erect, flat, bilobed, without a dense fringe of hairs; lower lip 3-lobed with the median lobe the largest, emarginate. *Stamens* 4, didynamous, ascending, the anterior pair longer and shortly exserted; filaments inserted near the throat, villous; anthers 2-thecous, cells diverging. *Style* subequally 2-lobed, shortly exserted. *Nutlets* ovoid-oblong, rounded at the apex, smooth. Fig. 13.

Calyx 5-toothed in some non-Southern African species.

About 33 species concentrated around the Mediterranean and adjoining Asia Minor, 1 of which is a fairly widespread weed; 4 species in Ethiopia-Somalia area and 1 indigenous in Southern Africa. The generic name is derived from *ballote,* the ancient Greek name for *B. nigra,* the Black Hoarhound.

Closely related to *Marrubium* (no. 6) but differs mainly in the spreading and short-toothed calyx limb, and the upper pair of stamens being exserted from the corolla tube.

Ballota africana *(L.) Benth.*, Lab. 594 (1834); in DC., Prodr. 12: 517 (1848); Skan in F.C. 5,1: 368 (1910); Salter in Fl. Cape Penins. 699 (1950); Patzak in Annln naturh. Mus. Wien 63: 62 (1959). Type: Cape, collector unknown (LINN).

Marrubium africanum L., Sp. Pl. 683 (1753); Thunb., Prodr. 96 (1800); Fl. Cap. edn Schult. 447 (1823). *Pseudodictamnus emarginatus* Moench, Meth. Pl. Suppl. 139 (1802), nom. illegit. Type: same as *M. africanum* L.

M. thouinii Schult. ex Weinm. in Ratisb. Syll. Pl. 2: 23 (1828). Type: a cultivated plant.

Soft, erect or spreading, greyish shrublet, 0,3−1,2 m tall. *Leaves* petiolate; blade orbicular to ovate, 15−50 × 15−45 mm, densely pubescent, soft to rugose, apex rounded to subacute, base cordate to rounded, margin irregularly crenate-dentate; petiole villous, 10−40 mm long. *Inflorescence* simple or branched, of few to many verticils, spaced below, more crowded above; verticils usually many-flowered, subglobose, about 20 mm in diam. *Calyx* densely hispid-villous, glandular; tube 6−8 mm long at flowering, enlarging slightly when mature; limb spreading, 9−11 mm in diam., 10−20-toothed; teeth ovate-deltoid, subulate or spine-tipped, 0,5−1,5 mm long, the additional intermediate teeth often smaller. *Corolla* purple or pinkish to pale mauve, 10−14 mm long; tube 7−9 mm long, exannulate; upper lip 3−5 mm long; lower lip 4−7 mm long.

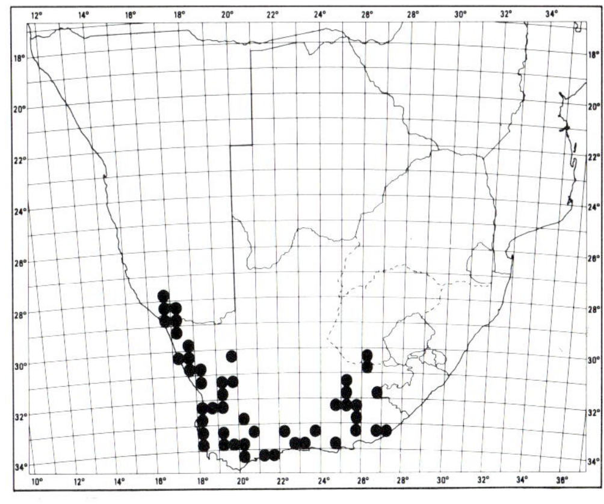

MAP 20. — **Ballota africana**

FIG. 13. — 1, **Ballota africana**, flowering stem, × 1; a, corolla, × 4; b, section through corolla, × 4; c, mature calyx, × 4; d, gynoecium, × 10; e, nutlet, × 9 (*Van Jaarsveld* 6609).

Found mainly in arid parts of the winter-rainfall area of the Cape Province as far north as the Orange River and southern S.W.A./Namibia, often along water-courses, in the shelter of rocks or bushes, and as a semi-weed of disturbed places. Map 20.

Vouchers: *Galpin* 10590; *Rodin* 1372; *Schlechter* 11242.

The common name Kattekrui (Catmint) refers to the not unpleasantly aromatic foliage.

Introduced to Europe before 1701 when it was illustrated by Commelin, Hort. med. Amst. t.90, under the phrase name Pseudodictamnus africanus foliis subrotundis subtus incanis.

Marrubium crispum L., Sp. Pl. edn 2.2: 1674 (1763), based on a plant reputedly from Europe, is included in synonymy under *B. africana* by Bentham (l.c.) and Skan (l.c.) but not by the most recent monographer of the genus, Patzak (l.c.). Linnaeus related it to his *M. africana*, but its identity is uncertain.

7281 ### 13. STACHYS

Stachys *L.*, Sp. Pl. 580 (1753); Gen. Pl. edn 5: 253 (1754); Benth., Lab. 525 (1834); in DC., Prodr. 12: 462 (1848); Benth. & Hook. f., Gen. Pl. 2,2: 1208 (1876); Briq. in Natürl. PflFam. 4,3a: 260 (1896); Bak. in F.T.A. 5: 465 (1900); Skan in F.C. 5,1: 337 (1910); R. A. Dyer, Gen. 1: 528 (1975). Type species: *S. sylvatica* L.

Sideritis sensu Thunb., Prodr. 95 (1800); Fl. Cap. edn Schult. 444 (1823).

Annual or perennial herbs, undershrubs, or sometimes shrubs, with various kinds of indumentum or sometimes nearly glabrous. *Leaves* sessile or petiolate, entire or toothed. *Inflorescence* a terminal or axillary spike or raceme; flowers in 2−many-flowered verticils, sessile or pedicellate; bracts leaf-like or reduced; bracteoles usually present, linear. *Calyx* subequally 5-toothed or rarely more or less bilabiate, 5−10-nerved; teeth usually shorter than the tube, ovate-acuminate to lanceolate-acuminate, sometimes ending in a short bristle-like point. *Corolla* bilabiate; tube straight or curved, sometimes pubescent without and usually annular-pilose near the base within; upper lip erect or ascending, usually concave or arched, entire or very shortly emarginate, usually shorter than the lower lip; lower lip spreading or deflexed, 3-lobed, with the middle lobe the largest. *Stamens* 4, didynamous, ascending under the upper lip, the lower pair the longer, usually shortly exserted from the corolla tube; anthers 2-celled, with the cells usually divergent and at length divaricate. *Style* terete, as long as the stamens, equally bifid. *Nutlets* ovoid or oblong in outline, often triquetrous, obtuse or rounded at the apex.

A genus of about 450 species occurring mainly in the subtropical and temperate regions of both hemispheres. A few species are attractive horticultural subjects and one of these, *S. byzantina* C. Koch, is grown in Southern African gardens. Of the 41 species dealt with below, *S. arvensis* L. is a cosmopolitan weed while the remainder are indigenous. Certain of the latter are used medicinally and are known as Wildetee, Boesmantee or Bushman Tea, while an infusion of the leaves of *S. linearis* Burch. ex Benth. is claimed to promote the flow of milk in nursing mothers.

1 Pubescence of simple hairs or plants subglabrous with no branched hairs on calyx or corolla: (second half of couplet on p. 4: 54)

  2 Corolla tube 12−20 mm long, often twice or more than twice as long as the calyx:

    3 Stem robust, somewhat prickly on the angles; calyx 10−14 mm long; corolla red to purple .... 1. *S. thunbergii*

    3 Stem slender, not prickly; calyx 6−8 mm long; corolla white to mauve, often flecked with deeper mauve .... 2. *S. tubulosa*

  2 Corolla tube less than 12 mm long, not twice as long as the calyx:

    4 Leaves ovate, cordate, large, 35−100 mm long and 25−70 mm broad:

      5 Leaves sparingly hispid or pilose beneath:

        6 Rhachis shortly retrorse-pubescent, not glandular; inflorescence of 1−4 verticils, compact to subcapitate .... 15. *S. graciliflora*

        6 Rhachis finely glandular-tomentose; inflorescence usually slender of few to several spaced verticils:

          7 Verticils 2-flowered .... 7. *S. rudatisii*

          7 Verticils 4−6-flowered .... 14. *S. aethiopica*

      5 Leaves densely and softly hairy beneath or, if pilose, then usually more than 15 pairs of marginal teeth:

        8 Calyx teeth not spreading, narrow:

          9 All leaves distinctly petiolate; calyx teeth more than 2 mm long:

            10 Upper bracts scarcely longer than the calyx; calyx more or less densely covered with short hairs .... 3. *S. grandifolia*

            10 Upper bracts distinctly longer than the calyx; calyx densely covered with long hairs .... 4. *S. bolusii*

          9 Upper leaves sessile or subsessile; calyx teeth up to 2 mm long............................... 5. *S. kuntzei*

8 Calyx teeth spreading, broad-based; leaves very large and freely gland-dotted beneath
.... 6. *S. albiflora*

4 Leaves oblong-lanceolate, linear-oblong to linear or, if ovate to ovate-oblong, then rarely up to 35
mm long or 25 mm broad (specimens of *S. simplex* may exceed this, but then leaves not
cordate-based; occasional abnormal specimens of *S. natalensis* may also exceed this, but then
verticils 2-flowered):

11 Verticils all 2-flowered:

12 Calyx villous; leaves subglabrous to densely villous above, sparsely to densely tomentose or
hispid beneath:

13 Leaves densely and softly tomentose beneath:

14 Leaves discolorous, sparingly pubescent and greenish brown above, white tomentose
and freely dotted with yellowish gland-dots beneath............................... 8. *S. arachnoidea*

14 Leaves concolorous, villous above, densely matted grey tomentose beneath, obscuring the
surface .... 12. *S. sessilifolia*

13 Leaves villous to hispid or sparingly strigose beneath:

15 Inflorescence fairly compact to lax, (40−) 60−150 mm long; corolla whitish, the lower lip
5−7 mm long, shorter than the tube .... 13. *S. natalensis*

15 Inflorescence compact, 30−60 mm long; corolla purple, the lower lip up to 8 mm long,
longer than the tube .... 19. *S. flexuosa*

12 Calyx subglabrous to hispid or glandular-puberulous; leaves glabrous to sparingly hispid
above, subglabrous to sparingly hispid or glandular-puberulous beneath:

16 Lower surface of leaf, calyx and rhachis densely and finely glandular-puberulous; leaves
20−45 × 15−25 mm, margin finely crenulate .... 7. *S. rudatisii*

16 Lower surface of leaf, calyx and rhachis glabrous to hispid, often with some glands or, if
glandular-puberulous, then leaves smaller than above or margin rather coarsely toothed:

17 Leaves petiolate, broadly ovate to ovate-deltoid or, if narrowly deltoid, then leaves small
with deeply crenate margins:

18 Stem glabrous to sparingly retrorse-scabrid; leaves eglandular, drying dark brown, hairs
on upper surface bulbous-based .... 16. *S. scabrida*

18 Stem variously pubescent or, if subglabrous, then leaves often glandular beneath and
hairs on upper surface not bulbous-based:

19 Leaves narrowly deltoid with deeply crenate margins........................... 18. *S. sublobata*

19 Leaves not as above:

20 Leaves small, often less than 10 mm long, broadly ovate to suborbicular; stems
short, subglabrous to glandular-puberulous, usually radiating from a central
taproot... 17. *S. cymbalaria*

20 Leaves usually exceeding 10 mm long, broadly ovate to ovate-deltoid; stems usually
long and straggling, up to 500 mm long, variously pubescent:

21 Calyx usually densely hispid, eglandular (Natal, Swaziland and Transvaal)
.... 13. *S. natalensis*

21 Calyx sparsely hispid, often glandular (Cape)............................... 14. *S. aethiopica*

17 Leaves sessile or subsessile, ovate to narrowly deltoid, margin not deeply crenate
.... 20. *S. humifusa*

11 Verticils 3−10-flowered or some 2-flowered and some more than 2-flowered on the same
inflorescence:

22 Leaves ovate to ovate-oblong, about as long as broad, up to twice as long as broad or, if more
than twice as long as broad, not usually exceeding 15 mm long:

23 Under-surface of leaf densely and softly tomentose:

24 Stems with appressed retrorse hairs; leaves usually reticulate-veined beneath
(Transvaal) .... 10. *S. reticulata*

24 Stems with spreading hairs; leaves not noticeably reticulate-veined beneath (eastern
Cape):

25 Leaves petiolate.. 11. *S. malacophylla*

25 Leaves sessile or subsessile .... 12. *S. sessilifolia*

23 Under-surface of leaf strigose, hispid or glandular-puberulous to subglabrous:

  26 Annual; corolla small, scarcely exceeding the calyx .... 28. *S. arvensis*

  26 Perennial; corolla distinctly longer than the calyx:

    27 Stem glabrous to sparingly retrorse-scabrid or shortly retrorse-pubescent; calyx eglandular:

      28 Leaves petiolate:

        29 Leaves thin to fairly firm in texture, thinly pilose, the hairs on the upper surface not bulbous-based; inflorescence of 1−4 (rarely more) verticils, usually somewhat subcapitate.. 15. *S. graciliflora*

        29 Leaves firm in texture, subglabrous to thinly hispid, drying dark brown, the hairs on the upper surface bulbous-based; stems and leaves with a rather varnished appearance.. 16. *S. scabrida*

      28 Leaves sessile to subsessile, ovate to narrowly deltoid .... 20. *S. humifusa*

    27 Stem variously pubescent, if shortly retrorse-pubescent then inflorescence usually slender of few to several spaced verticils; calyx glandular or eglandular:

      30 Leaf base distinctly cordate; under-surface of leaf densely glandular or leaf thin-textured and sparingly pubescent:

        31 Leaf thin-textured, glandular or sparingly pubescent beneath, margin crenate to crenate-serrate; stems slender, decumbent to spreading.................... 14. *S. aethiopica*

        31 Leaf thick-textured, glandular beneath, margin finely crenulate; stem usually erect, fairly stout.. 27. *S. tysonii*

      30 Leaf base rounded to subcordate; leaf relatively thick-textured, usually hispid-villous:

        32 Leaves placed mainly near the base of the stem; inflorescence slender, elongate .... 25. *S. simplex*

        32 Leaves placed along the length of the stem; inflorescence fairly compact:

          33 Stems very densely villous; lower leaves shortly petiolate, upper leaves subsessile, 17−35 × 10−20 mm.. 26. *S. obtusifolia*

          33 Stems shortly villous; leaves all petiolate, smaller than above, 10−20 × 6−15 mm... 19. *S. flexuosa*

22 Leaves oblong, oblong-lanceolate, ovate-lanceolate or deltoid-lanceolate, about 2,5 to several times longer than broad, rarely less than 18 mm long:

  34 Leaves sessile or subsessile:

    35 Stem subglabrous or with short scabrid hairs.. 20. *S. humifusa*

    35 Stem hispid to villous:

      36 Leaves blackish when dry, oblong-lanceolate to linear-lanceolate, 30−60 × 3−10 mm... 23. *S. nigricans*

      36 Leaves not or slightly blackish when dry, deltoid-lanceolate, oblong-lanceolate, oblong or ovate-deltoid, 15−50 × 5−18 mm;

        37 Leaves rounded to subtruncate at the base; calyx fairly densely hispid-villous .... 24. *S. sessilis*

        37 Leaves truncate to somewhat auricled at the base; calyx sparingly hispid .... 21. *S. rivularis*

  34 Leaves petiolate:

    38 Stem retrorse-hispid to scabrid or shortly glandular-pubescent; leaves cordate to subtruncate at the base:

      39 Stem retrorse-hispid to scabrid; leaves ovate-lanceolate to oblong-lanceolate, sparingly hispid.. 22. *S. erectiuscula*

      39 Stem glandular-tomentose; leaves ovate to ovate-lanceolate, glandular-pubescent .... 27. *S. tysonii*

    38 Stem villous; leaves rounded to subtruncate at the base, strigose to villous:

40 Leaves placed mainly near the base of the stem, often broadly elliptic............. 25. *S. simplex*

40 Leaves placed along the length of the stem, lanceolate-deltoid to ovate-deltoid .... 24. *S. sessilis*

1 (from p. 4: 51) Pubescence of stellate or branched hairs, often forming a dense velvety or felt-like tomentum or more or less floccose, rarely plants subglabrous but then some branched hairs on calyx or corolla:

41 Stems procumbent, herbaceous; leaves ovate, cordate; pubescence of greyish brown stellate hairs .... 9. *S. rehmannii*

41 Stems erect, woody or subherbaceous; leaves linear to lanceolate or obovate, rarely ovate (and then stems woody), not cordate at the base; pubescence usually of white, grey or yellowish stellate to branched hairs:

42 Calyx very thinly and minutely stellate-tomentulose or sometimes glabrescent:

43 Leaves lanceolate, pubescent, usually serrate; a soft, branched shrub 1−3 m tall .... 29. *S. caffra*

43 Leaves linear to oblong-lanceolate, often glabrescent, usually entire; stems 150−500 mm tall arising annually from a perennial rootstock.. 30. *S. hyssopoides*

42 Calyx markedly stellate-hispid or densely covered with a felt-like or wool-like indumentum:

44 Calyx stellate-hispid or covered with a short felt-like indumentum:

45 Leaves sparingly to fairly densely stellate-pilose with the leaf surfaces visible through the tomentum on both the upper and lower surfaces:

46 Leaves oblong-linear or spathulate to elliptic, usually not markedly cuneate at the base; stem sparingly to densely tomentose; flowers subsessile .... 31. *S. dregeana*

46 Leaves obovate, cuneate at the base; stem thickly white-felted; flowers in pedunculate cymes with peduncles 3−8 mm long and pedicels 2−3 mm long..................... 32. *S. dinteri*

45 Leaves with a dense felt-like tomentum at least on the lower surface, the upper surface sometimes *(S. cuneata)* less dense to thinly hispid:

47 Leaves small, usually less than 20 mm long (up to 25 mm long in *S. cuneata*), broadly ovate or obovate to oblanceolate, margin crenate:

48 Leaves obovate to oblanceolate, cuneate at the base................................ 33. *S. cuneata*

48 Leaves broadly ovate, base truncate .... 34. *S. zeyheri*

47 Leaves usually more than 20 mm long (sometimes shorter in *S. rugosa* but then base not or only slightly cuneate and margin not crenate), linear or spathulate to elliptic, elliptic-ovate, obovate- or ovate-elliptic; margin entire to faintly toothed or occasionally serrulate:

49 Plant greyish in the dried state or, if yellowish, bracteoles much shorter than the calyx:

50 Leaves narrowly linear to spathulate or oblanceolate-spathulate (rarely obovate), smooth, entire; verticils usually 2-(occasionally 3- or 4-) flowered:

51 Leaves linear-spathulate to oblanceolate-spathulate (rarely obovate), rounded and broadest at or near the apex.. 35. *S. spathulata*

51 Leaves narrowly linear, usually narrowed towards the apex.................. 36. *S. linearis*

50 Leaves oblong, oblong-elliptic or occasionally linear-elliptic to lanceolate, broadly elliptic or obovate, entire or toothed, often very rugose: verticils (2−) 4−20-flowered;

52 Calyx more or less distinctly 2-lipped .... 38. *S. burchelliana*

52 Calyx not 2-lipped .... 37. *S. rugosa*

49 Plant yellowish in the dried state; bracteoles strongly developed, subequal to the calyx in length.. 41. *S. flavescens*

44 Calyx densely covered with wool-like to almost plumose indumentum:

53 Leaves thick-textured, rugose, densely white-felted beneath, 20−45 × 10−20 mm .... 39. *S. lamarckii*

53 Leaves thin-textured, subglabrous to sparingly hispid, 10−20 × 3−5 mm .... 40. *S. aurea*

1. Stachys thunbergii *Benth.*, Lab. 540 (1834); in DC., Prodr. 12: 467 (1848); Skan in F.C. 5,1: 342 (1910); Salter in Fl. Cape Penins. 697 (1950). Type: "Hartequaskloof; in Duyvelsberg et prope Constantiam", *Thunberg* s.n. (UPS, holo., microfiche 566/13502!).

Galeopsis hispida Thunb., Prodr. 96 (1800); Fl. Cap. edn Schult. 446 (1823). *Stachys hispida* (Thunb.) Briq. in Natürl. PflFam. 4,3a: 263 (1897), nom. illegit., non *S. hispida* Pursh (1814). Type: as for *S. thunbergii* Benth.

Perennial herb with stout, 4-angled, erect or ascending softly woody stems up to 2 m long, sparingly branched, armed with strong retrorse prickles along the angles and at the nodes, otherwise glabrous. *Leaves* petiolate; blade rather thick in texture drying dark brown and somewhat rugose, ovate-deltoid to lanceolate-deltoid, 30—60 × 15—30 mm, sparingly hispid to glabrous, apex acute, base cordate, margin regularly and finely crenate; petiole 10—20 mm long. *Inflorescence* lax to fairly dense, 60—150 mm long; verticils 4—6-flowered. *Calyx* puberulous to hispidulous, 12—14 mm long at flowering; teeth lanceolate-acuminate, 5 mm long, usually spine-tipped. *Corolla* red, magenta or purple, minutely pubescent; tube 16—20 mm long, arcuate; upper lip 4—5 mm long; lower lip 6—7 mm long.

Common on south-eastern slopes of Devils Peak and adjoining mountains of the Cape Peninsula and again in forest margins in the George-Knysna-Humansdorp area, with a few scattered records in between. Map 21.

Vouchers: *Compton* 13015; *Hutchinson* 1283.

A distinctive species with long, often subscandent stems armed with retrorse prickles and long red to purple corolla.

2. Stachys tubulosa *MacOwan* in Kew Bull. 1893: 13 (1893); Skan in F.C. 5,1: 342 (1910); Ross, Fl. Natal 304 (1972); Compton, Fl. Swaziland 497 (1976). Lectotype: Griqualand East, *Tyson* sub Herb. Norm. Austr.-Afr. 1297 (K, lecto.; PRE!).

S. dolichodeira Briq. in Bull. Herb. Boissier sér. 2,3:1081 (1903). Type: Griqualand East, *Tyson* 2549 (K; PRE!; SAM!).

Soft straggling herb, probably perennial; stems weak, slender, sparingly branched, softly pilose, with long internodes. *Leaves* long petiolate; blade thin-textured, broadly ovate, 35—65 × 25—55 mm, softly pilose

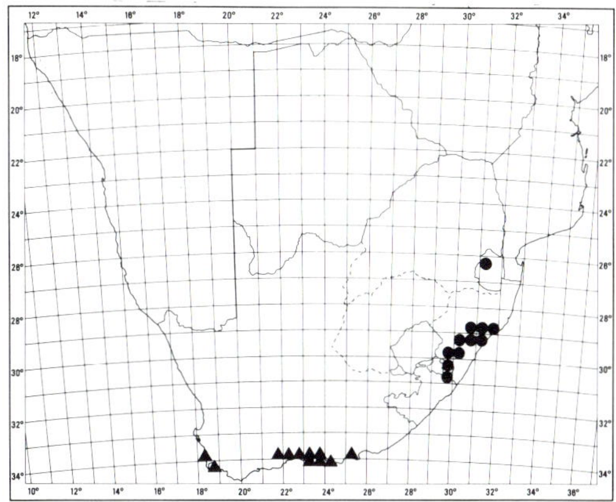

MAP 21. — ▲ **Stachys thunbergii**
● **S. tubulosa**

especially on the nerves, apex acute or subacute, base deeply cordate, margin regularly crenate; petiole 18—40 mm long. *Inflorescence* of 1—3 verticils, subcapitate or interrupted; verticils 4—6-flowered, flowers subsessile. *Calyx* softly hispid, 6—8 mm long at flowering. *Corolla* pinkish white flecked with mauve or deep mauve; tube 12—18 mm long, arcuate or nearly straight; upper lip ascending, 7 mm long; lower lip deflexed, 5 mm long.

A soft straggling herb found in moist, shady forest and forest margins in Swaziland, Natal coast and midlands, East Griqualand and Transkei. Map 21.

Vouchers: *Acocks* 13563; *Galpin* 12009.

Related to *S. graciliflora* (no. 15), from which it is distinguished by its longer corolla tube.

3. Stachys grandifolia *E. Mey. ex Benth.* in E. Mey., Comm. 239 (1838); Benth. in DC., Prodr. 12: 475 (1848); Skan. in F.C. 5,1: 342 (1910); Ross, Fl. Natal 303 (1972); Compton, Fl. Swaziland 496 (1976). Lectotype: Cape, between Umtata and Umzimvubu Rivers, *Drège* 4781a (K, lecto.!).

Straggling or much-branched perennial herb up to 1 m tall; stems densely and softly pubescent with longish, often crisped hairs but no glands. *Leaves* petiolate; blade broadly ovate, 35—80 × 28—65 mm, thinly to softly pubescent above, usually denser and sometimes softly grey-velvety beneath,

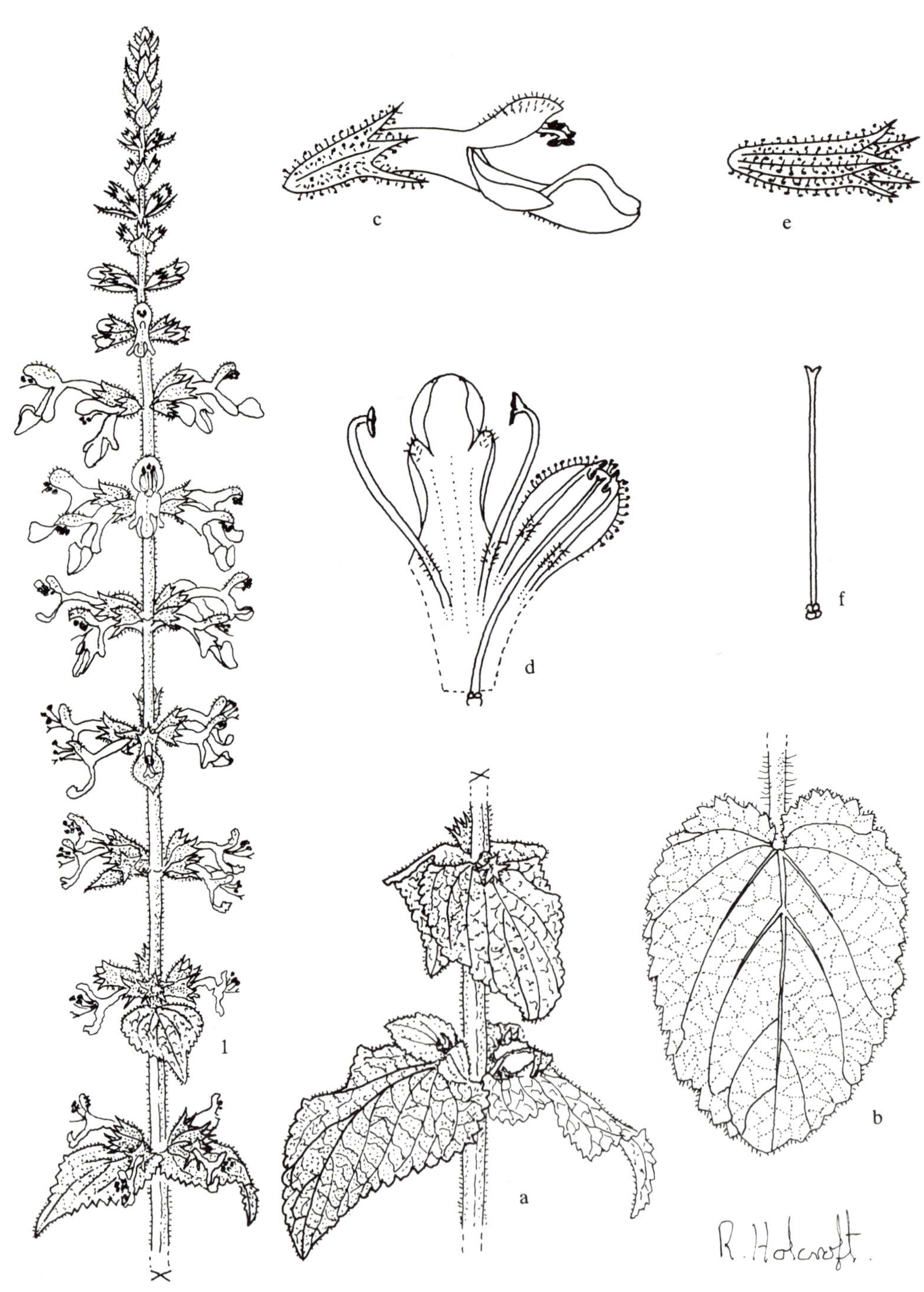

apex acute or subacute, base deeply cordate often with a wide sinus, margin regularly and rather finely crenate with about 18−25 teeth on each side; petiole 15−45 mm long. *Inflorescence* simple or often with a pair of branches near the base, tapering towards the apex, usually with 3−several spaced verticils; verticils (4−) 6-flowered; bracts reduced. *Calyx* softly pilose, 6 mm long. *Corolla* white usually with mauve spots on the lower lip; tube 7−11 mm long, straight; upper lip horizontal, 4−6 mm long; lower lip deflexed, 6−7 mm long. Fig. 14.

A bushy herb at forest margins and along mountain streams, distributed from Woodbush in the Transvaal to northern Swaziland, along the Natal Drakensberg escarpment and adjoining areas, and extending into the Transkei and eastern Cape Province. Map 22.

Vouchers: *Codd* 8540; *Pegler* 434; *Schlechter* 4741.

Resembles *S. graciliflora* (no. 15) but is a more robust plant with denser indumentum on the leaves and stems, more finely toothed leaf margins and more slender, elongate inflorescences. Also related to *S. kuntzei* (no. 5), in which the upper leaves tend to be sessile, the bracts are broader, the calyx tube is longer in relation to the teeth, which are markedly spine-tipped, and the rhachis and calyx are glandular-pubescent.

Flowers of *S. grandifolia* are freely visited by bees and other insects suggesting that it may be a good bee plant.

4. **Stachys bolusii** *Skan* in F.C. 5,1: 343 (1910). Lectotype: Cape, Malmesbury dis-

trict, near Hopefield and Saldanha Bay, *Bolus* 12809 (K, lecto.; BOL!; PRE!).

Perennial herb, spreading or ascending; stems branched, up to 0,45 m long, sparingly to fairly densely villous with long spreading to retrorse hairs and some gland-tipped hairs. *Leaves* petiolate; blade broadly ovate, the larger 30−55 × 25−40 mm, fairly densely appressed-pubescent on both surfaces, apex obtuse to rounded, base deeply cordate with a wide sinus, margin regularly and somewhat coarsely crenate with about 10−14 rounded teeth on each side; petiole 10−30 mm long. *Inflorescence* simple, scarcely tapering, up to 150 mm long, of several 6-flowered verticils; bracts densely villous, leaf-like, especially the lower, smaller above but longer than the corolla, broadly ovate; flowers subsessile. *Calyx* densely villous, 7 mm long. *Corolla* white with purple or pink markings on the lower lip; tube 7 mm long, widening slightly towards the mouth; upper lip ascending, 4−5 mm long; lower lip deflexed, 7−8 mm long.

Found among rocks in the Malmesbury district of the south-western Cape Province. Map 23.

Vouchers: *Boucher* 3173; *Galpin* 10711.

Apparently a rare species, widely separated from its nearest relative, *S. grandifolia* (no. 3), from which it differs in the usually smaller leaves with 10−14 teeth on each margin and the larger bracts which usually exceed the corolla in length.

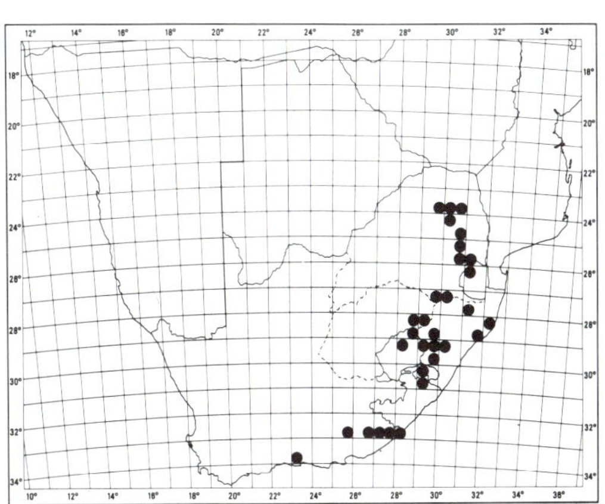

MAP 22. — **Stachys grandifolia**

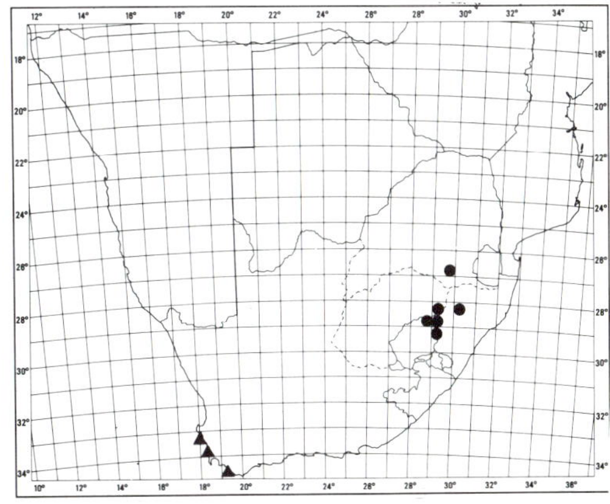

MAP 23. — ▲ **Stachys bolusii**
● **S. kuntzei**

FIG. 14. — 1, **Stachys grandifolia**, inflorescence, × 1; a, lower part of stem, × 1; b, leaf, × 1; c, flower, × 3; d, corolla opened longitudinally, × 3; e, mature calyx, × 3; f, gynoecium, × 3; (living plant, BRI garden).

Compton 23616 from Mossel River shore, Caledon district, resembles *S. bolusii* but is more densely hispid and the corolla tube is longer, up to 10 mm long. It may represent a distinct entity and should be investigated further.

5. Stachys kuntzei *Gürke* in Kuntze, Rev. Gen. 3,2: 262 (1898); Skan in F.C. 5,1: 344 (1910); Ross, Fl. Natal 303 (1972). Type: Natal, Van Reenen, *Kuntze* s.n. (NY, holo.; PRE, photo.!).

S. petrogenes Briq. in Bull. Herb. Boissier sér. 2,3: 1085 (1903). Type: Natal, Van Reenen, *Schlechter* 6969 (PRE!).

Perennial herb, decumbent or ascending, 0,45–1 m tall; stems stout, semi-succulent, densely pilose with long spreading hairs and shorter gland-tipped hairs. *Leaves* sessile above, petiolate below; blade fairly thick-textured, broadly ovate, 35–70 × 25–50 mm, densely appressed-tomentose on both surfaces, often with gland-tipped hairs beneath, apex obtuse to rounded, base cordate, margin finely and regularly crenate; petiole up to 30 mm long. *Inflorescence* simple or branched near the base, tapering and denser towards the apex, of many fairly closely spaced 6-flowered verticils; rhachis densely glandular-pubescent; bracts broadly ovate, broad-based, the lower pair somewhat leaf-like, the upper much reduced. *Calyx* densely glandular-pubescent, 6–7 mm long. *Corolla* white or tinged with mauve; tube 7–9 mm long; upper lip horizontal, 3–4 mm long; lower lip deflexed 4–5 mm long.

Found in grass among rocks and on sandstone ledges in mountain grassland in the Natal Midlands and Drakensberg region, extending to the adjacent eastern Orange Free State and south-eastern Transvaal. Map 23.

Vouchers: *Galpin* 9513; *Jacobsz* 213.

Sometimes confused with *S. grandifolia* (no. 3), but the two species may be distinguished on several character differences, as discussed under that species.

6. Stachys albiflora *N.E. Br.* in Kew Bull. 1901: 131 (1901); Skan in F.C. 5,1: 344 (1910); Ross, Fl. Natal 303 (1972). Type: Natal, Drakensberg, *Evans* 395 (K, holo.; NH!; PRE!).

A robust perennial herb, 0,6–1,3 m tall; stems stout, erect or ascending, branched above, densely glandular-pilose. *Leaves* very large, petiolate; blade rather thin-textured, broadly ovate, 80–110 × 50–75 mm, appressed-pilose, paler beneath and densely dotted with yellowish sessile glands, apex acute to subacute, base deeply cordate, margin finely and regularly crenate; petiole 20–40 mm long. *Inflorescence* simple or with a pair of branches near the base, tapering and denser towards the apex, of several spaced 6-flowered verticils; rhachis densely glandular-pubescent; bracts ovate, longer than the calyx. *Calyx* hispidulous and freely gland-dotted, 7 mm long; teeth distinctly spreading, 3 mm long, deltoid-subulate and markedly spine-tipped. *Corolla* white, tube 6–7 mm long; upper lip horizontal, 6 mm long; lower lip deflexed, 6 mm long.

A robust herb locally frequent in *Leucosidea sericea* communities at altitudes of 2 000–2 200 m in a restricted area in the Drakensberg, in Natal and the eastern Orange Free State. Map 24.

Vouchers: *Killick* 1329; *Killick & Vahrmeijer* 3788.

Resembles *S. kuntzei* (no. 5) but may readily be distinguished by the spreading calyx teeth, by the narrower bracts which taper towards the base, and by the freely gland-dotted undersides of the leaves.

7. Stachys rudatisii *Skan* in F.C. 5,1: 347 (1910); Ross, Fl. Natal 305 (1972); Codd in Bothalia 12: 182 (1977). Type: Natal, Dumisa, *Rudatis* 405 (K, holo.; NH!; PRE!).

Apparently a prostrate or decumbent, branched, perennial herb; stems fairly stout, branched, up to 0,3 m or more long, deeply 4-furrowed, glandular-hispidulous, with retrorse multicellular hairs and copious short glandular hairs. *Leaves* petiolate; blade ovate, (20–) 30–45 × (15–) 20–25 mm, densely glandular-hispidulous, apex obtuse to rounded, base cordate, margin regularly and shallowly crenulate; petiole 10–25 mm long. *Inflorescence* of few to several 2-flowered verticils in the axils of leaf-like bracts; bracts scarcely differentiated or becoming smaller towards the apex. *Calyx* glandular-hispid, 7–8 mm long. *Corolla* white; tube 8–9 mm long; upper lip ascending, 4 mm long and equally broad; lower lip horizontal, 7–8 mm long.

In damp grassy places among rocks and in shady thickets in southern Natal. Map 24.

Voucher: *Hilliard & Burtt* 9040.

Differs from *S. natalensis* (no. 13) in the densely glandular stems and leaves; in *S. natalensis* the stems and leaves are hispid-villous and the undersides of the

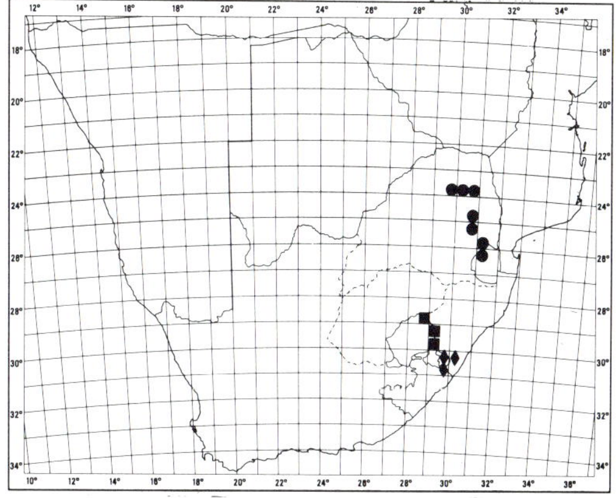

MAP 24. — ■ *Stachys albiflora*
♦ *S. rudatisii*
● *S. arachnoidea*

leaves often bear short gland-tipped hairs. *S. aethiopica* (no. 14) often has densely glandular stems and leaves but the leaves are usually smaller and broadly ovate, while the verticils (in Natal specimens) are normally 4—6-flowered, not 2-flowered as in *S. rudatisii*.

Rattray sub BOL 14275 from the Hogsback may belong in *S. rudatisii* but the leaves are very broadly ovate and the locality is widely separated from the two listed above. Further material is desirable.

8. **Stachys arachnoidea** *Codd* in Bothalia 12: 182 (1977). Type: Swaziland, near Mbabane, *Compton* 25890 (PRE, holo.!; K!; NBG!).

S. rehmannii sensu Compton, Fl. Swaz. 66 (1966). *S. nr. rudatisii* sensu Compton, l.c. (1966). *S. sp.* sensu Compton, l.c. (1966).

Perennial herb; stems procumbent to subscandent, branched, up to 1 m long, densely and softly white villous. *Leaves* subsessile to shortly petiolate; blade thin-textured, broadly ovate-deltoid to subrotund, 18—40 × 15—30 mm, discolorous, upper surface greenish to brown and thinly to fairly densely pubescent, under-surface with a dense white web-like tomentum and freely dotted with minute yellowish pustule-like gland-dots, apex obtuse to rounded, base cordate, margin regularly and shallowly crenulate; petiole up to 10 mm long. *Inflorescence* lax; verticils 2-flowered; bracts leaf-like, subsessile. *Calyx* densely and softly white tomentose with numerous minute gland-dots, 8—9 mm long. *Corolla*

white with mauve-purple upper lip and speckled lower lip; tube 8—9 mm long; upper lip ascending, concave, 3—4 mm long, lower lip horizontal, 8—9 mm long.

Found in moist places in forest margins and grassy slopes on the mountains of eastern Transvaal and northern Swaziland at altitudes of 1 300—2 000 m. Map 24.

Vouchers: *Codd* 9858; *Scheepers* 729.

Has been confused with other species with procumbent stems and 2-flowered verticils, such as *S. rudatisii* (no. 7), *S. rehmannii* (no. 9) and *S. natalensis* (no. 13), but differs in the discolorous leaves with white cobwebby tomentum and minute yellowish gland-dots on the under-surfaces. In addition, *S. rehmannii* is distinguished by the dense stellate tomentum on all parts of the plant, while the verticils are often 4—6-flowered.

9. **Stachys rehmannii** *Skan* in F.C. 5,1: 345 (1910); Codd in Bothalia 12: 183 (1977). Type: Transvaal, Houtbosch, *Rehmann* 6178 (K, holo.!).

Perennial herb; stems procumbent, branched, up to 0,5 m long, sparsely to densely stellate-hispid. *Leaves* petiolate; blade fairly thick-textured, broadly ovate-deltoid, 10—22 × 10—20 mm, reticulate, densely grey stellate-hispid on both surfaces, apex obtuse to rounded, base deeply cordate, margin regularly crenate; petiole 3—7 mm long. *Inflorescence* lax below, dense above; verticils usually 2-flowered, occasionally 4—6-flowered; bracts leaf-like below becoming rapidly smaller and eventually elliptic and shorter than the calyx above. *Calyx* densely and shortly stellate-villous, 7—9 mm long. *Corolla* white to rosy with a purplish blotch in the throat; tube 7—8 mm long; upper lip ascending, 3 mm long and equally broad; lower lip horizontal, 6—7 mm long.

Found among rocks in mountain grassland at altitudes of 1 300—2 200 m in the northern and north-eastern Transvaal. Map 25.

Vouchers: *Codd & Dyer* 9022; *Strey & Schlieben* 8515.

Readily distinguished from all other members of the *S. aethiopica* complex by the presence of dense stellate pubescence on all parts of the plant.

10. **Stachys reticulata** *Codd* in Bothalia 12: 183 (1977). Type: Transvaal, Mariepskop, *Werdermann & Oberdieck* 1868 (PRE, holo.!).

Perennial herb; stems decumbent to

procumbent, densely pilose, branched, up to 0,6 m long. *Leaves* petiolate; blade fairly firm-textured, ovate-deltoid to broadly ovate or subreniform, 10−30 × 8−25 mm, usually discolorous, upper surface brownish, hispid, under-surface paler, reticulate-veined, densely tomentose and gland-dotted, apex obtuse to subacute, base broadly cordate, margin crenulate; petiole 5−12 mm long. *Inflorescence* lax, of 2−4 verticils or occasionally subcapitate; verticils 4−6-flowered; bracts much reduced, the upper ones shorter than the calyx. *Calyx* fairly densely glandular-hispid, 7−8 mm long. *Corolla* white to pale mauve; tube 8−9 mm long; upper lip horizontal, oblong, concave, 4−5 mm long; lower lip deflexed, 6−7 mm long.

Found among rocks in exposed situations in mountain grassland in the Pilgrims Rest district of the eastern Transvaal at altitudes of 1 500 to 2 200 m. Map 25.

Vouchers: *Galpin* 14342; *Killick & Strey* 2391.

May be distinguished from *S. aethiopica* (no. 14) by the discolorous leaves which are reticulate-veined and densely tomentose below, by the spreading calyx teeth and by the characteristic appressed pilose tomentum of the stems. *S. natalensis* (no. 13) differs from it in having 2-flowered verticils, while in *S. rehmannii* (no. 9) the hairs are stellate. In *S. malacophylla* (below) the stem pubescence is softly spreading, the leaves are concolorous without glands and the calyx teeth are not spreading.

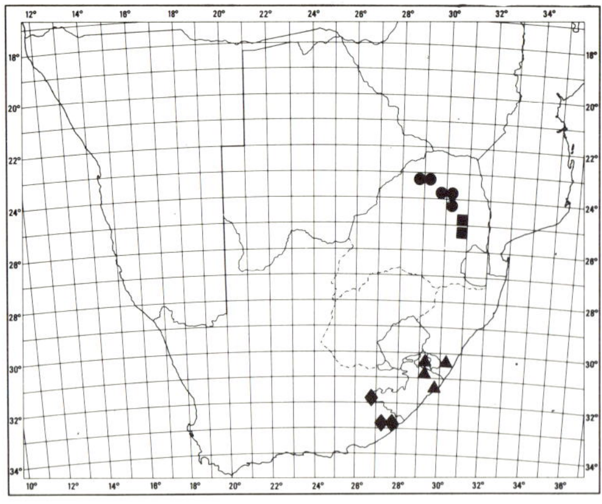

MAP 25. — ● **Stachys rehmannii**
■ **S. reticulata**
♦ **S. malacophylla**
▲ **S. sessilifolia**

11. **Stachys malacophylla** *Skan* in Kew Bull. 1909: 421 (1909); F.C. 5,1: 345 (1910); Codd in Bothalia 12: 184 (1977). Lectotype: Cape, Queenstown, *Galpin* 1955 (K, lecto.; PRE!).

Perennial herb, stems decumbent, sparingly branched, fairly densely pilose, up to 0,4 m long. *Leaves* petiolate; blade fairly firm-textured, broadly ovate, 14−30 × 10−22 mm, greyish, concolorous, densely appressed pilose above, matted velvety pilose beneath, glands not evident, apex obtuse to rounded, base broadly and deeply cordate, margin crenulate; petiole 5−14 mm long. *Inflorescence* lax, of 2−8 verticils; verticils (2−) 3−6-flowered; bracts reduced, the upper ones shorter than the calyx and narrowly elliptical. *Calyx* densely hispid-villous and finely gland-dotted, 6−7 mm long. *Corolla* mauve; tube 8 mm long; upper lip horizontal, 4 mm long; lower lip deflexed, 5−6 mm long.

A little-known species from mountains in the eastern Cape Province. Map 25.

Voucher: *Sim* 19590.

Closely related to *S. sessilifolia* (below) from which it is distinguished by the petiolate leaves and usually 4−6-flowered verticils, while the tomentum on stems and leaves is less densely woolly. In *S. reticulata* (no. 10) from the eastern Transvaal, which resembles *S. malacophylla* superficially, the stem pubescence is strongly retrorse, the leaves tend to be discolorous and noticeably reticulate below, while the calyx teeth are distinctly spreading.

12. **Stachys sessilifolia** *E. Mey. ex Benth.* in E. Mey., Comm. 239 (1838); Benth. in DC., Prodr. 12: 476 (1848); Skan in F.C. 5,1: 345 (1910); Codd in Bothalia 12: 184 (1977). Type: Cape, between Umzimvubu and Umsikaba Rivers, *Drège* 4752 (K, holo.!).

S. bachmannii Gürke in Bot. Jb. 26: 75 (1898). Type: Cape, Pondoland, near Dorking, *Bachmann* 1169.

Perennial herb; stem decumbent or ascending, branched, up to 0,7 m long, densely and softly villous. *Leaves* often shortly petiolate below, sessile above; blade ovate to ovate-deltoid or narrowly ovate, 10−22 × 7−11 mm, densely appressed villous above, densely matted-villous beneath, apex obtuse, base rounded to subcordate, margin finely crenulate. *Inflorescence* crowded at the apex, laxer below, of few to several verticils; verticils 2 (−6)-

flowered; bracts leaf-like, becoming smaller but longer than the calyx towards the apex. *Calyx* densely shaggy-villous, 7—8 mm long. *Corolla* "white with carmine on the lower lip" (fide Skan, l.c.); tube narrow, 7—8 mm long; upper lip ascending, 3—4 mm long; lower lip horizontal, 6—8 mm long.

Found in dense grassland in the Transkei and eastern Cape Province. Map 25.

Voucher: *Coleman* 834.

A little-known species closely related to *S. natalensis* (below) but distinguished from that species by the densely matted-villous lower surface of the leaves. Further material is required in order to determine how meaningful this distinction is. A specimen from northern Natal, near Luneburg, *Galpin* 9870, has this type of tomentum but differs in having petioles up to 7 mm long. Until more material is forthcoming, it is referred to *S. sessilifolia* with some hesitation; from *S. malacophylla* (no. 11) it differs in having 2-flowered, not 4—6-flowered verticils.

The type of *S. bachmannii* Briq. has not been seen; the species was included in *S. sessilifolia* by Skan and, according to its description, this decision appears to be correct.

13. Stachys natalensis *Hochst.* in Flora 28: 65 (1845); Codd in Bothalia 12: 185 (1977). Type: Natal, Table Mtn, *Krauss* 1139.

Perennial herb; stems several, erect, 0,12—0,2 m tall or few, decumbent to straggling, up to 0,5 m long, variously pubescent, glandular hairs usually absent. *Leaves* subsessile or petiolate; blade firm to thick-textured, ovate to ovate-deltoid, variable in size, 10—40 × 6—24 mm, shortly and sparingly appressed pubescent to densely strigose above, less dense and more spreading beneath, often with some glandular hairs but not conspicuously glandular, apex obtuse to rounded, base deeply cordate to rounded, margin regularly and shallowly crenate; petiole up to 12 mm long. *Inflorescence* fairly dense above, laxer below, of 5—20 verticils; verticils 2-flowered; bracts similar to the leaves below, becoming smaller towards the apex, eventually elliptic, equal to or shorter than the calyx. *Calyx* densely villous to densely and shortly pubescent, without glands, 7—10 mm long. *Corolla* white with a few lilac markings on the lower lip; tube 7—11 mm long; upper lip ascending, 2—3 mm long; lower lip deflexed, 5—7 mm long.

Found in grass on stony hillsides, or in semi-shady kloofs and wooded places in the mountains of northern, central and eastern Transvaal, Swaziland, northern and coastal Natal as far south as Durban, with an occasional record from the eastern Cape Province. Also recorded from Zimbabwe.

No material of the type, *Krauss* 1139 from Table Mtn, near Pietermaritzburg, has been traced but the description is considered adequate to identify it with the present concept. The specimen *Schlechter* 2894 from near Verulam, Natal may be regarded as representative.

Two varieties are recognized:

1 Pubescence on stem, leaves and calyx sparingly villous to shortly scabrid or to-mentose.............................. (a) var. *natalensis*

1 Pubescence on stem, leaves and calyx densely villous........................ (b) var. *galpinii*

(a) var. **natalensis.**

Codd in Bothalia 12: 185 (1977).

S. natalensis Hochst. in Flora 28: 65 (1845); Skan in F.C. 5,1: 367 (1910); Ross, Fl. Natal 303 (1972). Type: Natal, Table Mtn, *Krauss* 1139.

S. transvaalensis Gürke in Bot. Jb. 28: 316 (1901); Skan, l.c. 346 (1910). Type: Transvaal, Lydenburg district, *Wilms* 1136 (BM).

S. leptoclada Briq. in Bull. Herb. Boissier sér. 2,3: 1084 (1903); Skan, l.c. 351 (1910); Ross, Fl. Natal 303 (1972). Type: Natal, Bluekrantz River, *Schlechter* 6865 (Z, holo.!; BOL!).

S. aethiopica sensu Letty, Wild Flow. Transv. 284, t.141: 3 (1962); sensu Compton, Fl. Swaziland 496 (1976).

Stems decumbent to straggling, up to 0,5 m long; pubescence on stem, leaves and calyx shortly scabrid or tomentose to sparingly or fairly densely villous; leaves petiolate.

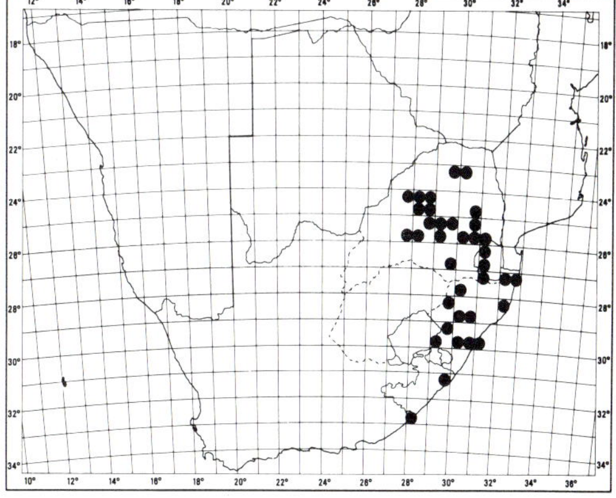

MAP 26. — **Stachys natalensis** var. **natalensis**

Distribution and ecology more or less as for the species, but not prevalent in the mountain grassland of the eastern Transvaal. Map 26.

Vouchers: *Codd* 8622; *Compton* 26835; *Junod* 123; *Schlechter* 2894; *Strey* 3947.

Specimens with scabrid or shortly tomentose pubescence are closely related to typical *S. aethiopica* (no. 14) but can be separated, where the two species overlap, by the 2-flowered verticils. The occurrence of long villous hairs varies from sparse to fairly dense, grading into var. *galpinii* with very dense villous hairs.

(b) var. **galpinii** *(Briq.) Codd* in Bothalia 12: 185 (1977).

S. galpinii Briq. in Bull. Herb. Boissier sér. 2,3: 1082 (1903); Skan, l.c. 346 (1910); Ross, Fl. Natal 303 (1972); Compton, Fl. Swaziland 496 (1976). Type: Transvaal, near Barberton, *Galpin* 681 (K; PRE!; SAM!).

S. lupulina Briq., l.c. 1082 (1903). Type: "Natal, near Claremont, *Schlechter* 4651" (see note below).

S. parilis N.E. Br. in Kew Bull. 1901: 131 (1901); Skan, l.c. 347 (1910); Ross, l.c. 303 (1972). Type: Natal, Drakensberg, Tiger Cave Valley, *Evans* 387 (K; NH!; PRE, photo.).

S. villosissima H.M. Forbes in Bothalia 4: 38 (1941); Ross, l.c. 304 (1972). Type: Natal, Entumeni, *Forbes* 783 (NH, holo.!; PRE!).

Stems erect, 0,12−0,2 m tall or decumbent to straggling, 0,3−0,4 m long; stem, leaves and calyx densely villous; leaves subsessile or petiolate.

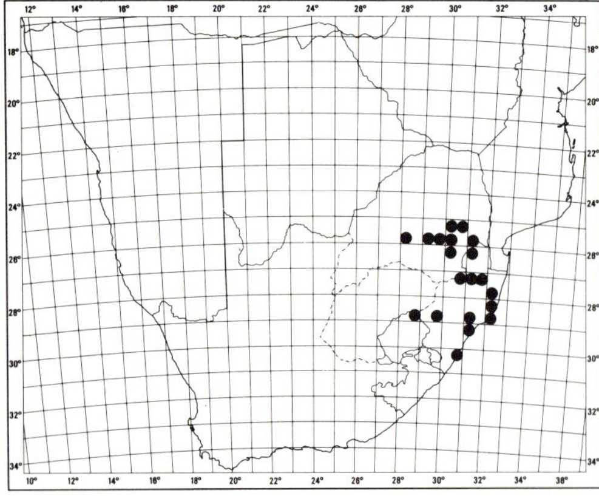

MAP 27. — **Stachys natalensis** var. **galpinii**

In dense grass, often among rocks, in central and eastern Transvaal and northern Swaziland; extending to Natal. Map 27.

Voucher: *Codd* 8063; *Galpin* 10183; *C.A. Smith* 3272.

This variety can be recognized by the combination of densely villous pubescence and 2-flowered verticils. The leaves may be petiolate or subsessile and the latter specimens come near to *S. sessilifolia* (no. 12) in which undersides of the leaves are densely matted-villous and the stems are softly tomentose. In *S. malacophylla* (no. 11) the undersides of the leaves are also densely pubescent and the verticils are usually 6-flowered.

Skan, l.c, draws attention to the confusion concerning the type of *S. lupulina*. Briquet cites the specimen as "Natal, Claremontplats prope Claremont, *Schlechter* 4651, ann. 1892." On the type sheet in Z and an isotype in BOL the label reads: "Claremont flats prope Cape Town, *Schlechter* 465, 9, III. 1892". It is undoubtedly conspecific with *S. natalensis* var. *galpinii*, which does not occur in the Cape, but could have been collected while Schlechter was in northern Natal or the eastern Transvaal. Skan concluded that it had probably been introduced at the Cape but there is no evidence to support this.

14. **Stachys aethiopica** L., Mant. 1: 82 (1767); Benth. in DC., Prodr. 12: 476 (1848); Skan in F.C. 5,1: 348 (1910); Marloth, Fl. S. Afr. 3,2: 180, t.47B (1932); Salter in Fl. Cape Penins. 697 (1950); Jacot Guill., Fl. Lesotho 237 (1971); Ross, Fl. Natal 303 (1972); Codd in Bothalia 12: 186 (1977). Type: Cape Province, LINN 736.13.

Betonica capensis Burm. f., Fl. Cap. Prodr. 16 (1768). Type: Pluk., Almagest. Bot. t.315, f.3 (1696).

S. pulchella Salisb., Prodr. 83 (1796), nom. illegit. Type: same as for *S. aethiopica* L.

S. serrulata Burch. ex Benth., Lab. 549 (1834); in DC., Prodr. 12: 477 (1848); Skan, l.c. 350 (1910). Type: Cape, near Knysna, *Burchell* 5155 (K, holo.!).

S. aethiopica var. *grandiflora* Burch. ex Benth. in E. Mey., Comm. 239 (1837). Type: Cape, Klein Winterhoek, *Drège* 75d (K, holo.!).

—var. *hispidissima* Benth., l.c. 239 (1837); Skan, l.c. 348 (1910). Type: Cape, Hex River Kloof, *Drège* 75h (K, holo.!).

S. capensis Presl, Bot. Bemerk. 100 (1844); Benth. in DC., Prodr. 12: 496 (1848); Skan, l.c. 366 (1910). Type: Cape, without locality, *Krebs* 273 (PRC, holo.!).

S. hispidula Hochst. in Flora 28: 66 (1845); Skan, l.c. 367 (1910). Type: Cape, Humansdorp District, *Krauss* 1125.

S. fruticetorum Briq. in Bull. Herb. Boissier sér. 2,3: 1083 (1903); Skan, l.c. 351 (1910). Type: Cape, Sir Lowrys Pass, *Schlechter* 1179 (Z, holo.!; BOL!).

S. aethiopica var. *glandulifera* Skan, l.c. 348 (1910); Jacot Guill., l.c. 237 (1971). Syntypes: several, incl. Zwartkei River, *Baur* s.n. (K; PRE!).

—var. *parviflora* Skan, l.c. 348 (1910); Salter, l.c. 697 (1950). Syntypes: several incl. Cape Peninsula, Signal Station, *Wolley-Dod* 3048 (K; BOL!).

S. attenuata Skan, l.c. 351 (1910). Syntypes: Cape, near Bains Kloof, *Bolus* 2896 (BOL!); Paarl Mtn, *Drège* 75b (K!).

S. harveyi Skan, l.c. 350 (1910). Type: Cape, near Cape Town, *Harvey* s.n. (TCD, holo.).

Perennial herb, sparingly to freely branched; stems decumbent or ascending up to 0,5 m long or more, variously pubescent with short antrorse hairs, longish spreading to retrorse hairs mixed with glandular hairs, or densely glandular. *Leaves* petiolate; blade thin to fairly firm in texture, broadly ovate to ovate-deltoid, 8−35 (−60) × 6−25 (−35) mm, sparingly and shortly hispid with or without glandular hairs, rarely almost glabrous, apex subacute to obtuse, base deeply to shallowly cordate, margin regularly crenate to crenate-serrate; petiole 3−30 mm long. *Inflorescence* lax below, dense above, occasionally subcapitate, of few to several verticils; verticils (2−) 4−6-flowered; bracts similar to the leaves below becoming smaller towards the apex, eventually elliptic and equal to or shorter than the calyx. *Calyx* sparingly to densely hispid and often glandular, 5−8 mm long. *Corolla* white or pink to deep mauve, usually with purplish flecks on the lower lip; tube 7−10 mm long; upper lip ascending, 2−5 mm long; lower lip deflexed, 5−8 mm long.

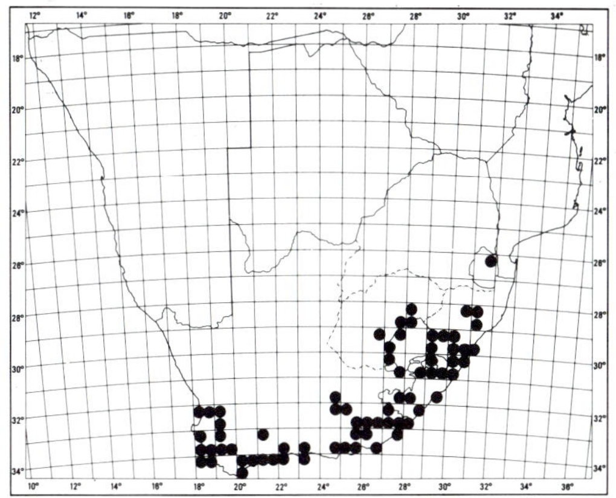

MAP 28. — **Stachys aethiopica**

Distributed from the Orange Free State, Lesotho, Natal and Transkei to the eastern Cape and more or less along the coast to the Peninsula and thence northwards to Clanwilliam district; found in a variety of habitats from mountain grassland, usually on sandstone formation, to dry woodland and coastal dune bush, and among rocks in fynbos. Map 28.

Vouchers: *Boucher* 2343; *Dieterlen* 101; *Hilliard & Burtt* 3194; *Pegler* 231; *Scheepers* 1412.

The variation in this widespread species is discussed more fully elsewhere (Codd, l.c.) and it is pointed out that *S. aethiopica* tends to be a repository for specimens which are not distinct enough to be satisfactorily segregated as separate species. Thus the key characters for separating it from other species may break down and it is often necessary to use a combination of characters or facies in allocating some specimens.

It is separated from the more northern species, *S. natalensis* (no. 13), mainly on the pubescence of leaves and calyx: *S. natalensis* usually has villous to densely villous and scarcely glandular leaves and calyx (glandular hairs rarely present), though some specimens in Natal are somewhat intermediate in having tomentose rather than villous pubescence. The stem pubescence in *S. aethiopica* is mainly hispid, either antrorse, retrorse or spreading, with sometimes scattered longish hairs and glandular hairs, while the leaves and calyx are often glandular-hispid, especially the lower surfaces of the leaves, which may be densely glandular-puberulous. However, there is a considerable intergrading of characters so that the varieties recognized in Flora Capensis are not upheld.

Usually *S. aethiopica* has 3−6-flowered verticils whereas in *S. natalensis* they are strictly 2-flowered. However, some depauperate specimens of *S. aethiopica* in the south-western Cape Province may have 2-flowered verticils (e.g. the type specimen and *S. attenuata*), while certain closely related segregate species such as *S. flexuosa* Skan (no. 19), *S. cymbalaria* Briq. (no. 17) and *S. sublobata* Skan (no. 18), also have 2-flowered verticils.

In *S. aethiopica* the inflorescence is usually fairly elongate, consisting of a number of verticils, but sometimes it is reduced to one or two verticils which give the inflorescence a subcapitate appearance. In such cases the distinction between *S. aethiopica* and *S. graciliflora* Presl (stems glabrous to sparingly retrorse-pubescent) becomes rather arbitrary (see below).

A large-leaved form with thin-textured, sparingly pilose leaves 30−60 × 20−40 mm occurs along the Natal coast from near Durban to Port Shepstone. It has glandular-tomentose stems and usually 6-flowered verticils, and appears to behave as a semi-weed. It appears to grade into *S. aethiopica* and no character could be found for separating it satisfactorily though the extremes look very different.

15. **Stachys graciliflora** *Presl*, Bot. Bemerk. 100 (1844); Benth. in DC., Prodr. 12: 496 (1848); Skan in F.C. 5,1: 366 (1910); Codd in Bothalia 12: 187 (1977). Type: Cape, without locality, *Krebs* s.n. (PRC, holo.!, as to left-hand specimen on sheet labelled *S. graciliflora* Presl; PRE, photo.).

S. cooperi Skan in Kew Bull. 1909: 420 (1909); in
F.C. 5,1: 343 (1910); Ross, Fl. Natal 303 (1972).
Syntypes: Cape, Albany Division, *Cooper* 15 (K!);
Kentani, *Pegler* 908, collected April 1909 (K!; PRE!).

Perennial, decumbent to prostrate or
subscandent herb; stems up to 0,4 m or
more, sparingly branched, shortly and often
sparingly retrorse-pubescent. *Leaves* petiolate, blade often thin-textured, ovate to
broadly ovate, 20−65 × 14−50 mm,
eglandular, sparingly hispidulous or shortly
and sparingly pilose, the hairs on the upper
surface soft and not bulbous-based, apex
subacute to obtuse, base deeply cordate
with a wide sinus and distant rounded
auricles, margin coarsely crenate; petiole
8−30 mm long. *Inflorescence* somewhat lax
below or often subcapitate, of 1−4 (rarely
more) verticils; verticils (2−) 4−6-flowered;
rhachis shortly retrorse-tomentulose; bracts
leaf-like below, smaller above and finally
lanceolate, shorter than the calyx. *Calyx*
softly pubescent to sparingly hispidulous,
6−8 mm long. *Corolla* white, sometimes
with mauve spots on the lower lip; tube
6−10 mm long, arcuate; upper lip ascending, 5−6 mm long; lower lip horizontal,
6−8 mm long.

A soft straggling herb of moist places in forest
margins, in grass, fynbos or coastal scrub from southern
Natal to Knysna in the Cape. Map 29.

Vouchers: *Galpin* 2069; *Pegler* 908; *Strey* 6169.

There is a gradation in leaf size from the specimens
with larger and softer leaves, described as *S. cooperi*
Skan, to those occurring further west with smaller and
firmer leaves, which match the type of *S. graciliflora*
Presl. The stem pubescence of the latter tends to be
slightly more scabrid and thus approaches the condition
found in *S. scabrida* Skan (below). However, in *S.
scabrida* the leaves are somewhat thicker in texture, dry
dark brown and the hairs on the upper leaf surface are
thicker with distinctly swollen bases. On these grounds
S. scabrida is kept distinct, but the two overlap from
southern Transkei to Knysna and further study in this
area is required.

As mentioned under the previous species, *S.
aethiopica* sometimes has few, fairly condensed
verticils, which resemble those of *S. graciliflora*.
However, these plants usually have a coarser pubescence on the stems, while gland-dots are often present
on the calyx and on the lower surface of the leaves. No
glandular hairs are found on *S. graciliflora*.

Specimens with large leaves resemble *S. tubulosa*
(no. 2) but may be separated when flowers are
available by the shorter corolla tube which does not
exceed 10 mm in length. In *S. tubulosa* the corolla tube
is 12−18 mm long and the species has a more northerly
distribution from East Griqualand to Swaziland.

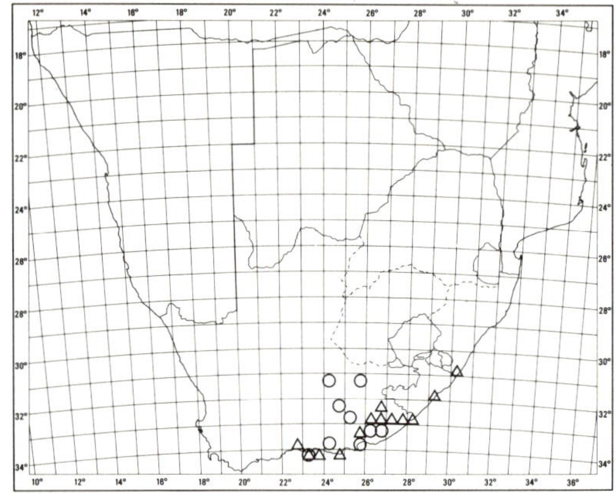

MAP 29. — △ **Stachys graciliflora**
○ **S. scabrida**

16. **Stachys scabrida** *Skan* in F.C. 5,1:
349 (1910); Codd in Bothalia 12: 188 (1977).
Lectotype: Cape, Bruintjieshoogte, *Burchell* 3037 (K, lecto.!; PRE!).

S. priorii Skan, l.c. 353 (1910). Type: Cape, Algoa
Bay, *Prior* s.n. (K., holo.!).

Perennial, decumbent to prostrate
herb; stems about 0,2−0,4 m long, sparingly
branched, subglabrous to scabrid with
strong retrorse-scabrid hairs or longer multicellular retrorse hairs. *Leaves* petiolate, drying dark brown; blade firm to coriaceous,
ovate to ovate-deltoid or deltoid, 12−30 ×
8−20 mm, eglandular, subglabrous to sparingly hispid, the upper surface usually with
scattered short to longish bulbous-based
hairs, apex usually acute, base deeply cordate with a wide sinus and distant rounded
auricles, margin regularly crenate to crenate-dentate, often somewhat thickened;
petiole 5−18 mm long. *Inflorescence* usually
slender, lax below, occasionally somewhat
condensed, of 2−6 verticils; verticils 2−6-
flowered; rhachis glabrous to retrorse-scabrid; bracts differentiated towards the apex.
Calyx subglabrous to sparingly hispid, 6−8
mm long. *Corolla* white, sometimes with
mauve spots on the lower lip; tube 6−8 mm
long, slightly arcuate; upper lip ascending,
5−6 mm long; lower lip horizontal, 6−9 mm
long.

A straggling herb in grass, fynbos or coastal scrub, extending from the southern Transkei to Knysna and, inland, to Steynsburg and Somerset East districts. Map 29.

Vouchers: *Bayliss* 8384; *Schonland* 3177.

S. scabrida appears to be related to *S. graciliflora* (no. 15) but the pubescence is coarser and more scabrid, while the leaves are thicker-textured and the hairs on the upper leaf surface tend to be bulbous-based. The inflorescences tend to be more slender, rather than subcapitate, as is the case in *S. graciliflora*. The distinction is by no means clear-cut, as indicated in the discussion of the latter species, and specimens such as *Story* 2445 tend to be somewhat intermediate. The type of *S. priorii* is somewhat intermediate between *S. scabrida* and *S. humifusa* (no. 20).

17. **Stachys cymbalaria** *Briq.* in Bull. Herb. Boissier sér. 2,3: 1088 (1903); Skan in F.C. 5,1: 352 (1910); Ross, Fl. Natal 303 (1972); Codd in Bothalia 12: 188 (1977). Type: Cape, Cradock, *Cooper* 516 (K, holo.!; W!).

S. aethiopica var. *tenella* Kuntze, Rev. Gen. 3,2: 262 (1898). Type: Cape, Cradock, *Kuntze* s.n.

S. cymbalaria var. *alba* Skan, l.c. 352 (1910). Type: Natal, Richmond, *Medley Wood* 1846 (K, holo.!; NH!).

Perennial prostrate herb; stems radiating from a central taproot, up to 0,3 m long, subglabrous or with few long slender spreading hairs or occasionally with a short sparse to fairly densely glandular pubescence. *Leaves* subsessile or shortly petiolate; blade broadly ovate-deltoid to suborbicular, 8−15 × 6−12 mm, subglabrous to thinly appressed-pubescent or glandular-puberulous, apex obtuse to rounded, base broadly cordate, margin crenate. *Inflorescence* lax, of 1−5 verticils; verticils 2-flowered; bracts elliptical, smaller than the calyx. *Calyx* puberulous to hispidulous, 5−6 mm long. *Corolla* purple, pink or white; tube 5−7 mm long; upper lip horizontal, 3−4 mm long; lower lip deflexed, 6 mm long.

Found among rocks in exposed mountain grassland at a few disjunct localities from Graaff-Reinet to Cradock through Transkei to the Kokstad area and in southern Natal. Map 30.

Vouchers: *Galpin* 10011; *Hilliard* 8106.

The species is characterized by the very small ovate leaves which are usually subglabrous to sparingly pubescent. Two gatherings from Mt Insizwa, *Schlechter* 6467 and *Hilliard & Burtt* 6568, are included with some hesitation as they are more markedly pubescent with somewhat narrower, subsessile leaves. These resemble *S. humifusa* (no. 20) but the calyx teeth are shorter and broader and the pubescence shorter than in that species.

18. **Stachys sublobata** *Skan* in F.C. 5,1: 354 (1910); Codd in Bothalia 12: 188 (1977). Lectotype: Cape, Swellendam district, Barrydale, *Galpin* 4425 (K, lecto.; PRE!).

Perennial many-stemmed herb; stems decumbent or ascending up to 0,3 m long, sparsely hispidulous with occasionally longish spreading hairs. *Leaves* shortly petiolate; blade small, often folded along the midrib, fairly thick in texture, narrowly triangular, 10−15 × 4−6 mm, sparingly hispidulous on both surfaces with gland-dots sometimes present beneath, apex acute, base openly cordate, margin deeply crenate with 4 or 5 lobe-like teeth up to 1,5 mm long on each side. *Inflorescence* lax, of a few 2 (−6)-flowered verticils; bracts leaf-like below becoming linear-lanceolate and subequal to the calyx above. *Calyx* thinly glandular-hispid, 6−7 mm long. *Corolla* mauve; tube 6−8 mm long; upper lip spreading to erect, 4−5 mm long; lower lip deflexed, 6−8 mm long.

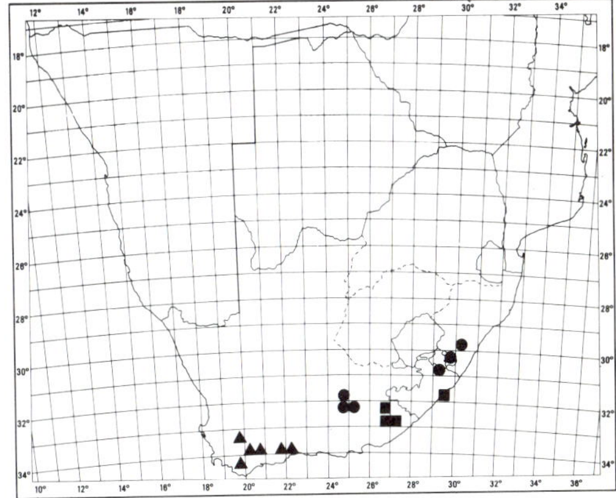

MAP 30. — ● Stachys cymbalaria
▲ S. sublobata
■ S. flexuosa

Found on hillsides in fynbos at altitudes of 300 to 900 m in the south-western Cape Province, recorded from Caledon to Mossel Bay districts and inland to Ladismith and Oudtshoorn districts. Map 30.

Vouchers: *Acocks* 20779; *Galpin* 4426.

Related to *S. cymbalaria* (no. 17) but may be distinguished by the deeply crenate, rather narrowly deltoid leaves and the more pronounced pubescence.

19. **Stachys flexuosa** *Skan* in F.C. 5,1: 352 (1910); Codd in Bothalia 12: 189 (1977). Type: Cape, Stockenstroom district, old Katberg Pass, *Galpin* 2393 (wrongly listed in F.C. as 2093) (K, holo.; PRE!; SAM!).

Perennial herb, branching at the base; stems slender, decumbent to ascending, sparingly branched, 0,15−0,25 m long, fairly densely hispid to villous-pilose with long spreading hairs and some short glandular hairs. *Leaves* petiolate; blade ovate, 10−20 × 6−15 mm, somewhat appressed villous with some bulbous-based hairs above, hispid mainly on the nerves beneath, apex obtuse to rounded, base truncate to subcordate, margin crenate; petiole 3−8 mm long. *Inflorescence* fairly dense, of few to several 2−6-flowered verticils; rhachis densely hispid with some glands; bracts leaf-like below, becoming smaller and subequal to the calyx above. *Calyx* hispid and with some glands, 7−8 mm long. *Corolla* purple; tube 6 mm long; upper lip erect, 4 mm long; lower lip deflexed, 8 mm long.

Known from a few localities, in the Transkei, eastern Cape Province, and Natal Drakensberg, among rocks in mountain grassland. Map 30.

Vouchers: *Galpin* 8379; *Rattray* 403.

The small ovate leaves with a rather wide sinus at the base are reminiscent of the *S. aethiopica* complex, but the relationship appears to be nearer to *S. obtusifolia* (no. 26) and *S. tysonii* (no. 27), from which it differs mainly in the smaller leaves, less densely villous stems and leaves, and in having few gland-dots on the leaves in contrast to the markedly glandular pubescence of *S. tysonii*.

Two gatherings, *Fourcade* 2281 and 4455 (both seen in STE), from near the mouth of the Krom River, Humansdorp district, can scarcely be separated from *S. flexuosa*. However, they are widely separated from the specimens cited above and grow under such different ecological conditions that further investigation is necessary before including them in *S. flexuosa*.

20. **Stachys humifusa** *Burch. ex Benth.*, Lab. 547 (1834); Benth. in DC., Prodr. 12: 476 (1848); Skan in F.C. 5,1: 358 (1910). Type: Cape, Bathurst district, Kowie, *Burchell* 3794 (K, holo.).

S. subsessilis Burch. ex Benth., Lab. 548 (1834); Benth. in E. Mey., Comm. 240 (1837); in DC., Prodr. 12: 476 (1848); Skan, l.c. 353 (1910). Type: Cape, near Port Elizabeth, *Burchell* 4326 (K, lecto.).

S. tenella Skan, l.c. 358 (1910). Type: Cape, Griqualand East, near Kokstad, *Tyson* 1790 (K, holo.; PRE!; SAM!).

Perennial (?) herb; stems procumbent to weakly ascending, branched or sometimes simple, up to 0,45 m long, often very slender, glabrous or with scattered short scabrid hairs, occasionally with short glandular hairs or slender long white hairs. *Leaves* sessile to shortly petiolate; blade usually drying brownish, somewhat leathery, ovate-deltoid to lanceolate-deltoid, 7−30 × 3−10 mm, often punctate, thinly strigose to subglabrous, apex obtuse to acute, base cordate, margin crenate, often slightly thickened. *Inflorescence* lax below, often compact above, of few to several 2−6-flowered verticils; bracts lanceolate-elliptic, entire, shorter than the calyx. *Calyx* glabrous to scabrid or sparingly hispid, 5−7 mm long. *Corolla* white to pale mauve; tube 6−7 mm long; upper lip ascending, 2−3 mm long; lower lip deflexed, 5−6 mm long.

Found in grassland or low-lying places from Knysna to King William's Town districts with one record in mountain grassland near Kokstad. Map 31.

Vouchers: *Bolus* 9139; *Galpin* 384.

More collecting is required in order that the correct limits of this species may be determined. *S. humifusa* and *S. subsessilis* grade into each other and the alleged difference in leaf shape does not hold good. The type of *S. priori* is somewhat intermediate between *S. humifusa* and *S. scabrida* (no. 16) and is included in synonymy under the latter; no other specimen exactly matching it has been seen. It is also not possible to separate the type of *S. tenella* from *S. humifusa* though its distance from the remainder of the distribution indicates that it requires closer study. As yet, no other gathering is known which matches it, though it also resembles some specimens now included in *S. cymbalaria* (no. 17). At the other end of the distribution range, near Knysna, two specimens collected by *Breyer* (TRV 23323, 23365), come near to *S. scabrida*, a species with larger, broader leaves and longer petioles, and this area should also be collected more thoroughly to determine whether there is a gradation between *S. scabrida* and *S. humifusa*.

21. **Stachys rivularis** *Wood & Evans* in J. Bot., Lond. 35: 489 (1897); Skan in F.C. 5,1: 358 (1910); Ross, Fl. Natal 303 (1972). Type: Natal, Mooi River, *Medley Wood* 6252 (NH, holo.!; K; PRE!).

S. schlechteri Gürke in Bot. Jb. 26: 74 (1898). Type: Natal, Mooi River, *Schlechter* 6837 (K; PRE!).

Perennial erect herb 0,2−0,3 m tall, branching near the base; stems several, simple, pilose with long spreading multicellular hairs and shorter gland-tipped hairs and with a hairy interpetiolar ridge at the nodes. *Leaves* sessile or the lower ones shortly petiolate; blade deltoid to ovate-deltoid, 15−30 × 5−12 mm, shortly appressed hispid, apex obtuse or subacute, base subcordate to truncate, margin crenulate, not or scarcely thickened. *Inflorescence* lax below, compact above, of several 4−6-flowered verticils; bracts ovate, similar to the leaves but smaller, usually longer than the calyx. *Calyx* sparingly hispid and with some short glandular hairs, 5−6 mm long. *Corolla* white with mauve spots on the lower lip; tube 5−6 mm long; upper lip ascending, 2,5−3 mm long; lower lip deflexed, 5−6 mm long.

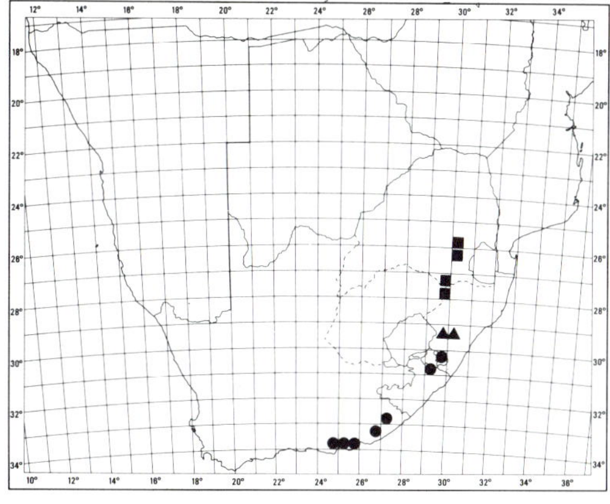

MAP 31. — ● **Stachys humifusa**
▲ **S. rivularis**
■ **S. erectiuscula**

A little-known grassland species found growing near Mooi River in the Natal Midlands at an altitude of about 1 500 m. Map 31.

Vouchers: *Mogg* 3331; *Schlechter* 6837.

Skan, l.c., considered *S. rivularis* to be possibly only a hairy variety of *S. humifusa* (no. 20), but the evidence indicates that the two are quite distinct, with *S. rivularis* having several stems, more rigidly erect, with shorter internodes, in addition to the more pronounced pubescence of stems, leaves and calyx. Its affinity is closer to *S. sessilis* (no. 24) and *S. obtusifolia* (no. 26), both of which can be separated by their even denser and stronger indumentum. Further material is desirable.

22. **Stachys erectiuscula** *Gürke* in Bot. Jb. 28: 315 (1900); Skan in F.C. 5,1: 357 (1910). Type: Transvaal, near Lydenburg, *Wilms* 1116.

S. erectiuscula var. *natalensis* Skan, l.c. 357 (1910); Ross, Fl. Natal 303 (1972). Syntypes: Natal, near Newcastle, *Medley Wood* 6349 (K, PRE!); 6795 (K).

Perennial, few-stemmed erect herb 0,25−0,5 m tall; stems sparingly branched, retrorse-hispid to scabrid and with a hairy interpetiolar ridge at the nodes. *Leaves* petiolate, drying brown or blackish; blade oblong-deltoid to ovate-deltoid, 20−40 × 5−15 mm, sparingly appressed pilose above with occasional bulbous-based hairs towards the margin, paler, reticulate and sparingly pilose beneath, especially on the nerves, apex obtuse to rounded, base truncate to subcordate, margin finely crenate; petiole 5−10 mm long. *Inflorescence* simple or with a pair of branches at the base, lax below, dense towards the apex; verticils 2−10-flowered; upper bracts subequal to the calyx or longer, hispidulous and gland-dotted below. *Calyx* somewhat sparingly hispidulous, 5−7 mm long. *Corolla* colour unknown; tube 5−7 mm long; upper lip ascending, 5−6 mm long; lower lip slightly deflexed, 8−10 mm long.

A little-known grassland species last collected in 1911, recorded from south-eastern Transvaal and northern Natal. Map 31.

Vouchers: *Burtt Davy* 7661; *Galpin* 13099.

Related to *S. nigricans* (below) but can be distinguished by the sparser and shorter pubescence of the calyx and leaves and by the retrorse hairs on the rhachis and stems. The record from Pretoria District requires confirmation.

23. **Stachys nigricans** *Benth.* in E. Mey., Comm. 238 (1838); in DC., Prodr. 12: 471 (1848); Wood, Natal Pl. 3: t.271 (1902); Skan in F.C. 5,1: 355 (1910); Ross, Fl. Natal, 303 (1972); Compton, Fl. Swaziland 496 (1976). Lectotype: Natal, between Umzimkulu and Umkomaas Rivers, *Drège* 4729b (K, lecto.; PRE, photo.!).

Perennial, erect, single- to few-stemmed herb, 0,4−1,2 m tall, with several horizontal fusiform roots; stems simple or sparingly branched above, villous to hispid with longish, spreading to antrorse multicellular hairs and with an interpetiolar ridge at the nodes. *Leaves* sessile or subsessile, drying dark brown to blackish; blade oblong-lanceolate to linear-lanceolate, 30—60 × 3—10 mm, strigose with usually bulbous-based appressed hairs, apex obtuse, base truncate to subcordate, margin crenate, slightly thickened. *Inflorescence* simple or occasionally with a pair of branches near the base, lax below, denser above; verticils usually 6-flowered; lower bracts leaf-like, becoming differentiated and equal to or shorter than the calyx above. *Calyx* fairly densely appressed-hispid, 5—6 mm long. *Corolla* usually white, occasionally speckled, or tinged with pink or mauve; tube 6—7 mm long; upper lip ascending, subrotund, concave, 3 mm long; lower lip deflexed, 6—7 mm long.

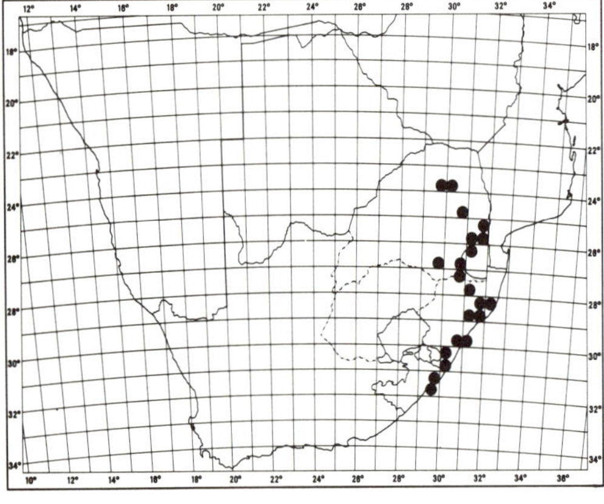

MAP 32. — **Stachys nigricans**

Found in grassland, usually subjected to frequent burning, in the mountains of north-eastern and eastern Transvaal and Swaziland at altitudes of 1 300—1 700 m, continuing through the semi-coastal area of Natal to Port St Johns in the Transkei. Map 32.

Vouchers: *Acocks* 13164; *Codd* 8122; *Galpin* 9610; *Medley Wood* 8320.

S. nigricans, *S. sessilis* (no. 24) and *S. simplex* (no. 25) form a closely related group with the same kind of pubescence on stems, leaves and calyx, consisting of strong, villous to hispid hairs. The hairs on the upper surfaces of the leaves are appressed and tend to be bulbous-based, especially towards the margin of the leaf. *S. nigricans* may be confused with *S. sessilis*, but the leaves of the former tend to be narrower and to dry a characteristic blackish brown colour. Although the two show a similar distribution pattern, they apparently

do not overlap, with *S. nigricans* having a semi-coastal distribution in Natal and Transkei while *S. sessilis* is found more inland at higher altitudes.

24. **Stachys sessilis** *Gürke* in Bot. Jb. 26: 74 (1898); Skan in F.C. 5,1: 355 (1910); Ross, Fl. Natal 304 (1972). Type: Natal, Inchanga, *Medley Wood* 4806 (K; NH!; PRE!; W!).

Perennial, erect, single- to few-stemmed herb, 0,15—1 m tall; stems arising annually at the end of a short horizontal rhizome bearing clusters of fusiform roots, usually simple, hispid-villous with longish spreading hairs and with an interpetiolar ridge at the nodes. *Leaves* sessile or the lower leaves shortly petiolate; blade drying greenish to brown, deltoid-lanceolate or oblong-lanceolate to oblong, or ovate-deltoid, broadest at the base to near the middle, 20—50 × 6—18 mm, strigose with the appressed hairs on the upper surface usually bulbous-based, apex rounded to obtuse, base truncate to rounded, margin crenate, slightly thickened or inrolled. *Inflorescence* simple, lax below, denser above; verticils 4—8-flowered; lower bracts leaf-like, becoming smaller, lanceolate and equal to or shorter than the calyx above. *Calyx* fairly densely appressed-hispid, 5—6 mm long. *Corolla* white, usually with purplish spots on the lower lip; tube 5—6 mm long; upper lip ascending, 2,5—3 mm long; lower lip deflexed, 7—8 mm long.

A species of dense grassland, usually subjected to frequent burning, found in southern Transvaal and extending on the Drakensberg to adjoining parts of the Orange Free State and Lesotho, to the Midlands of Natal and into Transkei as far as Umtata. Map 33.

Vouchers: *Devenish* 1570; *Hilliard & Burtt* 8938; *Killick* 1286.

Allied to *S. nigricans* (no. 23) and some specimens are difficult to assign with certainty. The differences are discussed under that species. In northern Natal it shows a tendency to grade into *S. simplex* (below) but the latter species can usually be separated by the larger, broader and fewer leaves which are placed low down on the stems. Although closely related, it is considered that the two can be maintained as distinct species. *S. obtusifolia* (no. 26) and *S. tysonii* (no. 27) are sometimes superficially similar but the hairs on the upper surfaces of the leaves are longer and more slender, and lack the thickened base which is characteristic of *S. nigricans*, *S. sessilis* and *S. simplex*, while the pubescence on the lower surfaces is softer and freely gland-dotted in *S. tysonii*.

25. **Stachys simplex** *Schltr.* in J. Bot., Lond. 35: 221 (1897); Skan in F.C. 5,1: 356

(1910); Ross, Fl. Natal 304 (1972); Compton, Fl. Swaziland 497 (1976). Type: Transvaal, Barberton, *Galpin* 1006 (K, holo.; PRE!; SAM!).

S. chrysotrichos Gürke in Bot. Jb. 28: 316 (1900). Type: Transvaal, between Middelburg and Crocodile River, *Wilms* 1137 (K).

S. pascuicola Briq. in Bull. Herb. Boissier sér. 2,3: 1086 (1903). Type: Transvaal, "Elandsspruitbergen" (Steenkampsberg), *Schlechter* 3844 (K; PRE!; SAM!; W!).

Perennial, usually single-stemmed herb; stems simple or sparingly branched, densely villous, decumbent to semi-prostrate with the inflorescence ascending to 0,2−0,4 m tall. *Leaves* in few pairs, rather close together on the lower half of the stem, the lower petiolate, the upper subsessile to shortly petiolate; blade rather thick-textured, drying brownish, ovate, ovate-oblong to ovate-lanceolate, 30−55 × 12−25 mm, upper surface fairly densely appressed villous with long often bulbous-based hairs, hairs shorter and mainly on the nerves beneath, apex obtuse to rounded, base rounded, margin regularly serrate-crenate. *Inflorescence* simple, lax below, dense above, verticils 4−6-flowered; bracts much smaller than the leaves, lanceolate, about as long as the calyx. *Calyx* densely villous, 6−7 mm long. *Corolla* white to mauve; tube 5−6 mm long; upper lip ascending, 2,5 mm long; lower lip deflexed, 4−5 mm long.

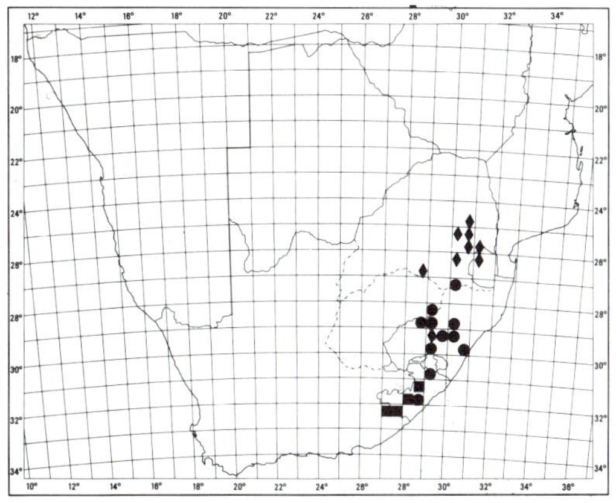

MAP 33. — ● **Stachys sessilis**
◆ **S. simplex**
■ **S. obtusifolia**

Found in mountain grassland subject to frequent burning, usually on stony slopes, at altitudes of 1 600 to 2 100 m, from the eastern and south-eastern Transvaal to the northern Drakensberg region of Natal. Map 33.

Vouchers: *Codd* 8136; 8307.

Allied to *S. sessilis* (no. 24) and, for distinguishing characters, see notes under that species.

26. **Stachys obtusifolia** *MacOwan* in Kew Bull. 1893: 13 (1893); Skan in F.C. 5,1: 356 (1910); partly, excluding *Tyson* 2561. Lectotype: Cape, Baziya, *Baur* 75 (K, lecto.; PRE!; SAM!).

S. obtusifolia var. *flanaganii* Skan, l.c. 356 (1910). Type: Cape, Stutterheim, Kabousie River, *Flanagan* 496 (K, holo.; PRE!).

Perennial several-stemmed or bushy herb, 0,2−0,3 m tall; stems erect often from a shortly creeping or decumbent base, sparingly branched, with long slender spreading hairs and shorter gland-tipped hairs. *Leaves* shortly petiolate below, subsessile above; blade drying brownish, broadly ovate to ovate-elliptic, 17−35 × 8−20 mm, upper surface fairly densely appressed pilose-villous with long slender multicellular hairs, not or scarcely bulbous-based, under-surface reticulate and villous on the nerves, not conspicuously gland-dotted, apex rounded, base cordate to subcordate, margin crenate, slightly thickened. *Inflorescence* fairly compact; verticils 4−6-flowered; bracts ovate, leaf-like, becoming smaller and subequal to the calyx upwards, densely villous and often glandular. *Calyx* densely pilose and often glandular, 6−7 mm long. *Corolla* colour unknown; tube 6−7 mm long; upper lip ascending, 3−4 mm long; lower lip deflexed, 6−7 mm long.

A somewhat rare species, sometimes locally frequent, on grassy slopes at altitudes of 700−1 200 m in the Transkei and eastern Cape Province. Map 33.

Vouchers: *Acocks* 9404; 12541.

A strongly aromatic, softly pubescent herb, sometimes confused with *S. sessilis* (no. 24) which has fewer to solitary, taller stems and harsher pubescence with bulbous-based hairs on the upper leaf surfaces while the bracts are smaller and more lanceolate than in *S. obtusifolia*. Also related to *S. tysonii* (below) which has shorter, softer and gland-dotted pubescence on the under-surfaces of the leaves, longer petioles and a more inland distribution at higher altitudes than *S. obtusifolia*.

Skan, l.c., upheld two varieties, of which var. *flanaganii* is not significantly distinct from *S. obtusifolia*. Var. *angustifolia* has markedly gland-dotted under-surfaces of the leaves and the leaves have longish petioles, so it is now included in *S. tysonii*, but more material of both *S. tysonii* and *S. obtusifolia* is required for further study.

27. Stachys tysonii *Skan* in F.C. 5,1: 357 (1910). Type: Cape, Griqualand East, near Clydesdale, *Tyson* 2561 (K, holo.!; BOL!; SAM!).

S. obtusifolia MacOwan var. *angustifolia* Skan, l.c. 356 (1910). Type: Orange Free State, Besters Vlei, Witzieshoek, *Bolus* 8240 (K, holo.; BOL!; PRE!).

Perennial, few- to several-stemmed herb, 0,15−0,3 m tall; stems erect often from a decumbent base, sparingly branched, glandular-pilose with long slender hairs and short gland-tipped hairs. *Leaves* petiolate; blade drying green to brownish, ovate to ovate-lanceolate or ovate-oblong, 15−35 × 8−20 mm, upper surface softly to stiffly appressed pubescent, under-surface usually with copious gland-dots, apex rounded, base usually deeply cordate, margin regularly crenate-serrate; petiole 4−12 mm long. *Inflorescence* lax below, fairly dense above; verticils 4−6-flowered; bracts leaf-like below becoming smaller, lanceolate and subequal to the calyx above. *Calyx* shortly glandular-pubescent to glandular-pilose, 6−7 mm long. *Corolla* whitish to mauve with darker flecks on the lower lip; upper lip ascending, 2−3 mm long; lower lip deflexed, 5−7 mm long.

Found in mountain grassland at altitudes of 1 000−2 800 m with scattered records from the Orange Free State, Natal, Lesotho and the Cape Province. Map 34.

Vouchers: *Acocks* 23875; *Jacot Guillarmod* 1255.

Related to *S. obtusifolia* (no. 26), *S. tysonii* is distinguished by the longer petioles and more oblong cordate-based leaves, and by the generally shorter and more glandular pubescence over the whole plant. For example, the lower surface of the leaf is usually shortly and softly pubescent and freely gland-dotted. On this basis, *S. obtusifolia* var. *angustifolia* Skan is now included in *S. tysonii*. However, much more material is required before the limits of *S. obtusifolia* and *S. tysonii* can be confidently assessed.

28. Stachys arvensis *L.*, Sp. Pl. edn. 2: 814 (1763); Benth. in DC., Prodr. 12: 477 (1848); Bolus & Wolley-Dod in Trans. S. Afr. phil. Soc. 14: 310 (1904); Skan in F.C. 5,1: 354 (1910); Salter in Fl. Cape Penins. 698 (1950). Type: from Europe, Hort. Cliff. Herb. (BM).

Annual herb branching from the base; stems erect or decumbent, simple or branched, 0,2−0,4 m long, hispid-pilose with long spreading hairs. *Leaves* petiolate; blade broadly ovate, 20−30 × 10−20 mm, appressed-pilose on both surfaces, apex rounded, base shallowly cordate to truncate, margin crenate to crenate-serrate; petiole up to 10 mm long. *Inflorescence* of several verticils, lax below, crowded towards the apex; verticils (2−) 4−6-flowered; bracts leaf-like below, becoming smaller and subequal to the calyx above. *Calyx* hirsute with long multicellular hairs and some gland-tipped hairs, 6−7 mm long. *Corolla* scarcely longer than the calyx, mauve; tube 5 mm long; upper lip 1,5 mm long; lower lip 2 mm long.

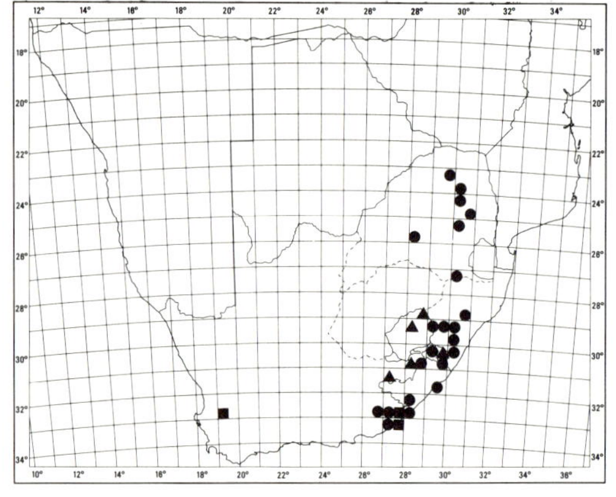

MAP 34. — ▲ **Stachys tysonii**
■ **S. arvensis**
● **S. caffra**

Indigenous in Europe, the Middle East, north Africa and the Atlantic Islands, now widespread throughout the World and introduced into South Africa before the end of the 17th Century (represented in Herb. Oldenland fide Burm. f., Fl. Cap. Prodr. 16, 1768); found as a garden weed mainly in the south-western Cape Province, with two records from East London. Map 34.

Vouchers: *Acocks* 23491; *Hanekom* 938.

Characterised by the very small flowers, up to 7 mm long, scarcely exceeding the calyx in length.

29. Stachys caffra *E. Mey. ex Benth.* in DC., Prodr. 12: 495 (1848); Skan in F.C. 5,1: 366 (1910); Ross, Fl. Natal 303 (1972).

Type: Transkei, between Umtata and Umzimvubu Rivers, *Drège* 4750 (K, lecto.).

Erect soft freely branched shrub, 1—3 m tall; stems slender, more or less stellate-tomentulose, occasionally glabrescent with age. *Leaves* petiolate; blade thin-textured, lanceolate, 30—90 × 10—30 mm, upper surface thinly hispidulous, lower surface paler and stellate-tomentose, apex acute to acuminate, base cuneate to obtuse, margin minutely serrulate except in the lower part; petiole 5—10 mm long. *Inflorescence* terminating the branchlets, of several spaced verticils, denser towards the apex; verticils 4—10-flowered, cymes occasionally pedunculate; bracts similar to the leaves, becoming gradually smaller upwards. *Calyx* stellate-tomentulose, 4—5 mm long. *Corolla* white, cream or greenish yellow; tube 5—6 mm long, slightly curved; upper lip 2,5—3 mm long; lower lip 5 mm long.

A soft understorey shrub of forest margins and shady stream banks in the northern, eastern and central Transvaal, the midlands and foothills of the Drakensberg in Natal, at altitudes of 1 300—2 000 m, extending to semi-coastal areas of the Transkei and eastern Cape Province, as far south as the Peddie district. Map 34.

Vouchers: *Junod* 4331; *Killick* 1895; *Schlechter* 6277.

Easily recognized among the species with stellate hairs by its soft, shrubby habit with slender branches and large, lanceolate, dark green leaves.

30. **Stachys hyssopoides** *Burch. ex Benth.*, Lab. 558 (1834); Benth. in E. Mey., Comm. 240 (1838); in DC., Prodr. 12: 495 (1848); Skan in F.C. 5,1: 365 (1910); Jacot Guill., Fl. Lesotho 237 (1971); Ross, Fl. Natal 303 (1972). Type: Cape, near Kuruman, *Burchell* 2653 (K, holo.).

S. coerulea Burch. ex Benth., Lab. 558 (1834); Benth. in DC., Prodr. 12: 495 (1848). Type: Cape, at junction of Vaal and Riet Rivers, *Burchell* 1775 (K, holo.).

S. macilenta E. Mey. ex Benth. in E. Mey. Comm. 240 (1838); Benth. in DC., Prodr. 12: 495 (1848). Type: Cape, near Shiloh, *Drège* (K, holo.; SAM!).

Perennial herb with a creeping rhizomatous rootstock; stems erect to decumbent, 0,2—0,6 m tall, simple or sparingly branched, glabrous or with scattered stellate hairs or sometimes stellate-tomentose. *Leaves* scarcely petiolate, sometimes fascicled, the base of the leaf somewhat clasping the stem and forming an interpetiolar ridge; blade somewhat coriaceous, linear to oblanceolate or linear-oblanceolate, 20—55 × 2—8 (—10) mm, glabrous to thinly stellate-pubescent, rarely stellate-tomentose, apex usually obtuse, often apiculate, base attenuate, margin entire or sparingly toothed towards the apex. *Inflorescence* of few to several spaced verticils; verticils usually 2-flowered, occasionally 4- or 6-flowered; bracts lanceolate, shorter or as long as the calyx. *Calyx* thinly to fairly densely grey stellate-tomentulose, usually with a fringe of white hairs along the margins of the teeth, 6—7 mm long. *Corolla* pink or mauve, often with darker flecks, to purple; tube 6—7 mm long; upper lip spreading to erect, 5 mm long; lower lip somewhat deflexed, 7 mm long.

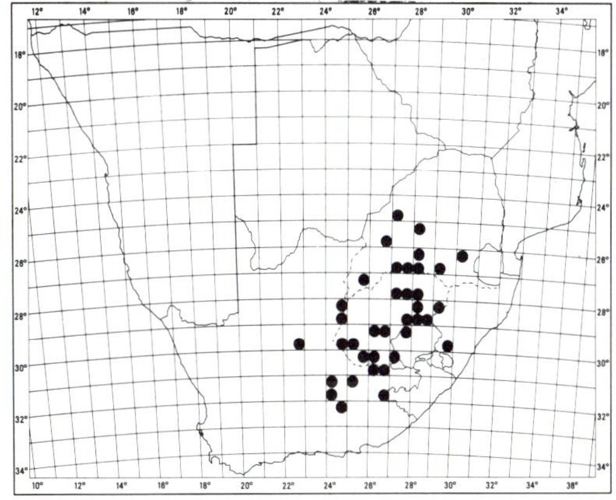

MAP 35. — **Stachys hyssopoides**

Often locally common on black clay or heavy loam soils, in depressions or on river banks, in the southern and south-western Transvaal, northern Natal, Lesotho, Orange Free State and northern, central and southern Cape Province. Map 35.

Vouchers: *Codd* 4463; *Galpin* 2251; *Medley Wood* 8253; *Schlechter* 3548.

Because of its underground rhizomes, it is occasionally recorded as a possible weed of disturbed areas. An infusion of the plant is used for chest complaints. Commonly known as "Pienksalie" (Pink Sage). Related to the following species, *S. dregeana*, from which it is readily separated by its narrower, firmer and more glabrous leaves.

31. Stachys dregeana *Benth.* in E. Mey., Comm. 240 (1838); in DC., Prodr. 12: 494 (1848); Skan in F.C. 5,1: 362 (1910); Jacot Guill., Fl. Lesotho 237 (1971). Lectotype: Cape, Wittebergen, *Drège* 7949c (K, lecto.).

S. foliosa Benth. in E. Mey., Comm. 241 (1838); in DC., Prodr. 12: 493 (1848); *S. rugosa* var. *foliosa* (Benth.) Skan, l.c. 359 (1910). Type: Cape, Sneeuwberg, *Drège* 3584b (K, holo.!).

S. lasiocalyx Schltr. in J. Bot., Lond. 36: 317 (1898). *S. dregeana* var. *lasiocalyx* (Schltr.) Skan, l.c. 362 (1910); Jacot Guill., l.c. (1971); Ross, Fl. Natal 303 (1972). Type: Mont-aux-Sources, *Thode* s.n.

S. dregeana var. *tenuior* Skan, l.c. 362 (1910). Lectotype: Cape, Andriesberg, *Galpin* 2031 (K, lecto.; PRE!).

Stems 1—several arising annually from a perennial taproot, erect, herbaceous, 0,1—0,4 m tall, sparingly to densely floccose-tomentose. *Leaves* sessile; blade rather thick-textured, oblong-linear or somewhat spathulate to elliptic-ovate, 15—60 × 3—17 mm, upper surface thinly to fairly densely stellate-pubescent, undersurface denser and often floccose, apex obtuse to rounded, base obtuse to somewhat narrowed, margin entire or shallowly crenate. *Inflorescence* of few to several verticils; verticils 2 (−4)-flowered; bracts leaf-like, becoming smaller upwards. *Calyx* densely and finely stellate-tomentose to floccose, 6—8 mm long. *Corolla* pink to mauve or purple; tube 6 mm long; upper lip 3—4 mm long; lower lip 6—8 mm long.

Found in subalpine grassland at altitudes of 2 000—3 000 m in the Drakensberg region of Lesotho and Natal and on adjacent mountain ranges in the north-eastern Cape Province. Map 36.

Vouchers: *Dieterlen* 905; *Galpin* 6817; *Killick & Marais* 2201.

A variable species which requires further study. At present the variation does not fall into a clear-cut pattern and so no purpose can be seen in upholding the varieties recognized in Flora Capensis. Its nearest affinity is with *S. hyssopoides* (no. 30) but it does not form rhizomes and the leaves are more markedly tomentose.

32. Stachys dinteri *Launert* in Mitt. bot. StSamml., Münch. 2: 313 (1957); Launert & Schreiber in F.S.W.A. 123: 30 (1969). Type: S.W.A./Namibia, Maltahöhe district, Tourlossie, *Dinter* 8285 (M, holo.; PRE!).

Shrublet 0,2—0,4 m tall, freely branching from a stout woody taproot; stems densely white stellate-floccose. *Leaves* sessile; blade fairly thick-textured, rugose, obovate to elliptic-obovate, 15—30 × 7—10 mm, upper surface greenish and thinly stellate-pilose, denser and greyish beneath, apex obtuse to rounded, base cuneate, margin somewhat coarsely crenate. *Inflorescence* of pedunculate, 2—5-flowered cymes in the axils of the upper leaves or sometimes starting low down. *Calyx* fairly densely stellate-pilose, 7—9 mm long. *Corolla* reddish purple; tube 5 mm long; upper lip concave, 3—4 mm long; lower lip horizontal, 5—6 mm long.

A species of restricted distribution in dry watercourses in the mountains of the Maltahöhe and Luderitz districts, S.W.A./Namibia. Map 36.

Vouchers: *Giess* 10446; *Strey* 2304; 2133.

Related to *S. cuneata* (below), but easily distinguished by the pedunculate cymes and the coarser, more floccose indumentum.

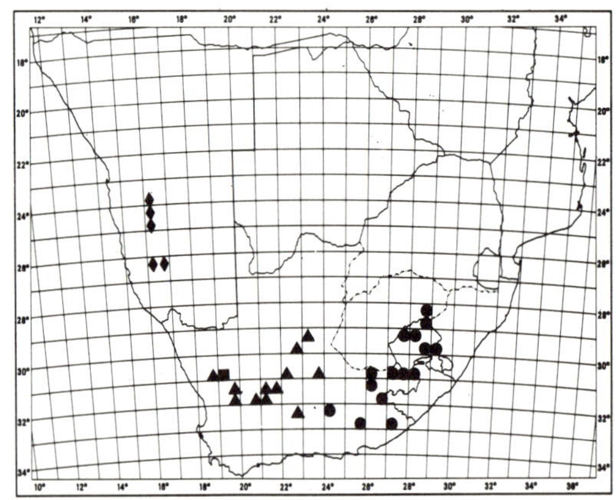

MAP 36. — ● **Stachys dregeana**
◆ **S. dinteri**
▲ **S. cuneata**
■ **S. zeyheri**

33. Stachys cuneata *Banks ex Benth.*, Lab. 560 (1834); Benth. in DC., Prodr. 12: 493 (1848); Skan in F.C. 5,1: 363 (1910). Type: Cape, *Masson* in Herb. Banks (BM, holo.; PRE, photo.!).

S. denticulata Burch. ex Benth., Lab. 560 (1834); Benth. in DC., Prodr. 12: 493 (1848). Type: Cape, Sutherland Div., Great Riet River, *Burchell* 1369 (K, holo.).

Shrub 0,6−1 m tall, branched; stems densely covered with a thick whitish felt-like tomentum, becoming grey with age. *Leaves* subsessile; blade thick-textured, obovate to oblanceolate, 10−30 × 5−10 mm, upper surface rugose, grey-green and finely (sometimes sparsely) stellate-tomentose to grey felted, paler grey-felted beneath, apex obtuse to rounded, base cuneate, margin finely to coarsely crenate, especially in the upper two-thirds. *Inflorescence* of several 2 (−6)-flowered verticils in the axils of the upper leaves, or upper leaves becoming smaller and bract-like. *Calyx* densely stellate-tomentose, strongly ribbed, 6−8 mm long. *Corolla* pink, mauve or purple; tube 6−8 mm long, slightly curved; upper lip concave, 3−4 mm long; lower lip horizontal, 6−7 mm long.

Found in dry watercourses on dolerite hills in the upper, central and western Karoo. Map 36.

Vouchers: *Acocks* 1740; 16401; 18875.

The distinction between *S. cuneata* and *S. dinteri* (no. 32) is discussed under the latter. *S. cuneata* is sometimes confused with *S. rugosa* (no. 37) but has usually shorter, more distinctly crenate leaves with the upper surface greenish yellow and often thinly pubescent. Occasional intermediates are found where the two overlap, for example, in the Victoria West district, which seems to indicate that the two may hybridize.

The plants are strongly and unpleasantly aromatic, being described as "foetid" and "like dead sheep"; nevertheless, they are browsed by stock and an infusion of the leaves is used as a medicine.

34. **Stachys zeyheri** *Skan* in F.C. 5,1: 363 (1910). Type: Cape, Calvinia district, *Zeyher* 1338 (K, holo.; BOL!; PRE!; SAM!).

A twiggy shrub, height unknown; branchlets at first densely white stellate-tomentose, glabrescent and reddish brown with age. *Leaves* very small, shortly petiolate; blade thick-textured, ovate to broadly ovate, 5−10 × 5−8 mm, densely and shortly white-tomentose on both surfaces, apex rounded, base obtuse to truncate, margin crenulate. *Inflorescence* of a few 2-flowered verticils in the terminal leaves of short shoots. *Calyx* densely stellate-tomentose, 5−6 mm long. *Corolla* colour not recorded; tube 5−6 mm long; upper lip 2 mm long, lower lip 3−4 mm long.

Known from only a few gatherings from mountain slopes in Namaqualand, Calvinia and Vanrhynsdorp districts. Map 36.

Vouchers: *Marloth* 12894; *Pearson* 3393 (BOL).

A little-known but distinctive, twiggy shrub with glabrescent branchlets and very small ovate, crenulate, densely tomentose leaves.

35. **Stachys spathulata** *Burch. ex Benth.*, Lab. 559 (1834); Benth. in E. Mey., Comm. 240 (1838); in DC., Prodr. 12: 494 (1848); Skan in F.C. 5,1: 362 (1910); Launert & Schreiber in F.S.W.A. 123: 31 (1969). Type: Cape, Griqualand West, *Burchell* 1738 (K, holo.!).

S. minima Gürke in Bull. Herb. Boissier 6: 550 (1898). Syntypes: Cape, Griqualand West, *Rehmann* 3360 (Z!); O.F.S., Olifantsfontein, *Rehmann* 3532.

S. pachycalamna Briq. in Bull. Herb. Boissier sér. 2,3: 1087 (1903). Type: S.W.A./Namibia, Auasberge, *Dinter* 814.

S. karasmontana Dinter in Fedde Reprium 17: 203 (1921). Type: S.W.A./Namibia, near Klein-Karas, *Schäfer* 316.

Perennial rhizomatous herb or sub-shrublet, branching near the base; stems erect or ascending, often decumbent at the base, 80−250 (−300) mm tall, whitish stellate-felted except on older stems. *Leaves* often crowded, sessile or subsessile; blade linear, linear-spathulate or linear-oblanceolate to rarely obovate-elliptic, often folded along the midrib, 15−50 × 2−12 mm, densely greyish white stellate-felted on both surfaces or more thinly and darker above, apex rounded, base attenuate to a

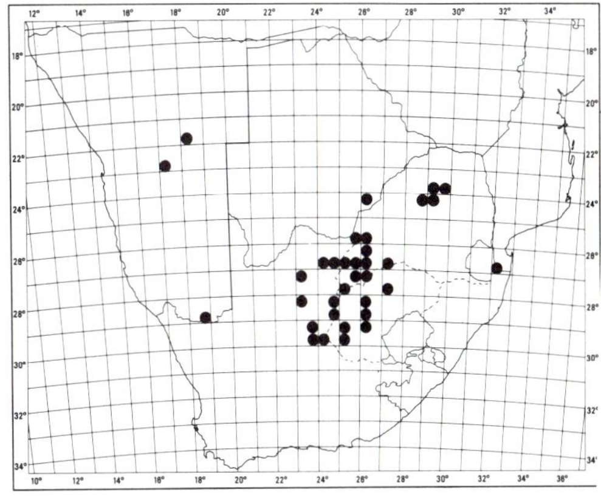

MAP 37. — **Stachys spathulata**

somewhat clasping base, forming an inter-petiolar ridge, margin entire. *Inflorescence* of few to several 2 (−4)-flowered verticils; bracts resembling the leaves below, becoming progressively shorter upwards. *Calyx* densely and shortly grey stellate-tomentose, 4−7 mm long. *Corolla* pink, mauve or rosy mauve; tube 4−5 mm long; upper lip ascending, 2,5−3 mm long; lower lip more or less horizontal, 4−5,5 mm long.

A widespread species from S.W.A./Namibia to Botswana, northern Cape, western Transvaal and western O.F.S., appearing again in northern KwaZulu and Mozambique. Often locally common on heavy soils in depressions, on river banks or in water courses, under arid to semi-arid conditions. Map 37.

Vouchers: *Leistner* 1016: *Merxmüller & Giess* 1271; *Schlechter* 4595; *Ward* 4516.

Similar to *S. hyssopoides* (no. 30) in habit and ecology and the two overlap to some extent in the western Transvaal, northern Cape Province and western Orange Free State. Like *S. hyssopoides*, *S. spathulata* tends to spread by rhizomes, often occupying disturbed places, but can be recognized by the dense, felt-like indumentum on all parts. Occasional specimens may be intermediate between the two. Some specimens of *S. spathulata* may have very narrow leaves and these may be confused with the next species, *S. linearis*, which has a distinctly shrubby habit, while the leaves are acute at the apex and the calyx teeth tend to be longer. The leaves of *S. spathulata* have a strong, unpleasant smell and are used medicinally, the plant being known as Teebossie or Boesmantee.

36. **Stachys linearis** *Burch. ex Benth.*, Lab. 559 (1834); in DC., Prodr. 12: 494 (1848). Type: near Phillipstown, *Burchell* 2717 (K, holo.!).

S. rugosa Ait. var. *linearis* (Burch. ex Benth.) Skan in F.C. 5,1: 359 (1910). Type: as above.

S. rosmarinifolia Benth., Lab. 559 (1834); in DC., Prodr. 12: 494 (1848). Type: a specimen in Herb. Vahl (C!).

— var. *burkei* Benth. in DC., Prodr. 12: 494 (1848). Type: near Grahamstown, *Burke* s.n. (K, holo.!).

S. recurva Gürke in Bull. Herb. Boissier 6: 549 (1898). Syntypes: Cape, Wittebergen, *Rehmann* 2883; Roggeveld, *Rehmann* 3196 (Z!).

A spreading or erect, branched shrublet 0,2−0,4 m tall with a strong woody taproot; stems decumbent to erect, whitish stellate-felted, occasionally glabrescent with age. *Leaves* sessile, forming an interpetiolar ridge; blade channelled above and tending to fold along the midrib, linear or rarely linear-lanceolate, 15−45 × 1,5−2,5 (−3,5) mm, densely and finely yellowish grey stellate-felted on both surfaces, rarely

greenish grey, narrowing gradually to base and apex, margin entire. *Inflorescence* of few to several 2 (rarely 4)-flowered verticils; bracts resembling the leaves, becoming progressively shorter upwards. *Calyx* densely and shortly grey stellate-felted, 5−8 mm long. *Corolla* pink to mauve or purplish; tube 6−8 mm long; upper lip ascending, 4 mm long; lower lip horizontal, 6−7 mm long.

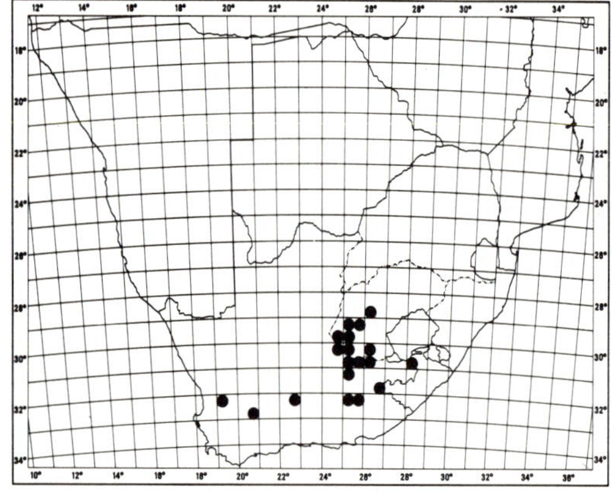

MAP 38. — **Stachys linearis**

Locally common on flats and especially on dolerite hills in the east central Karoo, extending to the south-western Orange Free State, the mountains of the north-eastern Cape Province and, westwards, with outliers on mountains in the Sutherland and Clanwilliam districts. Map 38.

Vouchers: *Acocks* 8714; *Bolus* 52; *Galpin* 2608; *Hutchinson* 3065.

Although some specimens resemble *S. spathulata* (no. 35; for differences see there), its relationship is closer to the following species, *S. rugosa*, and it was placed as a variety of *S. rugosa* in Flora Capensis. However, *S. linearis* is a smaller, more compact bush with narrower leaves and usually 2-flowered verticils; it can apparently be readily distinguished in the field from *S. rugosa*, so it is felt that *S. linearis* can be maintained as a separate species.

The leaves are strongly and unpleasantly aromatic and an infusion is taken medicinally, being known as Vaaltee, Boesmantee or Bushman Tea. It is said to stimulate the flow of milk in nursing mothers.

37. **Stachys rugosa** *Ait.*, Hort. Kew. 2: 303 (1789); Benth. in E. Mey., Comm. 241 (1838); in DC., Prodr. 12: 493 (1848); Skan in F.C. 5,1: 359 (1910), partly, excl. var.

linearis; Launert & Schreiber in F.S.W.A. 123: 31 (1969); Jacot Guill., Fl. Lesotho 237 (1971). Type: Hort. Kew., introduced from the Cape, *Masson*.

Sideritis pallida Thunb., Prodr. 95 (1800); Fl. Cap. edn Schult. 445 (1823). Type: Cape, *Thunberg* s.n. (UPS, microfiche 562/13407!).

S. rugosa Thunb., Prodr. 95 (1800); Fl. Cap. edn Schult. 445 (1823). Type: Cape, *Thunberg* s.n. (UPS, microfiche 562/13411!).

Stachys jugalis Burch. ex Benth., Lab. 562 (1834). Type: Cape, Juk River, *Burchell* 1233 (K, holo.).

S. rugosa Ait. var. *longiflora* Benth. in E. Mey., Comm. 241 (1837); in DC., Prodr. 12: 474 (1848). Type: Cape, Modderfontein, *Drège* (K, holo.).

S. desertii Benth. in DC., Prodr. 12: 494 (1848). Type: Karoo, *Ecklon* (K, holo.).

S. multiflora Benth. in DC., Prodr. 12: 492 (1848); Skan in F.C. 5,1: 361 (1910). Type: Cape, "inter Lekkersing et Noagas", *Drège* (K, holo.!).

S. crenulata Briq. in Bot. Jb. 19: 192 (1894); Bull. Herb. Boissier sér. 2,3: 1087 (1903). Type: S.W.A./ Namibia, *Steingröver* 8.

Shrub 0,3—1,2 m tall, freely branched; branches ascending, densely whitish stellate-tomentose, becoming greyish black with age. *Leaves* sessile or subsessile; blade linear-lanceolate to elliptic or ovate-lanceolate, variable in shape and size, 13—80 × 3—20 mm, densely stellate-tomentose to woolly, upper surface often greenish or drying blackish, whitish beneath, or more or less concolorous, usually conspicuously rugose, apex obtuse to acute, base obtuse to cuneate, margin usually finely and obscurely crenulate, occasionally distinctly crenate or entire. *Inflorescence* of several to many (2—) 6 (—8)-flowered verticils, well spaced below, denser above; bracts resembling the leaves below, becoming progressively smaller upwards. *Calyx* densely and coarsely stellate-felted, 6—8 mm long. *Corolla* yellow or shades of pink, mauve or purple, often mottled; tube 5—7 mm long; upper lip ascending, 3—6 mm long; lower lip horizontal, 6—8 mm long.

A variable species with an odd disjunct distribution, occurring on arid rocky formations in the western Karoo, Namaqualand and southern S.W.A./Namibia and again on rocky slopes in mountain grassland in Lesotho at altitudes of 2 500 to 3 000 m. Map 39.

Vouchers: *Dinter* 3633; *Killick* 1984; *Schlechter* 8245; 8639.

A good deal of variation is included in the above concept and further study may reveal that this

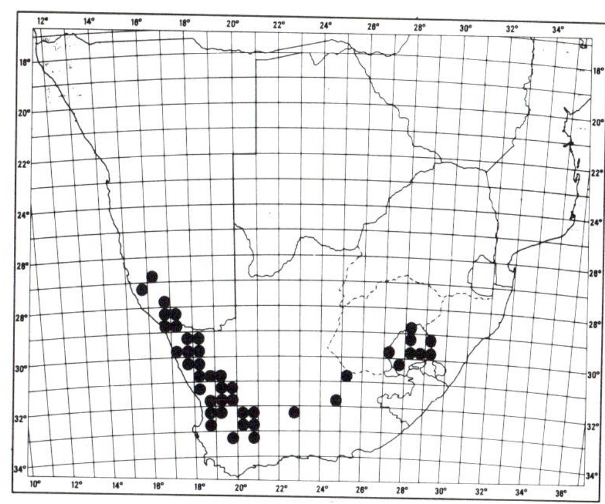

MAP 39. — **Stachys rugosa**

treatment is too broad. At this stage, however, it has not been found possible to classify the material into meaningful groups. The typical form with oblong-lanceolate obscurely crenulate leaves and yellow flowers occurs in Namaqualand. In the same area plants with purple to pink flowers occur and this is the common colour recorded to the south and east, while leaf shape varies from linear-lanceolate to broadly elliptic or ovate-lanceolate. In the Clanwilliam district plants occur with very narrow leaves which are somewhat intermediate with *S. linearis* (no. 36) while near Victoria West specimens with more markedly crenate leaves show a tendency to grade into *S. cuneata* (no. 33).

Further study is also required of the plants occurring at high altitudes in Lesotho which are now included in *S. rugosa*. The leaves of these plants tend to be darker and thinly hispid above, but the distinction is not constant. The habitat is very different from the arid fynbos of the western Cape but no reliable distinguishing characters can be found for separating the plants from *S. rugosa*.

38. **Stachys burchelliana** *Launert* in Mitt. bot. StSamml., Münch. 7: 301 (1968); Launert & Schreiber in F.S.W.A. 123: 30 (1969). Type: as for *Phlomis micrantha* Burch.

Phlomis micrantha Burch., Trav. 1: 340 (1822). Type: Cape, Asbestos Mts, *Burchell* 1672 (K, holo.).

Stachys burchellii Benth., Lab. 561 (1834); in DC., Prodr. 12: 493 (1848); Skan in F.C. 5,1: 360 (1910); nom. illegit. (see note below). Type: based on *Phlomis micrantha* Burch. and *Sideritis rugosa* Thunb.

S. rugosa sensu Marloth, Fl. S. Afr. 3,2: t.47A (1932).

Much branched shrub 0,4—1,2 m tall; branches ascending, white stellate-felted to almost floccose. *Leaves* subsessile; blade lanceolate to oblong-lanceolate, 25−50 × 5−10 mm, densely grey stellate-felted, often paler, floccose and rugose beneath, apex acute to obtuse, narrowed at the base, margin serrulate-crenulate, often somewhat obscurely. *Inflorescence* of several 6−10-flowered verticils usually rather close together in the upper leaf axils. *Calyx* 2-lipped, densely stellate-floccose, 6−7 mm long; upper lip slightly longer than the lower, shortly 3-toothed; lower lip deeply 2-toothed. *Corolla* yellow; tube 5−6 mm long; upper lip concave, spreading, 4 mm long; lower lip deflexed, 5−6 mm long.

Found on dry rocky hillsides and sandy soil overlying calcareous formations in the northern Cape Province and south-eastern S.W.A./Namibia. Map 40.

Vouchers: *Acocks* 16389; *De Winter* 3330.

A uniform species superficially resembling *S. rugosa* (no. 37) but may be distinguished by the characteristic bilabiate calyx. Its distribution does not overlap with that of *S. rugosa*.

The plants are unpleasantly aromatic and an infusion is taken for chest complaints. Like several allied species, it is commonly known as Boesmantee. On a specimen collected on the Asbestos Mts, Marloth records the vernacular names Dassiebos or Aas-voëlbos.

As pointed out by Launert, l.c., the name *S. burchellii* Benth. is illegitimate because Bentham included *Phlomis micrantha* Burch. in his protologue. Although the epithet *micrantha* was available to Bentham, it can no longer be taken up because of *Stachys micrantha* Koch (1848) and *S. micrantha* Griseb. (1879). Launert renamed the species *S. burchelliana*.

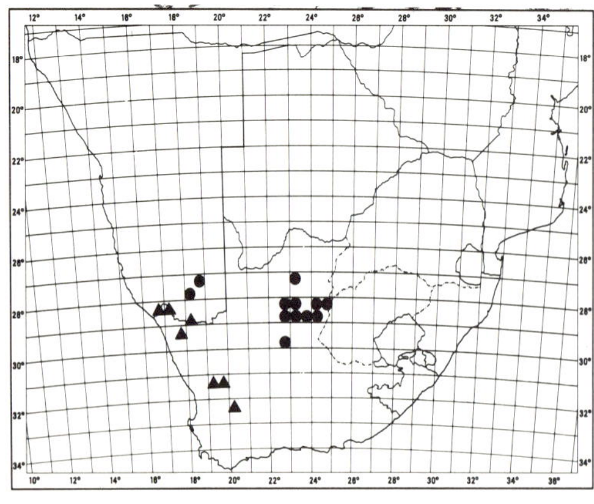

MAP 40. — ● **Stachys burchelliana**
▲ **S. lamarckii**

39. **Stachys lamarckii** *Benth.*, Lab. 562 (1834); in DC., Prodr. 12: 492 (1848); Skan in F.C. 5,1: 360 (1910). Lectotype: *Brugmans* in Herb. Vahl (C, lecto.!).

Sideritis decumbens Thunb., Prodr. 95 (1800); Fl. Cap. edn Schult. 444 (1823); nom. illegit., non Moench (1794). Type: Cape, "Bockland et Roggeveld", *Thunberg* (UPS, holo.).

Stachys rugosa sensu Lam., Tabl. Encycl. 3: 66, t.509, f.3 (1819).

S. nutans Benth., Lab. 561 (1834); in DC., Prodr. 12: 492 (1848); Skan in F.C., l.c. 367 (1910). Type: *Dahl* in Herb. Vahl (C, holo.!).

Shrub 0,2−1 m tall; branches ascending, white woolly stellate-tomentose on the younger parts, glabrescent and brownish purple with age. *Leaves* subsessile or shortly petiolate; blade rugose, fairly thick-textured, elliptic or lanceolate-elliptic to ovate-elliptic, 20−50 × 6−20 mm, upper surface subglabrous or thinly to fairly densely stellate-tomentose, more dense and somewhat woolly beneath, apex obtuse to rounded, base obtuse to cuneate, margin crenate. *Inflorescence* of few to several 6−many-flowered verticils, denser towards the apex. *Calyx* very densely white to creamy woolly-tomentose, 8−11 mm long. *Corolla* yellow; tube 6−7 mm long; upper lip ascending, 3 mm long; lower lip horizontal, 6 mm long.

Found among rocks at relatively high altitudes of 600−1 500 m in the mountains of the western Karoo and Namaqualand, extending from the Vanrhynsdorp district to the Richtersveld. Map 40.

Vouchers: *Acocks* 19533; *Hardy* 655; *Marloth* 12360.

A distinctive species characterized by the woolly stellate indumentum of the calyx and young vegetative parts and the glabrescent brownish and often shiny older parts of the stems, while the leaves are usually thinly stellate-pilose to glabrescent on the upper surface.

40. **Stachys aurea** *Benth.* in DC., Prodr. 12: 492 (1848). Type: Cape, Cedarberg, *Drège* 3098 (K, holo.!).

Betonica heraclea L., Mant. 83 (1767). Type: LINN 735.7.

Sideritis plumosa Thunb., Prodr. 95 (1800); Fl. Cap. edn Schult. 445 (1823). Type: Cape, *Thunberg* s.n. (UPS, Microfiche 563/13412!).

Phlomis parvifolia Burch., Trav. 1: 225 (1822). Type: Cape, near the Juk River, *Burchell* 1232 (K, holo.!).

Stachys integrifolia Vahl ex Benth., Lab. 562 (1834); in DC., Prodr. 12: 492 (1848); Skan in F.C. 5,1: 364 (1910); nom. illegit. Type: see note below.

S. hantamensis Vatke in Bot. Ztg 33: 462 (1875); Skan, l.c. 364 (1910). Type: Cape, Hantam Mts, *Meyer.*

S. teres Skan, l.c., 364 (1910). Syntypes: Cape, near the Juk River, *Burchell* 1232 (K!); 1276 (K!).

Freely branched shrub 0,3−1 m tall; branches spreading to ascending, densely yellowish white stellate-tomentulose. *Leaves* small, subsessile or shortly petiolate; blade green, thin-textured, obovate or oblanceolate to narrowly elliptic, 10−20 × 2−5 mm, sparingly to freely stellate-hispid on both surfaces or subglabrous above, apex acute to obtuse, base cuneate, margin entire or few-toothed near the apex. *Inflorescence* produced at the ends of slender branches, of few to several 4−6-flowered verticils; bracts resembling the leaves, stellate-hispid. *Calyx* densely yellowish woolly-tomentose, 6−9 mm long. *Corolla* yellow; tube 5−6 mm long; upper lip ascending, 2−3 mm long; lower lip deflexed, 5−6 mm long.

Found on rocky situations in the south-western Karoo and southern Namaqualand. Map. 41.

Vouchers: *Acocks* 18567; *Esterhuysen* 1298.

An unpleasantly aromatic plant but evidently well grazed. It is easily recognized by the small green leaves and the dense yellowish woolly covering of the calyces.

The nomenclature of this species is complicated partly by the fact that Bentham, when describing *S. integrifolia*, included three earlier names in the protologue, namely, *Betonica heraclea* L. (1767), *Sideritis plumosa* Thunb. (1800) and *Phlomis parvifolia* Burch. (1822). While the epithet *heraclea* could not be transferred because of *Stachys heraclea* Col. ex All. (1785), the other two epithets could have been taken up at that stage. Thus *S. integrifolia* Vahl ex Benth. (1834) is a superfluous name. However, since 1834 both the earlier epithets have been used in *Stachys* for different species *(S. plumosa* Griseb, 1844, and *S. parvifolia* Mart. 1844), so that neither *Sideritis plumosa* Thunb. nor *Phlomis parvifolia* Burch. can now be transferred to *Stachys.* The correct name for the species is, therefore, *S. aurea* Benth. (1848).

41. **Stachys flavescens** *Benth.* in E. Mey., Comm. 241 (1838); in DC., Prodr. 12: 493 (1848); Skan in F.C. 5,1: 361 (1910). Lectotype: Cape, between Pedroskloof and Leliefontein, *Drège* 3097 (K, lecto.!).

S. gariepina Benth. in DC., Prodr. 12: 493 (1848); Skan in F.C. 5,1: 362 (1910). Type: Namaqualand, *Ecklon* s.n. (S!, SAM!).

A rigid, branched shrub 0,6−1 m tall; branches erect or ascending, densely yellowish-felted when young, greyish with

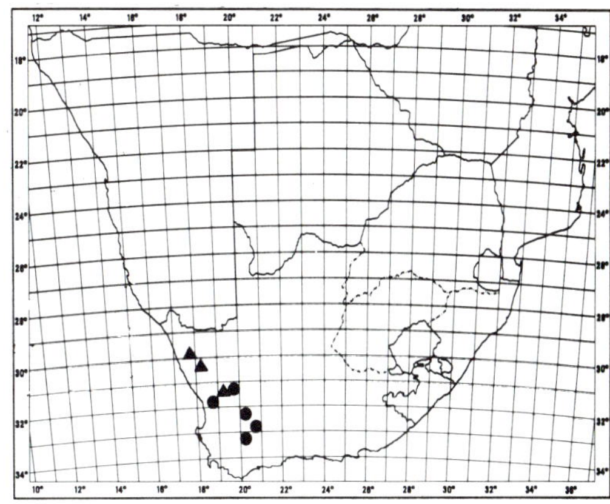

MAP 41. — ▲ **Stachys aurea**
● **S. flavescens**

age. *Leaves* sessile, crowded, thick-textured, lanceolate to oblong, 20−30 (−40) × 3−4 (−10) mm, densely yellowish-felted on both surfaces, apex acute, sometimes mucronate, base narrowed to truncate, margin entire. *Inflorescence* dense, of several closely placed 2−6-flowered verticils; bracts like the leaves but slightly smaller. *Calyx* densely stellate-felted, yellowish, 7−8 mm long. *Corolla* yellow; tube 5−6 mm long; upper lip ascending 2,5−3 mm long; lower lip spreading, 4,5−6 mm long.

Known from a few gatherings at 800−1 300 m altitude in the Bokkeveld Mts and Kamiesberg and reported from as far north as the Orange River, usually among rocks. Map 41.

Vouchers: *Acocks* 13214; 14204.

Characterized by the dense yellowish felt-like tomentum on all parts of the plant, the narrow to oblong, rigid, sessile leaves with a conspicuous midrib and obscure to inconspicuous secondary veins, and the well-developed bracteoles which are subequal to the calyx in length. Occasional specimens of *S. rugosa* (no. 37) dry with a somewhat yellowish colour and, if the leaves are narrow, they may be confused with *S. flavescens.* In such cases the bracteoles, which are slender and much shorter than the calyx in *S. rugosa*, should be diagnostic.

S. gariepina is stated by Bentham to be based on a specimen in Ecklon's herbarium, said to have been collected near the mouth of the Gariep (Orange) River. Ecklon's herbarium was acquired by Sonder, the greater part of whose herbarium went in turn to the Naturhistoriska Riksmuseet in Stockholm. There is a specimen in S which may well by the type. It is annotated: "Stachys gariepina n.sp. Iter ad montem Kamisberg, in terra Boshesmansland, et ad fluminis ostium, Gariep, Namaqualand, E. & Z". What appears to be a duplicate has been seen in SAM. The leaves are somewhat larger than those on plants from the Kamiesberg, and the dimensions given in brackets in the above description are derived from these specimens. In no other way do they differ from typical *S. flavescens* and they are included in the latter species without hesitation. It may be noted that Ecklon did not travel to Namaqualand so that the above-mentioned specimens attributed to Ecklon or Ecklon & Zeyher were no doubt collected by Zeyher.

7290   **14. SALVIA ***

Salvia *L.*, Sp. Pl. 23 (1753); Gen. Pl. edn 5: 15 (1754); Benth., Lab. 190 (1833); in DC., Prodr. 12: 262 (1848); Benth. & Hook. f., Gen. Pl. 2,2: 1194 (1876); Briq. in Natürl. PflFam. 4,3a: 270 (1896); Bak. in F.T.A. 5: 456 (1900); Skan in F.C. 5,1: 307 (1910); Hedge in Notes R. bot. Gdn Edinb. 33: 1 (1974); R.A. Dyer, Gen. 1: 529 (1975). Type species: *S. officinalis* L.

Annual or perennial herbs, undershrubs or shrubs with various kinds of indumentum. *Leaves* entire, toothed, or more or less deeply lobed. *Inflorescence* a spike, raceme or panicle; flowers in 2—many-flowered verticils; bracts usually reduced, rarely similar to the upper leaves, sometimes showy, deciduous or persistent; bracteoles usually present. *Calyx* 2-lipped, about as long as the corolla tube, sometimes accrescent, variously hairy and often glandular; upper lip entire or 3-toothed, median tooth often shorter or obsolete; lower lip equally 2-toothed, longer than the upper. *Corolla* 2-lipped; tube straight or curved, usually enlarging towards the throat, annular-pilose or exannulate within, invaginated with a plate of internal tissue or not; upper lip usually longer than the lower lip, straight or falcate, usually concave and ± compressed, entire or bifid; lower lip spreading, 3-lobed, the median lobe usually much larger than the two lateral. *Stamens* 2, curved; the connective produced and lying within the upper lip and bearing an oblong or linear anther-theca; the other part of the connective produced into a variously shaped appendage, sterile or bearing a much reduced anther-theca, the appendages cohering or not; staminodes 2, small and usually inconspicuous. *Style* included or exserted from the corolla, usually exceeding the stamens, unequally 2-lobed. *Nutlets* triquetrous to compressed, ovoid to subglobose, smooth, mucilaginous on wetting or not.

Probably between 800 and 900 species, widely distributed in the temperate and tropical regions of both hemispheres; 22 species occur naturally in Southern Africa while 1 European species and 3 from tropical America have become naturalized and are included in the key.

Several species are cultivated in Southern Africa, including the European species *S. officinalis* L., commonly known as Sage, which is grown as a culinary herb. The more commonly grown ornamental species may be divided into two groups on the basis of corolla colour:

1. *Corolla blue, lilac, violet, purplish blue or white: S. azurea* Lam., from the south-eastern United States, a soft shrub to 1,2 m tall with linear-lanceolate leaves and sky-blue corolla; *S. farinacea* Benth., from Mexico and Texas, a soft shrub up to 1,2 m with densely blue-tomentose calyx and white, blue or purple corolla; *S. leucantha* Cav., from Mexico, a soft shrub about 0,9 m tall, leaves densely white-tomentose below, calyx densely purplish-lanate and corolla white-tomentose; *S. patens* Cav., from Mexico, a rhizomatous plant with herbaceous stems 0,6—1 m tall and large deep blue flowers; *S. pratensis* L., from Europe, a leafy herb 0,6—1 m tall with large oblong-ovate leaves and blue-mauve flowers; *S. sclarea* L., from Europe and Asia Minor, with herbaceous stems up to 0,9 m tall, a rosette of large ovate leaves and conspicuous large persistent whitish mauve bracts; and *S. uliginosa* Benth., from South America, a soft shrub 0,6—1,5 m tall with dense slender racemes of blue flowers.

2. *Corolla pink, reddish or scarlet: S. coccinea* Etlinger, from tropical America, probably Brazil, annual or perennial, 0,3—0,5 m tall, small bracts and red flowers, frequently recorded as a garden escape and therefore included in the key below; *S. involucrata* Cav., from Mexico, a soft shrub to 1,5 m with large reddish bracts and corolla; *S. microphylla* Kunth, from Mexico, a soft shrub to 1,2 m with smallish leaves and pale pink to deep red flowers, planted sometimes as a low hedge and occasionally found as a garden escape; and *S. splendens* Sellow ex Roem. & Schult., the commonly cultivated Red Salvia, usually treated as an annual in gardens, with normally red flowers, though variously coloured cultivars have been developed. Further species have been introduced and grown in South Africa but do not appear to be widely cultivated.

1 Shrubs with campanulate-infundibuliform fruiting calyces with widely diverging lips and usually enlarging
   distinctly from flower to fruit:

 2 Calyx densely villous:

  3 Stems acutely quadrangular, white, with antrorse hairs .... 9. *S. albicaulis*

  3 Stems round-quadrangular, not white, with spreading hairs .... 2. *S. africana-coerulea*

---

* Adapted from Hedge, l.c.

2 Calyx pilose:

  4 Corolla 35—40 mm long:

    5 Bracts persistent; upper corolla lip c. 25 mm long; leaves greenish, canescent .... 2. *S. africana-lutea*

    5 Bracts soon deciduous; upper corolla lip c. 17 mm long; leaves greenish white with a dense appressed tomentum .... 3. *S. lanceolata*

  4 Corolla shorter than 30 mm long:

    6 Both leaf surfaces with a prominent indumentum of eglandular or glandular hairs:

      7 Fruiting calyces up to 16 mm long; leaf margins crenate-dentate or eroso-dentate, rarely entire:

        8 Stems with eglandular hairs; leaves coriaceous; flowers purplish blue .... 5. *S. dentata*

        8 Stems with glandular hairs; leaves herbaceous; flowers whitish .... 7. *S. garipensis*

      7 Fruiting calyces up to 25 mm long; leaf margins entire .... 6. *S. dolomitica*

    6 Both leaf surfaces with numerous gland-dots but otherwise ± glabrous; stems with short antrorse hairs only .... 8. *S. chamelaeagnea*

1 Shrubs, subshrubs, perennial or annual herbs with campanulate to tubular calyces with not or somewhat diverging lips, not or little enlarging in fruit:

  9 Shrubs or subshrubs with woody stems; corolla white, mauve or blue:

    10 Leaves obovate-elliptic, up to 13 × 8 mm .... 1. *S. muirii*

    10 Leaves irregularly lyrate-pinnatifid, up to 40 × 23 mm .... 10. *S. namaensis*

  9 Perennial or annual herbs with herbaceous to softly woody stems, sometimes with a woody rootstock, or shrubs with corolla red or pink:

    11 Upper lip of calyx 2- or 3-toothed (mainly indigenous species):

      12 Upper lip of corolla ± straight:

        13 Upper lip of fruiting calyx with closely connivent short teeth; corolla tube exannulate, 8—12 mm .... 21. *S. verbenaca*

        13 Upper lip of fruiting calyx with distinct ± spreading teeth separated by rather broad truncate sinuses; corolla tube annulate or exannulate:

          14 Lower parts of stem and upper leaf surface glabrous; upper leaves petiolate; corolla 20—25 mm long .... 15. *S. obtusata*

          14 Lower parts of stem and upper leaf surface pilose; upper leaves petiolate or sessile:

            15 Leaves simple, ovate-triangular:

              16 Leaf blade usually not exceeding 30 × 20 mm; calyx enlarging to 12 mm or more in fruit with a wide sinus between upper and lower lips .... 11. *S. triangularis*

              16 Leaf blade up to 60 × 40 mm or more; calyx enlarging to 10 mm long in fruit ...... 12. *S. aurita*

            15 Leaves simple to pinnate, not ovate-triangular:

              17 Corolla 25—40 mm long with a ± straight tube; calyx tubular, 10—14 mm long; leaves lyrate-pinnatifid .... 14. *S. scabra*

              17 Corolla 12—26 mm long; tube narrow and straight or broad and widened above; calyx tubular, tubular-campanulate or ovate-campanulate, 4—13 mm long; leaves various:

                18 Leaves simple, lyrate or runcinate:

                  19 Leaves runcinate with a terminal segment of up to 95 × 50 mm; verticils up to 24-flowered; corolla tube exannulate .... 13. *S. tysonii*

                  19 Leaves simple or runcinate or lyrate with a terminal segment of less than 50 × 30 mm; verticils 4—12-flowered; corolla tube annulate:

                    20 Calyx upper lip with lateral teeth c. 2,5 mm long and median clearly shorter .... 12. *S. aurita*

                    20 Calyx upper lip with subequal teeth, 0,5—1,5 mm long:

                      21 Fruiting calyces 7—13 mm long; corolla 10—26 mm long; plants usually rhizomatous .... 16. *S. repens*

                      21 Fruiting calyces 4—7 mm long; corolla 7—14 mm long; plants not rhizomatous:

22 Stems with a distinct indumentum of short to long eglandular hairs; leaves oblong-lanceolate to obovate; calyx ovate-campanulate .... 17. *S. runcinata*

22 Stems almost glabrous with few scattered eglandular hairs; leaves narrowly linear-oblong to oblong-lanceolate; calyx ovate .... 18. *S. stenophylla*

18 Leaves pinnatifid to pinnate with linear segments:

23 Calyx 4—7 mm long; corolla 7—14 mm long:

24 Stems with a distinct indumentum of short to long eglandular hairs; leaves oblong-lanceolate to obovate; calyx ovate-campanulate .... 17. *S. runcinata*

24 Stems almost glabrous with few scattered short eglandular hairs; leaves narrow, linear-oblong to oblong-lanceolate; calyx ovate .... 18. *S. stenophylla*

23 Calyx c. 9 mm long, tubular; corolla c. 25 mm long .... 19. *S. schlechteri*

12 Upper lip of corolla distinctly falcate:

25 Verticils 2-flowered; leaves linear to linear-oblanceolate, 3—10 mm wide, sparingly pilose, base attenuate .... 20. *S. granitica*

25 Verticils 6—8-flowered; leaves linear-oblong to ovate, 15—70 mm wide, rugose above, scabrid to lanate beneath, base obtuse to subcordate:

26 Mature calyx 7—9 mm long; corolla 8—12 mm long .... 21. *S. verbenaca*

26 Mature calyx 10—15 mm long; corolla 15—30 mm long:

27 Lower stem indumentum with numerous capitate glandular hairs; leaves not white-lanate beneath .... 22. *S. disermas*

27 Lower stem indumentum without capitate glandular hairs; leaves ± white-lanate beneath .... 23. *S. radula*

11 Upper lip of calyx entire, consisting of a single ovate tooth (introduced and naturalized species):

28 Annual herbs up to 0,6 m tall; corolla blue, up to 8 mm long:

29 Leaves ovate, 40—60 × 35—50 mm .... 24. *S. tiliifolia*

29 Leaves lanceolate to linear-lanceolate, 30—60 × 5—10 mm .... 25. *S. reflexa*

28 Perennial soft shrub 0,6—1,2 m tall; corolla red, 20—30 mm long .... 26. *S. coccinea*

1. **Salvia muirii** *L. Bol.* in J. Bot., Lond. 68: 103 (1930); Hedge in Notes R. bot. Gdn Edinb. 33: 35 (1974). Type: Mossel Bay, near Cloetes Pass, *Muir* 2025 (BOL, holo.; PRE!).

S. muirii var. *grandiflora* L. Bol., l.c. (1930). Type: Cape, George, hills E. of Great Brak River, *Fourcade* 3854 (BOL, holo.; K).

Stiff twiggy erect shrub up to 0,3—0,6 m tall; stems with a dense indumentum of short eglandular hairs and gland-dots, eventually glabrescent. *Leaves* shortly petiolate; blade simple, thick-textured, obovate-elliptic, 6—13 × 3—8 mm, grey-velutinous with dense short appressed hairs on both surfaces, gland-dotted, apex rounded to obtuse or abruptly acute, base cuneate, margin entire. *Inflorescence* slender, lax below, with up to 7 verticils; verticils 2(—3)-flowered. *Calyx* densely and shortly antrorse pubescent, dotted with orange-red glands, ciliate on the margins, c. 10 mm long, not or slightly enlarging in fruit. *Corolla* blue with white throat, up to 26 mm long; tube 14—18 mm long; upper lip more or less straight, c. 8 mm long; lower lip as long as or longer than the upper. Fig. 15: 1.

A little-known species from rocky hillsides on the northern foothills of the Langeberg in the Riversdale district to the Great Brak River in Mossel Bay district, Cape Province. Map 42.

Voucher: *Fourcade* 3854 (BOL).

Characterized by the woody habit, the small grey-velutinous, thick-textured leaves, the long-tubed blue corolla and the ciliate fringe on the calyx margins.

2. **Salvia africana-lutea** *L.*, Sp. Pl. 1: 26 (1753); Codd in Flower. Pl. Afr. 37: 1461 (1966). Type: Cape, collector unknown (LINN 42/38).

S. aurea L., Sp. Pl. edn 2,1: 38 (1762); Curtis in Curtis's bot. Mag. t.182 (1792); Thunb., Fl. Cap. edn Schult. 448 (1823); Benth. in E. Mey., Comm. 233 (1838); in DC., Prodr. 12: 273 (1848); Skan in F.C. 5,1: 313 (1910); Marloth, Fl. S. Afr. 3,2: t.46B (1932); Salter in Fl. Cape Penins. 697 (1950); Rice & Compton,

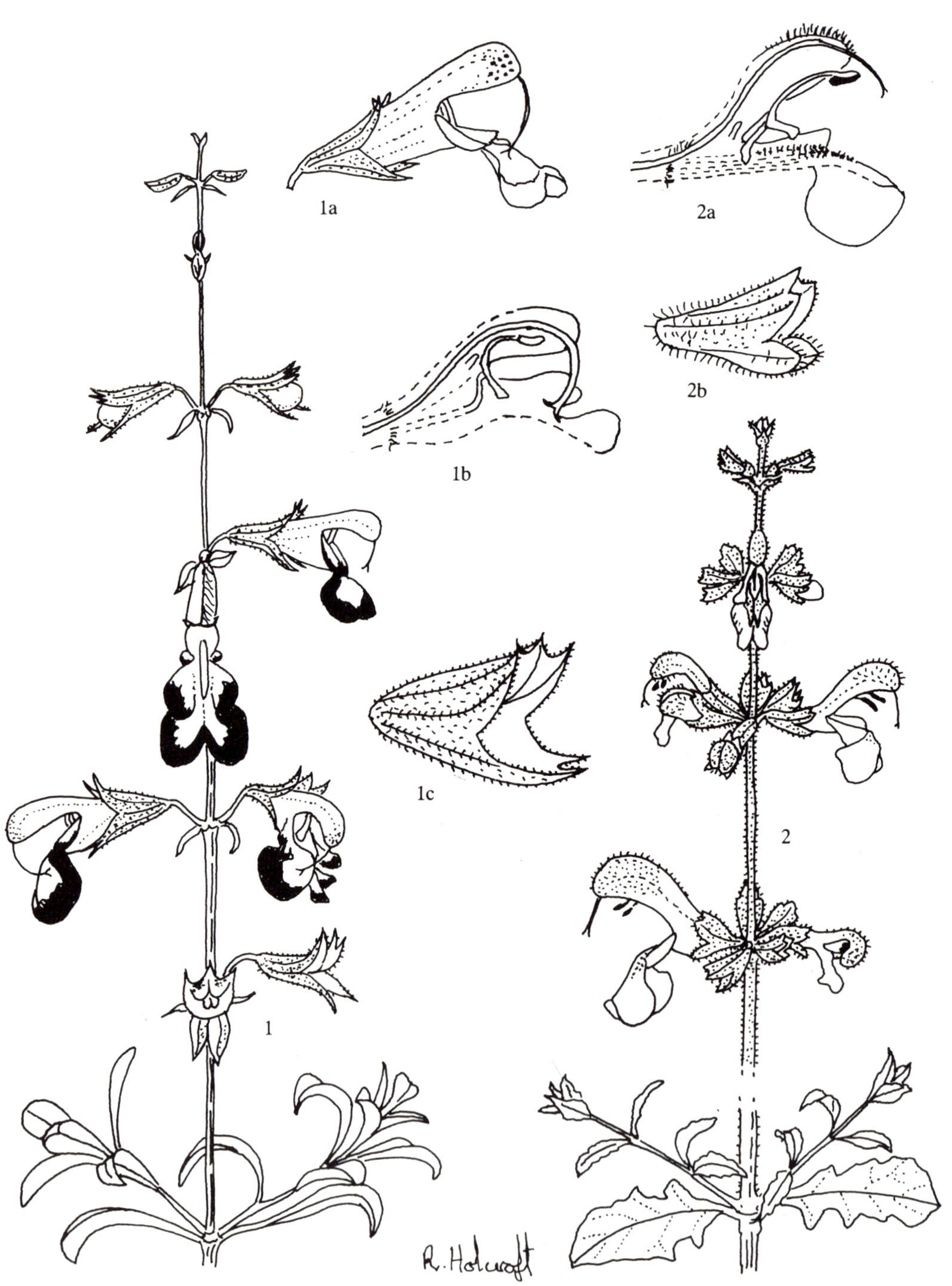
1a
2a
1b
2b
1c
1
2
R. Holcroft

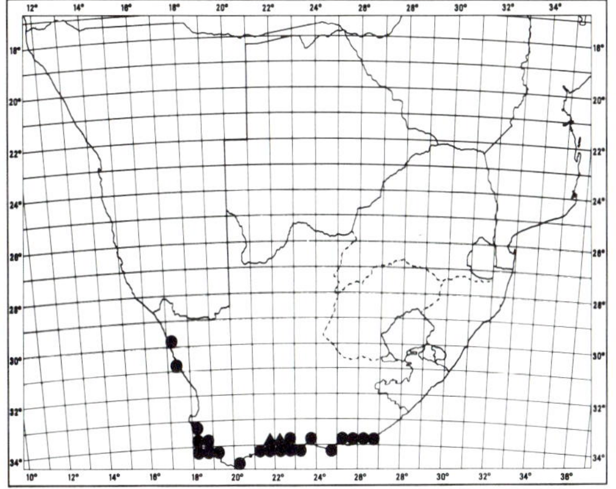

MAP 42. — ▲ **Salvia muirii**
       ● **S. africana-lutea**

occasionally purplish, (30−) 35−45 (−50) mm long; tube c. 20 mm long, upper lip c. 25 mm long, slightly falcate; lower lip c. 12 mm long.

Distributed from Namaqualand to the Cape Peninsula and, eastwards, to Port Alfred, on coastal sand dunes and in arid fynbos on rocky slopes to 800 m altitude. Map 42.

Vouchers: *Boucher* 463; *Galpin* 315; *Hutchinson* 141.

With its shrubby habit and large golden to brownish flowers it is sometimes confused with *S. lanceolata* (no. 3), but may be separated by the unbranched inflorescences, persistent bracts and longer upper lip of the corolla.

A good deal of variation in leaf size, shape and texture is included in the present concept of *S. africana-lutea*. The typical form has relatively thick-textured, small leaves, elliptic or oblong to obovate-elliptic or oblanceolate-oblong, 12−35 × 6−15 (−20) mm, densely grey-tomentose, with apex obtuse to rounded, base cuneate to obtuse, margin entire and petioles 2−7(−12) mm long. It occurs along the Cape coast from about Kleinsee in Namaqualand to Port Elizabeth in the east, on coastal dunes and adjoining hills.

In addition, there are a few gatherings from the Piketberg-Clanwilliam-Citrusdal area (*Schlechter* 4976; 8376; *Marloth* 11494; *Hanekom* 1186) at altitudes up to 800 m, with thin-textured, fairly large leaves, broadly ovate-elliptic to subrotund, 25−55 × 22−40 mm, sparingly glandular pubescent, apex rounded, base rounded to truncate or subcordate, sometimes auriculate, margin eroso-crenulate and petioles 5−15 mm long. The type of *S. eckloniana, Ecklon* s.n. from Clanwilliam district, is such a specimen. Hedge (l.c.) considers this form to represent a juvenile condition but this seems unlikely, and separate varietal status would appear to be justified. However, it is felt that more study is required, particularly in the field, especially as intermediate specimens have been seen from the Peninsula, at altitudes of up to 550m (*Marloth* 273; *Goldblatt* 2662). It may be noted that leaf shape and leaf margin in the closely related species, *S. lanceolata,* show a similar range of variation, but not as marked as in *S. africana-lutea.*

In the first edition of the Species Plantarum (1753), Linnaeus described two South African species under the names *S. afr. lutea* and *S. afr. caerulea*. In edn 2 (1762) he altered these names to *S. aurea* and *S. africana*. The latter two names have largely been used in subsequent literature. The earlier names were also rejected by Hedge, l.c., who considered that they were out of context with Linneaus's nomenclatural thinking at the time, and that the names in the second edition were preferable. This may be so, but they are not the only deviations in edn 1 and such cases are dealt with in

Wild Flow. Cape G.H. t.126 (1951); Bailey, Cycl. Hort. edn 20, 3: 3059 (1963); Hedge in Notes R. bot. Gdn Edinb. 33: 42 (1974); nom. illegit. (see notes below). Type: as above.

S. colorata L., Syst. Nat. edn 12,2: 66 (1767). Type: Cape, collector unknown (LINN 42/39).

S. eckloniana Benth. in DC., Prodr. 12: 273 (1848); Skan in F.C. 5,1: 313 (1910). Type: Cape, Clanwilliam, *Ecklon* s.n. (K, holo.!).

Much branched shrub up to 2 m tall; stems densely leafy, sparse to densely appressed white-tomentose, dotted with orange-red gland-dots. *Leaves* petiolate; blade simple, thickish or thin-textured, suborbicular to elliptic or narrowly obovate, 15−35 (−55) × 6−20 (−40) mm, grey-tomentose, gland-dotted, apex rounded to obtuse, base cuneate to cordate or shortly lobed, margin usually entire but sometimes crenate-dentate to eroso-crenate (the larger leaves); petiole 2−15 mm long. *Inflorescence* simple, usually dense, of 3−12 verticils; verticils 2(−4)-flowered; bracts broadly ovate or obovate, persistent. *Calyx* broad-campanulate, expanding to 30 mm long in fruit, purplish and membranous, with short spreading glandular and eglandular hairs, dotted with orange-red gland-dots. *Corolla* golden brown, reddish brown, khaki or

---

FIG. 15. — 1, **Salvia muirii**, flowering stem, × 1; 1a, flower, × 1,5; 1b, section through corolla, × 1,5; 1c, mature calyx, × 2 (*Van Jaarsveld & Mitchell* s.n.). 2, **Salvia africana-caerulea**, flowering stem, × 1; 2a, section through corolla, × 1,5; 2b, mature calyx, × 1,5 (*Codd* s.n.).

Art. 23 (1972) of the Code which states that if an epithet consists of two or more words these are to be united or hyphenated, and should not be rejected. Introduced to Europe before 1701 when it was illustrated in Commelin, Hort. med. Amst. t.92.

3. **Salvia lanceolata** *Lam.*, Tabl. Encycl. 1: 72 (1791); Hedge in Notes R. bot. Gdn Edinb. 33: 44 (1974). Type: Cape, without locality, *Sonnerat* (P).

S. nivea Thunb., Prodr. 96 (1800); Fl. Cap. edn Schult. 450 (1823); Benth. in E. Mey., Comm. 233 (1838); in DC., Prodr. 12: 273 (1848); Skan in F.C. 5,1: 314 (1910); Salter in Fl. Cape Penins. 697 (1950); Mason, W. Cape Sandv. Flow. t.71, f. 2. (1972). Type: Cape, without exact locality, *Thunberg* (UPS).

S. hastifolia Benth. in E. Mey., Comm. 233 (1837); in DC., Prodr. 12: 274 (1848); Skan, l.c. 314 (1910). Type: Cape, Clanwilliam area, Boschkloof, *Drège* 7934 (K, holo.).

S. nitida Drège, Zwei Pfl. Doc. 103 (1843), nom. nud., in error for *S. nivea*, see index p. 218.

S. diversifolia Benth. in DC., Prodr. 12: 274 (1848), in syn.

Much branched shrub 1−2 m tall; stems finely tomentulose, glabrescent, often reddish brown. *Leaves* petiolate; blade simple, thick-textured, linear-elliptic to ovate-oblong or oblanceolate, 10−35 × 5−20 mm, with a short dense greyish tomentum or with a sparse indumentum of short broad hairs mainly on the veins and leaf margin, gland-dotted, apex acute, base cuneate to hastate-auriculate, margin entire or irregularly crenate-dentate. *Inflorescence* usually branched, each branch with 3−5 verticils; verticils 2 (−4)-flowered; bracts obovate, acuminate, soon deciduous. *Calyx* fairly densely glandular-hispid, expanding to broad-campanulate, 25 mm long and purplish in fruit. *Corolla* dull rose to brownish crimson or grey-blue, 25−35 mm long; upper lip straight or slightly falcate, c. 17 mm long; lower lip c. 13 mm long.

Distributed from Namaqualand to the Cape Peninsula and eastwards to Montagu, in coastal sandveld and arid fynbos, on sandy soil and rocky hillsides at altitudes of 0−300 m. Map 43.

Vouchers: *Acocks* 17216; *Marloth* 7005; *Schlieben & Ellis* 12430.

Related to the previous species, *S. africana-lutea*, and the differences are noted under that species. *S. lanceolata* also varies a good deal in leaf size, indumentum and margin, with some specimens having larger and broader leaves with thinner indumentum, crenate-dentate margins and hastate base, in contrast to the normal densely grey-velvety entire leaves.

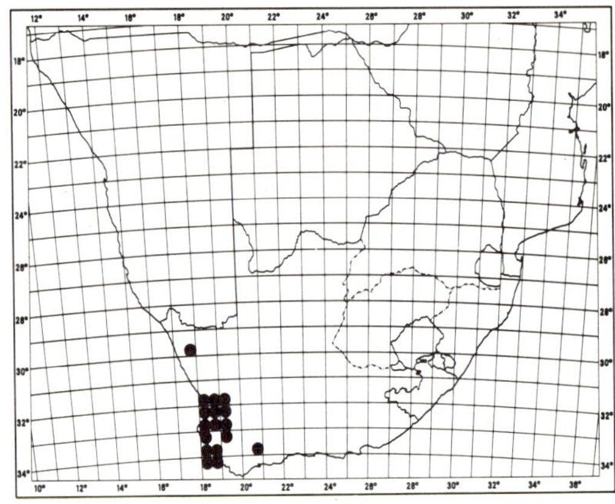

MAP 43. — **Salvia lanceolata**

4. **Salvia africana-caerulea** *L.*, Sp. Pl. 1: 26 (1753). Type: Cape, ex Hort. Cliff. (BM).

S. africana L., Sp. Pl. edn 2,1: 38 (1762); Thunb., Prodr. 96 (1800); Fl. Cap. edn Schult. 449 (1823); Benth. in E. Mey., Comm. 234 (1838); in DC., Prodr. 12: 274 (1848); Skan in F.C. 5,1: 315 (1910); Marloth, Fl. S. Afr. 3,2: 179, t.46A (1932); Salter, Fl. Cape Penins. 696 (1951); Rice & Compton, Wild Flow. Cape G.H. t.125 (1951); Mason, W. Cape Sandv. Flow. t.71, f. 4 (1972); Hedge, in Notes R. bot. Gdn Edinb. 33: 45 (1974); nom. illegit. *S. rotundifolia* Salisb., Prodr. 74 (1796), nom. illegit. Type: as for *S. africana-caerulea*.

S. lanuginosa Burm. f., Fl. Cap. Prodr. 1 (March 1768); Skan, l.c. 333 (1910). Type: Cape, *Oldenland* ex Hb. Burman (G).

S. integerrima Mill., Gard. Dict. edn 8: Salvia No. 12 (16 April 1768). Type: a cultivated plant.

S. barbata Lam., Tabl. Encycl. 1: 72 (1791). Type: Cape, without locality, *Sonnerat* (P).

S. colorata sensu Vahl, Enum. 1: 230 (1804).

S. africana var. *obtusa* Benth. in E. Mey., Comm. 234 (1837), nom. nud.

S. undulata Benth. in DC., Prodr. 12: 275 (1848); Skan, l.c. 316 (1910). Type: Cape, Clanwilliam district, *Ecklon* s.n. (K, holo.).

S. subspathulata Lehm. in Hamburg. Gart. Blumenzeit. 6: 457 (1850). Type: a cultivated plant.

Shrub 0,6−1,5 (−2) m tall, often branching at the base with several erect, usually sparingly branched stems; stems greyish-tomentulose to hispidulous, gland-dotted and occasionally with glandular hairs. *Leaves* petiolate; blade simple, subcoriaceous, obovate-elliptic to broadly obovate, 8−25 (−35) × 4−15 (−25) mm, greenish and somewhat rugose above with short eglandular hairs, greyish-tomentulose and gland-dotted beneath, rarely almost glabrous, apex subacute to rounded, base cuneate, margin subentire to eroso-denticulate, occasionally auriculate at the base. *Inflorescence* often dense or spaced below with 5−12 verticils; verticils 2−6-flowered; bracts ovate, cuspidate, up to 10 × 9 mm, persistent. *Calyx* somewhat funnel-shaped, glandular-villous, expanding to 14 mm in fruit, purple-tinged. *Corolla* light blue to bluish purple or pinkish, the lower lip usually with a paler blue margin and white to yellowish in the centre, 16−28 mm long; tube 8−10 mm long; upper lip falcate, 8−18 mm long; lower lip as long as the upper with a broad reflexed median lobe. Fig. 15:2.

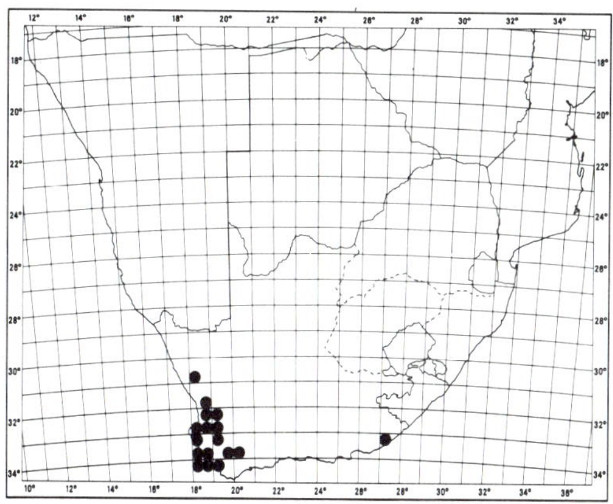

MAP 44. — **Salvia africana-coerulea**

Distributed from Vanrhynsdorp district to Cape Town and eastwards to Montagu and Caledon with an odd record from Peddie district in the eastern Cape Province; in coastal fynbos and on rocky slopes. Map 44.

Vouchers: *Bayliss* 3351; *Galpin* 4424; *Schlechter* 5221.

There is a good deal of variation in size and shape of leaves, density of indumentum and size of corolla. It is sometimes confused with *S. chamelaeagnea* (no. 8) which has a coarser, glandular pubescence, shorter hairs on the calyx and deciduous bracts.

See note under *S. africana-lutea* (no. 2) regarding the use of the name *S. africana-caerulea* L. (1753) instead of *S. africana* L. (1762).

5. **Salvia dentata** *Ait.*, Hort. Kew. 1: 37 (1789); Benth. in DC., Prodr. 12: 275 (1848); Skan in F.C. 5,1: 315 (1910); Hedge in Notes R. bot. Gdn Edinb. 33: 47 (1974). Type: Cape, without locality, *Masson* (BM).

S. angustifolia Salisb., Prodr. 73 (1796), nom. illegit. Type: as for *S. dentata* Ait.

S. rigida Thunb., Prodr. 96 (1800); Fl. Cap. edn Schult. 451 (1823). Type: Cape, *Thunberg* s.n. (UPS, holo., microfiche 27/634!).

S. crispula Benth. in E. Mey., Comm. 1: 234 (1837); in DC., Prodr. 12: 274 (1848). Syntypes: Cape, Modderfonteinsberg, *Drège* (BM); "Camiesbergen", *Drège* 3113 (K).

Twiggy erect shrub 0,6−1,5 (−2) m tall; stems greyish-tomentulose, gland-dotted. *Leaves* often crowded, shortly petiolate; blade simple, thick-textured, spathulate or obovate to narrowly elliptic or lanceolate, 8−22 × 4−12 mm, greenish above, densely greyish-tomentulose beneath, gland-dotted, rarely almost glabrous, apex obtuse, base cuneate, margin usually crenate-dentate to pinnatifid. *Inflorescence* of 2−9 spaced or crowded verticils; verticils 2−6-flowered. *Calyx* somewhat funnel-shaped, hispid and usually copiously red gland-dotted, occasionally hispid-villous, expanding to c. 15 mm long in fruit. *Corolla* light blue or whitish to violet-blue or purple, 16−25 mm long; tube 8−10 mm long; upper lip slightly falcate, 8−12 mm long; lower lip usually longer than the upper.

Distributed from Namaqualand to Clanwilliam and Calvinia districts in broken veld of rocky hillsides and water-courses, at altitudes of 700−1 700 m. Map 45.

Vouchers: *Acocks* 14923, 18456; *Galpin* 11166; *Rodin* 1445.

Related to *S. africana-caerulea* (no. 4) and occasionally specimens are somewhat intermediate, but can usually be distinguished by the undulate to pinnatifid leaves, the shorter pubescence on the calyx and the broader bracteoles. It occurs more to the north-west at somewhat higher altitudes than *S. africana-caerulea*.

6. **Salvia dolomitica** *Codd* in Flower. Pl. Afr. 32: t.1248 (1957); Letty, Wild Flow. Transv. 289 (1962); Hedge in Notes R. bot. Gdn Edinb. 33: 48 (1974). Type: Transvaal, Pilgrims Rest (cult. in Pretoria), *Codd* 8848 (PRE, holo.!).

Shrub 1−2 m tall, branched from the base; stems terete, ascending, densely

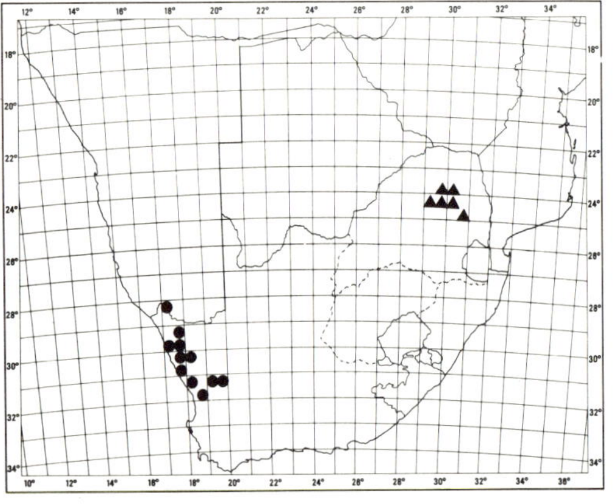

MAP 45. — ● Salvia dentata
▲ S. dolomitica

covered with a short whitish crisped tomentum. *Leaves* petiolate; blade simple, elliptic to obovate, 25—50 (—65) × 12—20 (—30) mm, densely greyish appressed tomentose on both surfaces, reticulate beneath and gland-dotted, apex obtuse, base obtuse to cuneate, margin entire. *Inflorescence* compact, of several 2-flowered verticils. *Calyx* broadly campanulate, often purple-tinged, glandular hirsute, enlarging to 25 mm long in fruit. *Corolla* light pink or lilac with cream or yellow markings on the lower lip, 20—28 mm long; tube c. 10 mm long; upper lip falcate, 10—14 mm long; lower lip 12—18 mm long.

Restricted to the eastern and north-eastern Transvaal, usually on dolomitic outcrops between 1 000 and 1 500 m altitude. Map 45.

Vouchers: *Codd* 10400; *Codd & De Winter* 3089; *Maguire* 2531.

A distinct species, geographically separated from the allied shrubby species, i.e. those with expanding fruiting calyces, all of which are found in the south-western to south-eastern Cape Province.

7. **Salvia garipensis** E. Mey. ex Benth.

in E. Mey., Comm. 1: 232 (1838); in DC., Prodr. 12: 273 (1848); Skan in F.C. 5,1: 311 (1910); Launert & Schreiber in F.S.W.A. 123: 28 (1969); Hedge in Notes R. bot. Gdn Edinb. 33: 51 (1974). Type: Cape, Namaqualand, *Drège* 3112 (K, holo.).

S. steingroeveri Briq. in Bot. Jb. 19: 191 (1894). Type: S.W.A./Namibia, without locality, *Steingröver* 55.

S. dinteri Briq. in Bull. Herb. Boissier sér. 2,3: 1075 (1903). Type: S.W.A./Namibia, Gubub, *Dinter* 1111 (Z, holo.).

Much branched shrub 0,6—1,2 m tall; stems glandular-pubescent with longish eglandular hairs, dense short gland-tipped hairs and gland-dots. *Leaves* petiolate; blade thin- or thick-textured, ovate to ovate-deltoid or subrotund, 12—50 × 10—30 mm, green, glandular-hispid and gland-dotted, upper surface smooth to rugose, under-surface often markedly reticulate, apex obtuse to rounded, base rounded to cordate, margin irregularly eroso-dentate. *Inflorescence* of few to several verticils, spaced below, close together above; verticils 2 (—4)-flowered. *Calyx* glandular-pubescent, enlarging to 16 mm long in fruit with widely spreading lips. *Corolla* white or pale blue to mauve, 20—25 mm long; tube c. 10 mm long; upper lip falcate, 10—15 mm long; lower lip subequal to or slightly shorter than the upper.

Distributed from the southern half of S.W.A./Namibia to the adjoining Cape Province, on stony hillsides and watercourses. Map 46.

Vouchers: *Dinter* 3547; *Giess & Müller* 11887.

The leaves vary considerably in size, texture and degree of crenation but the species is readily distinguished from the other shrubby South African species by the leaves being truncate to cordate at the base. Its nearest ally appears to be *S. dominica* L. of the eastern Mediterranean area.

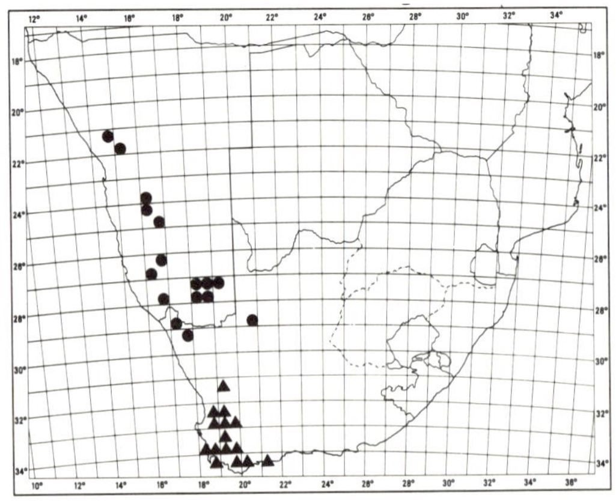

MAP 46. — ● **Salvia garipensis**
▲ S. chamelaeagnea

8. **Salvia chamelaeagnea** *Berg.*, Descr. Pl. Cap. 3 (Sept. 1767); Salter in Fl. Cape Penins. 696 (1950); Codd in Flower. Pl. Afr. 31: t.1219 (1956); Hedge in Notes R. bot. Gdn Edinb. 33: 54 (1974). Type: Cape, without locality, *Ekeberg* s.n. (STB).

S. paniculata L., Mant. 25 (Oct. 1767); Ait., Hort. Kew. 1: 45 (1789); Thunb., Fl. Cap. edn Schult. 450 (1823); Benth. in E. Mey., Comm. 1: 235 (1838); in DC., Prodr. 12: 275 (1848); Hook. f. in Curtis's bot. Mag. t.6790 (1884); Skan in F.C. 5, 316 (1910). Type: Cape, (LINN, holo.).

Much branched shrub 0,6—2 m tall; stems scabrid to pilose, gland-dotted. *Leaves* petiolate; blade simple, coriaceous, obovate to oblanceolate or broadly elliptical, 15—30 (—35) × 5—20 mm, green, subglabrous to slightly scabrid or appressed-pubescent, reticulate below, freely gland-dotted on both surfaces, apex obtuse, apiculate, base cuneate, margin subentire to denticulate. *Inflorescence* a large panicle 100—300 mm long; verticils 2-flowered. *Calyx* reddish purple, glandular-hispid and gland-dotted, enlarging to 12 mm long in fruit. *Corolla* blue or purplish blue often with white on the lower lip, 18—25 (—30) mm long; tube 6—8 mm long, not or scarcely exceeding the calyx; upper lip slightly falcate, 12—20 mm long; lower lip 10—15 mm long.

Distributed from Clanwilliam to Cape Town and eastwards to Ladismith and Riversdale districts, in fynbos along watercourses, in sandy soil among rocks and along roadsides, often locally common. Map 46.

Vouchers: *Acocks* 8584; *Rodin* 3060; *Schlechter* 9866.

A relatively constant, strongly aromatic species, occasionally confused with *S. africana-caerulea* (no. 4).

9. **Salvia albicaulis** *Benth.* in E. Mey., Comm. 1: 234 (1838); in DC., Prodr. 12: 274 (1848); Skan in F.C. 5,1: 317 (1910); Hedge in Notes R. bot. Gdn Edinb. 33: 57 (1974). Type: Cape, Tulbagh, *Ecklon* 7937 (K, holo.).

S. dregeana Benth. in E. Mey., Comm. 1: 234 (1838); in DC., Prodr. 12: 274 (1848). *S. albicaulis* var. *dregeana* (Benth.) Skan, l.c. 317 (1910). Type: Cape, between Pakhuis and Biedow, *Drège* 3114 (K, holo.).

Shrub or woody herb 0,3—0,6 m (or more) tall, branched from the base; stems erect, rather sparingly branched, sharply 4-angled, densely and shortly tomentose with, in addition, occasional long multicellular hairs. *Leaves* petiolate; blade simple, coriaceous, obovate-spathulate or obovate to broadly elliptic, 15—30 × 8—20 mm, subglabrous and somewhat varnished to shortly hispid above, reticulate and white-tomentose between the veins below, the veins often with short or long hispid hairs, apex subacute to rounded, base cuneate, margin crenate-dentate to irregularly and occasionally pinnately lobed. *Inflorescence* a panicle 70—200 mm long; verticils lax to fairly dense, 2—3-flowered. *Calyx* densely villous with long white hairs mainly along the nerves, 10—12 mm long, scarcely expanding in fruit. *Corolla* purplish, 18—24 mm long; tube 8—12 mm long; upper lip falcate, 10—12 mm long; lower lip about 10 mm long.

Distributed from Clanwilliam southwards to Ceres and Wellington districts, in fynbos on rocky slopes. Map 47.

Vouchers: *Leach & Carp* 11352; *Schlechter* 9970; *Taylor* 4799.

A distinct species characterized by the acutely quadrangular white stems and the villous calyces which expand only slightly when in fruit.

10. **Salvia namaensis** *Schinz* in Verh. bot. Ver. Prov. Brandenb. 31: 208 (1890); Skan in F.C. 5,1: 325 (1910); Wilman, Check List Griq. West 228 (1946); Launert & Schreiber in F.S.W.A. 123: 28 (1969); Hedge in Notes R. bot. Gdn Edinb. 33: 59 (1974); Type: S.W.A./Namibia, Tiras, *Schinz* 30 (K).

S. burchellii N.E.Br. in Kew Bull. 1901: 130 (1901); Skan, l.c. 325 (1910). Type: Cape, Richmond, Rhenoster Poort, *Burchell* (K, holo.).

— var. *hispidula* Skan, l.c. 326 (1910). Syntypes: without locality, *Thom* 209 (K); *Ecklon* 77 (K).

Much branched shrublet or bushy shrub 0,3—1,2 m tall, often herbaceous above and woody below, yellow-green or grey-green, strongly aromatic and somewhat viscid; stems shortly and often crisply tomentose and gland-dotted, eventually glabrescent. *Leaves* shortly petiolate; blade irregularly lyrate-pinnatifid, coriaceous, (10—) 15—40 × (5—) 10—20 mm, markedly rugose, shortly hispid above, denser to crisped, strongly reticulate and freely gland-dotted beneath. *Inflorescence* simple, of up to 14 verticils, spaced below, more crowded above, verticils 4 (—6)-flowered. *Calyx* glandular-hispidulous, up to 8 mm long in fruit. *Corolla* white, mauve or blue, 8—12 mm long; tube 5—8 mm long; upper lip straight, 3—6 mm long; lower lip usually longer than the upper.

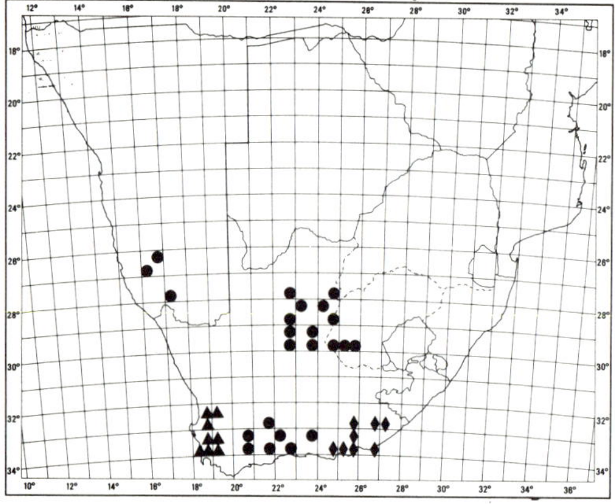

MAP 47. — ▲ **Salvia albicaulis**
     ● **S. namaensis**
     ♦ **S. triangularis**

Found in southern S.W.A./Namibia, northern and central Cape Province as far south as Oudtshoorn and Willowmore, and western Orange Free State, on rocky slopes below krantzes, in watercourses and on surface limestone. Map 47.

Vouchers: *Acocks* 20511; *Leistner* 1558; *Merxmüller* 2416; *Rodin* 3645.

Although shrubby in habit, the floral characters suggest a closer relationship to the herbaceous species which follow rather than to the preceding shrubby species. The leaves resemble some species of *S. stenophylla* (no. 18) but the latter has herbaceous, subglabrous to sparingly pubescent stems.

Dinter 4109 is an odd specimen with leaves somewhat like *S. namaensis* but the calyx, with its widely spreading lips, resembles *S. garipensis* (no. 7). Launert & Schreiber in F.S.W.A. 123: 28 (1969) suggest that it represents a hybrid between these two species.

11. **Salvia triangularis** *Thunb.*, Prodr. 96 (1800); Fl. Cap. edn Schult. 451 (1823); Benth. in E. Mey., Comm. 236 (1838); in DC., Prodr. 12: 351 (1848); Skan in F.C. 5,1: 323 (1910); Jacot Guill., Fl. Lesotho 238 (1971); Hedge in Notes R. bot. Gdn Edinb. 33: 70 (1974). Type: Cape, without locality, *Thunberg* s.n. (UPS, holo., microfiche 28/652!).

S. tenuifolia Burch. ex Benth., Lab. 304 (1833); in DC., Prodr. 12: 350 (1848). Type: Cape, without locality, *Burchell* s.n. (K, holo.).

Perennial herb branching from the scarcely woody base; stems sparingly branched, decumbent-ascending, 0,15—0,5 m long, hispid-villous. *Leaves* petiolate; blade simple, ovate-triangular, 20—40 × 15—30 mm, shortly villous, apex obtuse to rounded, base truncate to subcordate, occasionally shortly auriculate, margin irregularly crenate. *Inflorescence* of up to 10 verticils, spaced below, denser above; verticils (2—) 4—6-flowered. *Calyx* shortly villous, up to 13 mm long in fruit, campanulate, with a wide sinus between the upper and lower lips; upper lip with 3 subequal acuminate teeth 2—2,5 mm long, the middle tooth often slightly shorter than the outer two. *Corolla* pale blue, mauve or purple, 12—16 mm long; tube 8—11 mm long; upper lip straight, 2,5—3 mm long; lower lip slightly longer.

Found in south-eastern Cape Province from about King William's Town to near Humansdorp and inland to Somerset East and Keiskammahoek, in open grassy places between bushes and at forest margins. Map 47.

Vouchers: *Dahlstrand* 814; *Long* 817; *Marsh* 1369.

Related to the next species, *S. aurita*, but has smaller, more densely pubescent leaves, and a somewhat longer fruiting calyx in which there is a relatively wide sinus between the upper and lower lips.

12. **Salvia aurita** *L.f.*, Suppl. 88 (1781). Type: Cape, without locality, *Thunberg* s.n. (UPS, microfiche 23/547!).

Perennial herb with few to several stems from a subwoody base; stems ascending or straggling to 1,2 m long, pilose, usually with longish multicellular hairs, sometimes dense, or short and dense, gland-dotted and occasionally with capitate glandular hairs. *Leaves* petiolate, or the upper ones subsessile; blade simple to lyrate or runcinate, subcoriaceous, variable in shape from broadly ovate to ovate or oblong, 40—80 × 25—50 mm, suglabrous to fairly densely pilose, gland-dotted, apex obtuse to rounded, base truncate to auriculate, margin dentate or crenate, sometimes pinnatipartite or with distinct basal lobes (var. *galpinii*). *Inflorescence* of up to 15 verticils, the lower ones often widely spaced, denser above; verticils 6—8 (—12)-flowered. *Calyx* pilose, tubular-campanulate, up to 10 mm long in fruit; upper lip with 3 acuminate teeth, the 2 outer teeth c. 2,5 mm long and the central tooth shorter, 1,5−2 mm long. *Corolla* pale blue, lilac, white or pinkish, 16−20 mm long; tube 10−15 mm long; upper lip straight, 2,5−3 mm long; lower lip up to 4 mm long.

Distributed from southern and south-eastern Cape through Transkei and Natal to the Soutpansberg in Transvaal, on grassy slopes, stream banks and wooded places.

Two varieties are recognised:

1 Leaves simple or with a tendency towards indistinct basal lobing, ovate......... (a) var. *aurita*

1 Leaves pinnatipartite with distinct basal lobes, oblong to broadly oblong in outline .... (b) var. *galpinii*

(a) var. **aurita.**

Hedge in Notes R. bot. Gdn Edinb. 33: 65 (1974).

S. aurita L.f., Suppl. 88 (1781); Ait. f., Hort. Kew. edn 2,1: 62 (1810); Thunb., Fl. Cap. edn Schult. 451 (1823); Skan in F.C. 5,1: 322 (1910). Type: Cape, without locality, *Thunberg* (UPS).

S. sylvicola Burch. ex Benth., Lab. 304 (1833); in E. Mey., Comm. 236 (1838); in DC., Prodr. 12: 350 (1848). Type: Cape, without locality, *Burchell* (K, holo.).

S. lasiostachys Benth. in DC., Prodr. 12: 350 (1848); Skan. l.c. 324 (1910). Type: Cape, Uitenhage, *Ecklon* 62 (K, holo.).

S. pallidiflora Skan, l.c. 323 (1910). Syntypes: incl. Cape, Somerset East, *Burchell* 3165 (K).

S. peglerae Skan, l.c. 331 (1910); Jacot Guill., Fl. Lesotho 238 (1971). Syntypes: Cape, East London, Fort Pato, *Galpin* 7830 (K; PRE!); Cape, Kentani, *Pegler* 196 (K; PRE!).

Leaves simple, ovate-triangular to broadly ovate, base truncate to auriculate, or with a tendency towards indistinct basal lobing.

Distribution as for the species. Map 48.

Vouchers: *Bayliss* 6921; *Flanagan* 1288; *Scheepers* 1109.

See note after var. 'galpinii (below) and after *S. triangularis* (no. 11).

(b) var. **galpinii** *(Skan) Hedge* in Notes R. bot. Gdn Edinb. 33: 67 (1974). Type: Cape, near Queenstown, *Galpin* 1956 (BOL, holo.; PRE!).

S. galpinii Skan, l.c. 321 (1910); Ross, Fl. Natal 304 (1972); Compton, Fl. Swaziland 497 (1976).

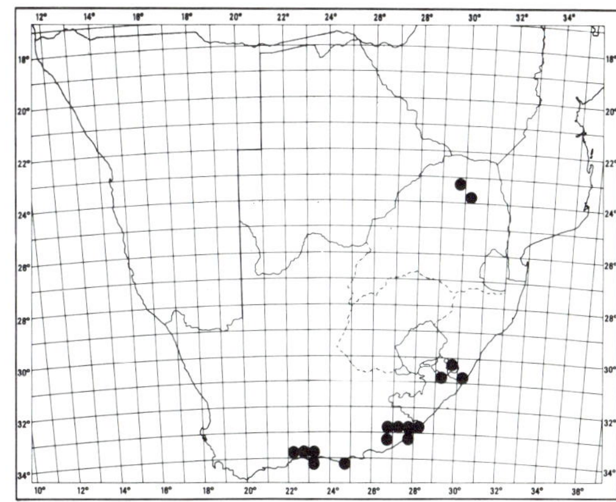

MAP 48. — **Salvia aurita** var. **aurita**

Leaves lyrate or runcinate, oblong to broadly oblong in outline, with distinct basal lobes.

Found in eastern Cape Province, Transkei, Natal, Swaziland and southern Transvaal. Map 49.

Vouchers: *Acocks* 9248; 22046; *Flanagan* 1213.

As stated by Hedge, l.c., the variation in this species appears to fall into two groups which are worth recognition as varieties. There are intermediates and both forms may occur together, e.g. *Galpin* 8164 (var. *aurita*) and 8164A (var. *galpinii*), both from Gatwyn, near Queenstown.

Related to *S. scabra* (no. 14) and *S. triangularis* (no. 11) but these may be distinguished by the longer corolla of *S. scabra* and by the smaller and more pubescent leaves of *S. triangularis*.

Bentham in E. Mey., Comm. 1: 237 (1837) and in DC., Prodr. 12: 351 (1848) appears to have confused *S. aurita* and *S. scabra* (no. 14). Thus *S. aurita* sensu Benth. is *S. scabra* and *S. scabra* sensu Benth. is a mixture of *S. aurita*, *S. namaensis* (no. 10), *S. repens* (no. 16) and *S. runcinata* (no. 17).

13. **Salvia tysonii** *Skan* in F.C. 5,1: 320 (1910); Ross, Fl. Natal 304 (1972); Hedge in Notes R. bot. Gdn Edinb. 33: 71 (1974). Lectotype: Cape, Griqualand East, near Clydesdale, *Tyson* 2171 (K, lecto.).

Robust erect herb up to 1,4 m tall from a creeping rootstock; stems usually simple, coarsely tomentose, gland-dotted. *Leaves:* the lower shortly petiolate, the upper subsessile; blade runcinate or pinnatifid, 50−95 × 40−50 mm, with a terminal segment of c. 60 × 50 mm, shortly appressed pubescent, apex subacute to obtuse,

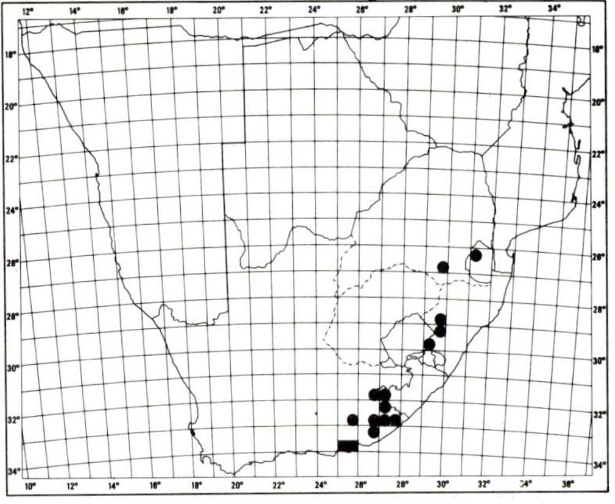

MAP 49. — ● **Salvia aurita** var. **galpinii**
■ **S. obtusata**

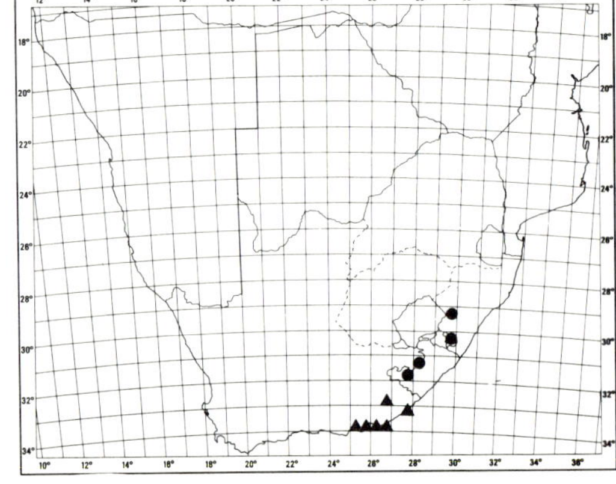

MAP 50. — ● **Salvia tysonii**
▲ **S. scabra**

margin irregularly and coarsely serrate. *Inflorescence* of many verticils, spaced below, denser above; verticils 8 (−24)-flowered. *Calyx* shortly hispid, tubular-campanulate, up to 10 mm long in fruit, upper lip with 3 acuminate teeth, the 2 outer teeth c. 2,5 mm long and the central tooth shorter, 1,5−2 mm long. *Corolla* blue, mauve or reddish, c. 13 mm long; tube straight c. 10 mm long; upper lip straight, 3 mm long; lower lip slightly shorter.

Known from a few scattered localities in eastern Cape, Transkei and Natal, beside streams in mountain grassland at altitudes of 750−1 450 m. Map 50.

Vouchers: *Flanagan* 2797; *Tyson* 1770.

Distinguished by the sturdy stems, rather coarse foliage and many-flowered verticils. More collections of this species with good field notes are needed.

14. **Salvia scabra** *L.f.*, Suppl. 89 (1781); Ait., Hort. Kew. 1: 41 (1789); Thunb., Fl. Cap. edn Schult. 452 (1823); Skan in F.C. 5,1: 321 (1910); Hedge in Notes R. bot. Gdn Edinb. 33: 67, t.18 (1974). Type: Cape, without locality, *Thunberg* s.n. (UPS, microfiche 27/639!).

S. graciliflora Avé-Lall. in Ind. Sem. Hort. Petrop. 10: 57 (1844). Syntypes: *S. aurita* sensu Benth. in E. Mey., Comm. 237 (1838), as to *Drège* b; and a cultivated plant.

Perennial erect herb from a subwoody rootstock, 0,3−1 m tall; stems several, branched, villous. *Leaves* petiolate or the upper ones subsessile; blade lyrate-pinnatifid, obovate to oblong in outline, 30−50 × 20−30 mm, scabrid-pilose above, hispid-pilose and gland-dotted beneath. *Inflorescence* of up to 12 verticils, spaced below, denser above; verticils 4−6-flowered. *Calyx* shortly villous, often tinged purple, tubular-campanulate, up to 14 mm long in fruit; upper lip with 3 acuminate subequal teeth 1,5−2 mm long. *Corolla* mauve, lilac or purple, 25−40 mm long; tube straight, 20−35 mm long; upper lip 3−4 mm long; lower lip 5−6 mm long. Fig. 16:1.

Limited to the eastern Cape Province coastal area from East London to near Humansdorp, in bush groups, coastal dunes and forest margins from near sea level to 180 m altitude. Map 50.

Vouchers: *Galpin* 3045; 10695; 10743.

A distinct species characterized by the lyrate-pinnatifid, scabrid leaves, and long corolla tube. See also note at end of *S. aurita* (no. 12).

15. **Salvia obtusata** *Thunb.*, Prodr. 97 (1800); Fl. Cap. edn Schult. 451 (1823); Benth. in DC., Prodr. 12: 351 (1848); Skan in F.C. 5,1: 324 (1910); Hedge in Notes R.

FIG. 16. — 1, **Salvia scabra,** flowering stem, × 0,7; 1a, mature calyx, × 3 (after Hedge in Notes R. bot. Gdn Edinb. 33: t.18, 1974, with his permission and that of Her Majesty's Stationery Office). 2, **S. runcinata,** upper part of plant, × 1; 2a, mature calyx, × 3; 2b, corolla opened longitudinally, × 3 (*Mrs Jenkins* s.n.).

bot. Gdn Edinb. 33: 69 (1974). Type: Cape, without locality, *Thunberg* s.n. (UPS, holo., microfiche 26/615!).

S. *marginata* Benth. in E. Mey., Comm. 236 (1838); in DC., Prodr. 12: 351 (1848). Syntypes: Cape, between Coega and Sundays River, *Drège* 7944a (K); Addo, *Drège* 7944 (K).

Perennial herb, somewhat woody at the base, with ascending stems up to 0,5 m long or more; stems glabrous below, glabrous to sparingly pubescent above. *Leaves* petiolate; blade often drying dark brown, subentire or lyrate-pinnatifid with a large terminal segment and one or two pairs of basal lobes, broadly elliptic to ovate in outline, 25−50 × 15−30 mm, upper surface subglabrous, under-surface sparingly pubescent on the nerves and margins, margin coarsely crenate; petiole up to 30 mm long, glabrous or with a few long stiff hairs. *Inflorescence* of up to 10 verticils, spaced below, denser above; verticils 2−8-flowered. *Calyx* sparingly hispid, purple tinged, tubular-campanulate, up to 10 mm long in fruit; upper lip with 3 acuminate subequal teeth 1−1,5 mm long. *Corolla* 20−25 mm long; tube c. 18 mm long; upper lip straight, 3 mm long; lower lip c. 5 mm long.

A little-known species apparently restricted to the south-eastern Cape Province from about Uitenhage to Albany district. Map 49.

Vouchers: *Germishuizen* 1418; *Zeyher* 3533.

Related to S. *scabra* (above) and S. *repens* (below) but differs from both in the stems and upper leaf surfaces being glabrous or nearly so. It also differs from S. *scabra* in having a shorter corolla, and from S. *repens* in lacking rhizomes.

16. **Salvia repens** *Burch. ex Benth.,* Lab. 306 (1833). Type: Cape, without locality, *Burchell* s.n. (K, holo.).

Perennial herb with few to several stems usually arising from a creeping rhizome; stems ascending, simple or branched, 0,25−0,6 (−0,8) m tall, shortly pilose to tomentose. *Leaves* usually crowded and larger at the base of the plant, the lower petiolate, the upper subsessile; blade simple to sublyrate or rarely runcinate, oblong to broadly obovate in outline, 30−100 × 8−45 mm, subglabrous to pilose, gland-dotted beneath, apex obtuse to rounded, base truncate to auriculate, margin irregularly crenate-dentate. *Inflores-*

cence of several to many verticils, widely spaced below, denser above; verticils 6 (−8)-flowered. *Calyx* shortly hispid, tubular-campanulate, up to 13 mm long in fruit; upper lip with 3 acuminate teeth 0,5−2 mm long, subequal or the central tooth slightly shorter than the outer two. *Corolla* pale blue or mauve to purple, rarely white, (10−) 14−26 mm long; tube (6−) 7−18 mm long; upper lip straight, 4−5 mm long; lower lip 4,5−7 mm long.

A very variable species extending from northern Transvaal, Orange Free State, Natal, Lesotho and north-eastern Cape Province to the southern and eastern Cape Province and Transkei.

More field observations are required for a better understanding of variation in leaf shape, size and indumentum, and calyx and corolla length. At present the material is classified into 3 varieties on the following lines:

1 Corolla less than 20 mm long; leaves 30−80 × 8−35 mm:

  2 Leaves elliptic to obovate, simple to runcinate, sparsely to freely gland-dotted; stems ascending, much or little branched; corolla 14−20 mm long .... (a) var. *repens*

  2 Leaves narrowly oblong, simple, densely gland-dotted; stems erect, much branched; corolla 10−15 mm long .... (c) var. *transvaalensis*

1 Corolla 20−26 mm long; leaves up to 100 × 45 mm.................................... (b) var. *keiensis*

(a) var. **repens.**

Hedge in Notes R. bot. Gdn Edinb. 33: 74 (1974).

S. *repens* Burch. ex Benth., Lab. 306 (1833); in DC., Prodr. 12: 353 (1848); Skan in F.C. 5,1: 328 (1910); Jacot Guill., Fl. Lesotho 238 (1971); Ross, Fl. Natal 304 (1972). Type: Cape, without locality, *Burchell* s.n. (K, holo.).

S. *subsessilis* Benth. in E. Mey., Comm. 237 (1838); in DC., Prodr. 12: 352 (1848). Syntypes: Zuurberg, *Drège* 4761 b (K); near Umzimvubu River, *Drège* (K).

S. *rudis* Benth, in E. Mey., Comm. 235 (1838); in DC., Prodr. 12: 350 (1848); Skan l.c. 331 (1910); Ross, Fl. Natal 304 (1972). Type: Cape, "Uitenhage", *Ecklon* s.n. (K, holo.).

S. *raphanifolia* Benth. in E. Mey., Comm. 237 (1838); in DC., Prodr. 12: 352 (1848); Skan, l.c. 330 (1910). Type: Cape, near Windvogelberg, *Drège* 7943 (K, holo.).

S. *incisa* Benth. in DC., Prodr. 12: 352 (1848). Type: Cape, "Karoo", Wodehouse, *Ecklon* 112 (K, holo.).

S. *woodii* Gürke in Bot. Jb. 26: 76 (1898); Skan, l.c. 332 (1910); Ross, Fl. Natal 304 (1972). Syntypes: several incl. Natal, Weenen County, *Medley Wood*

3621 (NH!); Transvaal, Standerton, *Rehmann* 6781 (PRE!).

S. natalensis Briq. & Schinz in Bull. Herb. Boissier sér. 2,3: 1078 (1903). Syntypes: Orange Free State, Harrismith, *Medley Wood* 4972 (NH, Z); Cape, near Kei River, *Schlechter* 6232 (Z).

S. schenckii Briq. in Bull. Herb. Boissier sér. 2,3: 1079 (1903). Type: Orange Free State, between Harrismith and Vaal River, *Schenck* 732 (Z).

S. cooperi Skan, l.c. 332 (1910); Ross, Fl. Natal 304 (1972). Syntypes: several, incl. Natal, *Cooper* 1279 (K); Orange Free State, near Witzies Hoek, *Bolus* 8237 (BOL, K).

Stems ascending, much or little branched; leaves simple to runcinate, sparingly to freely gland-dotted, elliptic to obovate, 30–80 × 14–35 mm; corolla 14–20 mm long.

Distribution as for the species, in mountain grassland, on river banks, in open woodland or karroid veld, often on heavy clay or clay-loam soils, sometimes a weed in gardens or disturbed places. Map 51.

Vouchers: *Codd* 10415; *Dieterlen* 958; *Flanagan* 1406; *Medley Wood* 5187; *Repton* 6260; *Schlechter* 3818.

As may be seen from the synonymy, a great deal of variation is included in this variety and the distinction between it and the next species, *S. runcinata*, is not always clear (see note after the latter).

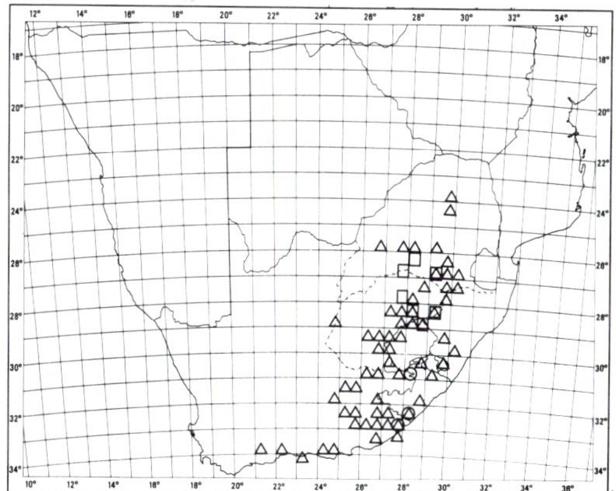

Map 51. — △ **Salvia repens** var. **repens**
    ○ **S. repens** var. **keiensis**
    □ **S. repens** var. **transvaalensis**

(b) var. **keiensis** *Hedge* in Notes R. bot. Gdn Edinb. 33: 75 (1974). Type: Cape, Komga, near Kei River, *Schlechter* 6232 (Z, holo.).

Stems ascending, not much branched; leaves simple or sometimes lobed near the base, broadly elliptic to obovate, 70–100 × 35–45 mm; corolla 20–26 mm long.

Restricted to the eastern Cape Province and Transkei, in grassland or open woodland. Map 51.

Vouchers: *Acocks* 22106; *Flanagan* 475.

Differs from var. *repens* in the larger leaves and longer corolla, though some specimens are somewhat intermediate, e.g. *Codd* 9243 from near Butterworth.

(c) var. **transvaalensis** *Hedge* in Notes R. bot. Gdn Edinb. 33: 75 (1974). Type: Transvaal, Vereeniging, *Burtt Davy* 17135 (BOL, holo.).

Stems erect, much branched, leafy; leaves usually simple, occasionally lobed towards the base, freely gland-dotted, narrowly oblong or elliptic, 30–50 × 8–15 (–20) mm; corolla 10–15 mm long.

Restricted to the southern Transvaal and northern Orange Free State; in grassland. Map 51.

Vouchers: *Acocks* 21023; *Burtt Davy* 9110.

Characterized by the rather dwarf, branched habit, leafy stems, small leaves densely covered with gland-dots and the small flowers.

17. **Salvia runcinata** *L.f.*, Suppl. 89 (1781); Thunb., Fl. Cap. edn Schult. 452 (1823); Benth. in E. Mey., Comm 237 (1838); in DC., Prodr. 12: 352 (1848); Skan in F.C. 5,1: 327 (1910); Wilman, Check List Griq. West 229 (1946); Jacot Guill., Fl. Lesotho 238 (1971); Hedge in Notes R. bot. Gdn Edinb. 33: 75 (1974). Type: Cape, without locality, *Thunberg* s.n. (UPS, microfiche 27/636!).

S. scabra sensu Benth., Lab. 305 (1833), partly.

S. monticola Benth. in E. Mey., Comm. 238 (1838); in DC., Prodr. 12: 353 (1848); Skan, l.c. 329 (1910); Ross, Fl. Natal 304 (1972). Syntypes: several, incl. Cape, near Windvogelberg, *Drège* 7946a (K).

S. runcinata var. *major* Benth. in DC., Prodr. 12: 352 (1848). Type: Cape, Uitenhage, *Ecklon* s.n. (K, holo.).

— var. *grandiflora* Skan, l.c. 327 (1910). Syntypes: several, incl. Cape, Victoria West. Div., *Drège* 4750c (K); Albert Div., *Drège* 7945 (K).

— var. *nana* Skan, l.c. 327 (1910). Syntypes: Transvaal, near Pretoria, *Burtt Davy* 606 (K; PRE!); *Leendertz* 965 (K; PRE!).

S. sisymbrifolia Skan, l.c. 328 (1910); Jacot Guill., Fl. Lesotho 238 (1971); Ross, Fl. Natal 304 (1972). Syntypes: several incl. Transvaal, Pretoria, *Burtt Davy* 7079 (K; PRE!).

Perennial erect herb 0,15–0,5 (–0,7) m tall with 1–several stems from a taproot or, occasionally, from a creeping rootstock;

stems hispid or crisped pilose, gland-dotted. *Leaves* shortly petiolate or the upper ones sessile; blade runcinate-pinnatipartite to lyrate, rarely almost entire, oblong-lanceolate to obovate in outline, 30—90 (—120) × 15—30 (—50) mm, hispid-scabrid, gland-dotted, lobes rounded to triangular, sometimes oblong and pinnatifid. *Inflorescence* of several to many verticils, widely spaced below, denser above; verticils 4—8-flowered. *Calyx* hispid-scabrid, gland-dotted, 5—8 mm long. *Corolla* white or pale blue to mauve or purplish, 7—14 mm long; tube 4,5—9 mm long; upper lip straight 3—4 mm long; lower lip often slightly longer. Fig. 16:2.

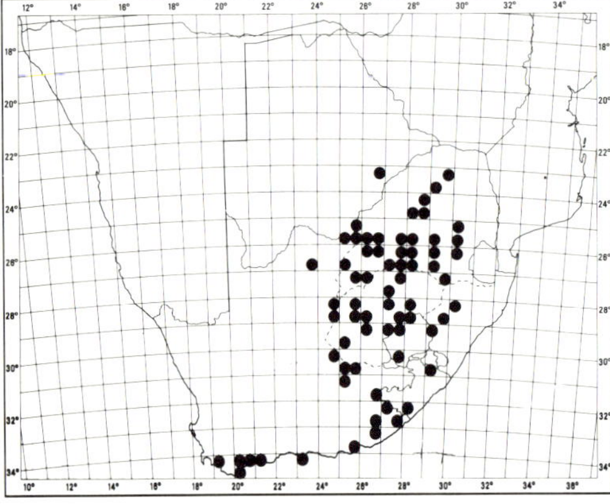

MAP 52. — **Salvia runcinata**

A very variable species extending from northern Transvaal and Botswana to northern Cape Province, Orange Free State and eastern and southern Cape Province as far south as Bredasdorp district but rare in Transkei, Natal and Lesotho; in a variety of habitats, but usually on heavy soils, sometimes spreading and locally common on disturbed places or overgrazed veld, for example under thorn trees. Also in Zimbabwe. Map 52.

Vouchers: *Acocks* 20984; *Galpin* M601; *Muir* 2652; *Schlechter* 3691.

The variation consists of intergrading forms which do not warrant taxonomic recognition. The limits of the species are also far from clear. In general, the calyx and corolla are shorter than in *S. repens* (above), the mature calyx has a wider sinus between the upper and lower lips and the leaves are more dissected, but intermediates may be found; ecologically the two are very similar, though *S. runcinata* seems to occupy more arid situations and usually lacks the creeping rootstock characteristic of *S. repens*.

S. stenophylla (below), which is probably the closest ally of *S. runcinata*, can usually be distinguished by the narrower leaves with narrower segments, and the almost glabrous stems, but some specimens are difficult to place.

It is probable that hybridization and introgression have contributed to the confusion but this can be confirmed only by field work in areas where one species overlaps with another. In the meantime it is reasonable to maintain the three species as distinct.

18. **Salvia stenophylla** *Burch. ex Benth.*, Lab. 306 (1833); Benth. in E. Mey., Comm. 238 (1838); in DC., Prodr. 12: 353 (1848); Skan in F.C. 5,1: 326 (1910); Wilman, Check List Griq. West 229 (1946); Launert and Schreiber in F.S.W.A. 123: 28 (1969); Jacot Guill., Fl. Lesotho 238 (1971); Ross, Fl. Natal 304 (1972); Hedge in Notes R. bot. Gdn Edinb. 33: 77, t.20 (1974). Type: Cape, Griquatown, *Burchell* 1881 (K, holo.).

S. xerobia Briq. in Bull. Herb. Boissier sér. 2,3: 1076 (1903). Type: Cape, near Keiskamma, *Schlechter* 6115 (Z, holo.).

S. chlorophylla Briq., l.c. 1080 (1903). Type: S.W.A./Namibia, Windhoek, *Dinter* 316 (Z, holo.).

S. stenophylla var. *subintegra* Skan, l.c. 326 (1910). Type: Botswana, "Batlapin Territory," *Holub* s.n. (K, holo.).

S. pallida Dinter ex Engl., Pflanzenw. Afr. 1,2: 570 (1910), nom. nud.

Perennial erect bushy herb 0,2—0,4 (—0,6) m tall, usually much branched from a woody taproot; stems subglabrous or with few hairs, usually with orange-red gland-dots. *Leaves* shortly petiolate or subsessile; blade often pinnatifid or pinnatisect, occasionally simple, linear-oblong to oblong-lanceolate in outline, (20—) 25—80 × (4—) 6—20 mm, sparingly pubescent on the nerves below and gland-dotted, margin often finely crenate-dentate. *Inflorescence* of several to many spaced verticils; verticils 6 (—8)-flowered. *Calyx* minutely hispidulous, gland-dotted, 4—5 mm long. *Corolla* pale blue or mauve, c. 12 mm long; tube c. 7 mm long; upper lip straight, 4 mm long; lower lip equal to or slightly longer than the upper.

Distributed from central S.W.A./Namibia and southern Botswana to Transvaal, Orange Free State, Lesotho, and northern, north-eastern and eastern Cape Province, rare in Natal and Transkei; in grassland, open woodland and semi-arid shrub, often on calcareous or brackish soil, sandy soil in watercourses or damp places, sometimes a semi-weed of disturbed places, for example at roadsides etc. Map 53.

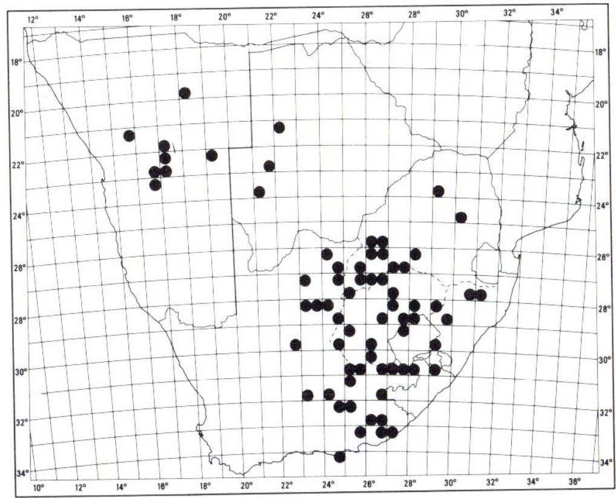

MAP 53. — **Salvia stenophylla**

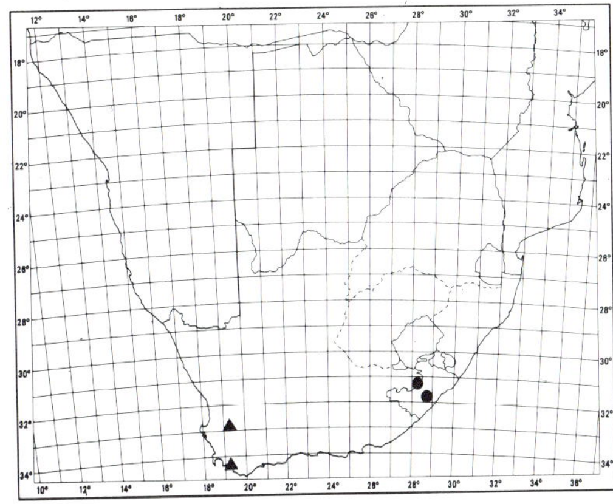

MAP 54. — ● **Salvia schlechteri**
▲ **S. granitica**

Vouchers: *Acocks* 15832; *De Winter* 7432; *Galpin* 1645; *Scheepers* 1342; *Strey* 2431.

The leaves vary from simple (the form described as *S. chlorophylla*) to pinnatifid (the typical form), and some specimens appear to grade into *S. runcinata* (above). The main differences are discussed under that species. A characteristic of the species is that the small calyces are more or less appressed to the rhachis.

19. **Salvia schlechteri** Briq. in Bull. Herb. Boissier sér. 2, 3: 1077 (1903); Hedge in Notes R. bot. Gdn Edinb. 33: 80 (1974). Type: Transkei, Umtata, *Schlechter* 6330 (Z, holo.).

S. monticola Benth. var. *angustiloba* Skan in F.C. 5,1: 330 (1910). Type: Transkei, "between Gekau and Bashee River", *Drège* 4751 (K, holo.).

Perennial herb up to 0,3 m tall, usually branched from a somewhat woody base; stems subglabrous or with some short stiff hairs and gland-dots. *Leaves* sessile or subsessile; blade finely pinnatifid or pinnatisect, 40—60 mm long, subglabrous, gland-dotted, with 4—6 pairs of short narrow irregularly dentate lateral segments 1 mm wide and a slightly larger terminal segment. *Inflorescence* of several to many verticils, spaced below, closer above; verticils 4—6-flowered. *Calyx* sparingly hispidulous, gland-dotted, c. 9 mm long. *Corolla* pale blue and white, up to 25 mm long; tube c. 14 mm long, widening to 8 mm wide at the throat; upper lip straight, 7—8 mm long; lower lip somewhat deflexed, 8—9 mm long.

Known from a restricted area in the Transkei. Map 54.

Voucher: *Van Breda* 865A.

Although so little known, this appears to be a distinct species recognizable by its finely pinnatisect or pinnatifid leaves and relatively large, wide-throated corolla.

20. **Salvia granitica** Hochst. in Flora 28: 65 (1845); Benth. in DC., Prodr. 12: 358 (1848); Skan in F.C. 5,1: 333 (1910); Hedge in Notes R. bot. Gdn Edinb. 33: 81, t.21 (1974). Type: Cape, near Caledon, Babylons Tower, *Krauss* 1120 (MB; W; BAS).

Stoloniferous perennial with a woody rootstock; stems erect-ascending up to 0,6 m long, not or little branched, subglabrous to sparingly pubescent. *Leaves* shortly petiolate; blade simple, linear to linear-oblanceolate, 30—50 × 3—6 (−10) mm, glabrous above, pilose beneath and gland-dotted, apex acute, base tapering narrowly, margin subentire or rarely irregularly dentate, often ciliate. *Inflorescence* of several spaced 2-flowered verticils. *Calyx* glandular-villous, 10—12 mm long. *Corolla* mauve-pink, c. 20 mm long; tube c. 14 mm long; upper lip falcate, 5—6 mm long; lower lip subequal to the upper. Fig. 17.

Recorded as yet only from the Clanwilliam and Caledon districts, on stony slopes. Map 54.

Vouchers: *Esterhuysen* 17911 (BOL); *Pillans* 8694 (BOL).

A species of doubtful affinity, characterized by the narrow, simple leaves, the 2-flowered verticils, and the falcate upper lip of the corolla.

21. **Salvia verbenaca** L., Sp. Pl. 25 (1753); Benth. in DC., Prodr. 12: 294 (1848); Launert & Schreiber in F.S.W.A. 123: 29 (1969); Hedge in Notes R. bot. Gdn Edinb. 33: 95 (1974). Type: from Europe (LINN, holo.).

Horminum verbenaca (L.) Mill., Gard. Dict. edn 8: Horminum No. 1 (1768). *S. verbenaefolia* Salisb., Prodr. 73 (1796), nom. illegit. Type: as for *S. verbenaca* L.

S. clandestina L., Sp. Pl. edn 2: 36 (1762). *S. verbenaca* subsp. *clandestina* (L.) Briq. var. *clandestina* (L.) Briq., Lab. Alpes Marit. 518 (1891). Type: from India (LINN, holo.).

S. controversa Ten., Syll. Fl. Neap. 18 (1831). *S. verbenaca* var. *controversa* (Ten.) Briq., l.c. 516 (1891). Type: from Italy.

S. clandestina var. *angustifolia* Benth. in DC., Prodr. 12: 295 (1848); Skan in F.C. 5,1: 319 (1910); Wilman, Check List Griq. West 228 (1946); Jacot Guill., Fl. Lesotho 238 (1971). *S. verbenaca* var. *angustifolia* (Benth.) Pugsley in J. Bot., Lond. 46: 144 (1908). Syntypes: several, incl. Cape, *Ecklon* (K); *Drège* (K).

S. cleistogama De Bary & Paul, Ind. Sem. Hort. Halens. 6 (1867). Type: a cultivated plant of South African provenance.

For further synonymy see Pugsley in J. Bot., Lond. 46: 144 (1908) and Hedge, l.c.

Perennial, probably short-lived, with stems arising from a woody taproot; stems erect, 0,15—0,4 m tall, densely glandular-villous with long spreading hairs and short-er gland-tipped hairs. *Leaves* mainly a dense basal rosette, shortly petiolate to subsessile, blade irregularly to deeply pinnatifid or sometimes almost entire, usually oblong to ovate-oblong in outline, 40—130 × 10—30 mm, rugose and usually densely pubescent and gland-dotted below, margin often eroso-dentate. *Inflorescence* usually branched with many verticils, spaced below and denser above; verticils usually 6-flow-ered. *Calyx* pilose and gland-dotted, densely villous in the throat, up to 8 mm long in fruit. *Corolla* light blue to purple, 8—12 mm long; tube 4—8 mm long; upper lip often slightly falcate, 2–4 mm long, lower lip usually slightly shorter.

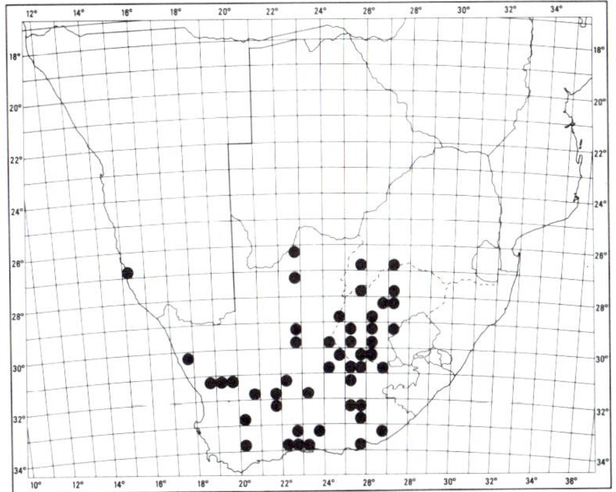

MAP 55. — **Salvia verbenaca**

Probably indigenous in the countries around the Mediterranean and on the Canary Islands and has spread further afield in Europe and Asia. If, as has been stated, it is an introduced plant in our Flora area, it is now widely distributed, mainly in the drier, western half of South Africa, in southern S.W.A./Namibia, northern, central and western Cape Province and western Orange Free State, with outliers reaching south-western Transvaal, Lesotho, southern and south-western Cape Province. In August 1811 Burchell collected it (No. 1454) between "Quaggasfontein and Dwaalfontein" (i.e. north of Fraserburg), and this would indicate that it may be indigenous in those parts. It has become naturalized in Australia and the United States. Map 55.

Vouchers: *Acocks* 2427; *Henrici* 2770; *Schlieben* 8702.

Plants in Southern Africa have somewhat nar-rower leaves than those occurring around the Mediterranean. Cleistogamous flowers (self-fertilized), which are smaller than normal flowers, are frequently found. Briquet, l.c., and Pugsley, l.c., have upheld several subspecies but it is impossible to key them out satisfactorily.

S. verbenaca differs from *S. runcinata* and related species (nos 16—19) in the shape of the upper lip of the calyx, with its teeth conniving into an apex which has 3 closely placed minute (0,5 mm) teeth. This is in contrast to *S. runcinata* and its allies in which the 3 teeth of the upper lip are somewhat spreading, acuminate, and 0,5—2,5 mm long (the central tooth often shorter than the outer 2).

Its nearest affinity is with *S. disermas* (below) which, together with *S. radula* (no. 23), has the upper calyx lip with the 3 teeth connivent as in *S. verbenaca*. *S. verbenaca* tends to be smaller in stature with shorter calyx and corolla (see key to species), and the upper lip

FIG. 17. — 1, **Salvia granitica**, habit, × 0,7; a, corolla, × 1,5; b, stamen, × 3; c, calyx, × 1,3 (from Hedge, Notes R. bot. Gdn Edinb. 33: t.21, 1974, with his permission and that of Her Majesty's Stationery Office).

of the corolla is not as distinctly falcate (often almost straight). Also, the leaves of *S. verbenaca* often have more deeply dissected margins than those of *S. disermas,* but depauperate specimens of the latter are sometimes difficult to identify with certainty.

22. **Salvia disermas** *L.*, Sp. Pl. edn 2: 36 (1762); Benth. in DC., Prodr. 12: 291 (1848); Skan in F.C. 5,1: 319 (1910); Hedge in Notes R. bot. Gdn Edinb. 33: 105 (1974). Type: "Syria" (LINN, holo.).

S. rugosa Ait., Hort. Kew. 1: 42 (1789). Type: Cape, without locality, *Masson* ex Hort. Kew. (BM, holo.).

S. rugosa Thunb., Prodr. 97 (1800); Fl. Cap. edn Schult. 451 (1823); Benth. in E. Mey., Comm. 235 (1838); in DC., Prodr. 12: 291 (1848); Skan, l.c. 318 (1910); Wilman, Check List Griq. West 228 (1946); Letty, Wild Flow. Transv. 288, t.144 (1962); Launert & Schreiber in F.S.W.A. 123: 28 (1969); nom. illegit. Type: Cape, without locality, *Thunberg* (UPS).

— var. *angustifolia* Benth. in DC., Prodr. 12: 291 (1848). Syntypes: Cape, without locality. *Burchell* 1801 (K); near Swellendam, *Ecklon* s.n.

S. fleckii Gürke in Bull. Herb. Boissier 6: 551 (1898); Bak. in F.T.A. 5: 457 (1900). Syntypes: S.W.A./Namibia, *Fleck* 168a; *J. Graf Pfeil* 78.

Soft shrub or perennial herb 0,3−1 (−1,2) m tall with one or more stems from a woody rootstock; stems glandular-villous. *Leaves* often crowded and larger near the base of the plant, petiolate or the upper ones subsessile; blade broadly ovate to oblong-lanceolate, 40−160 × 15−70 mm, rugose, scabrid, glandular-pubescent, lower surface markedly reticulate, apex acute to rounded, base cordate to obtuse, margin irregularly crenate to eroso-dentate. *Inflorescence* of 15 or more verticils, spaced below, denser above; verticils usually 6-flowered. *Calyx* glandular-hispid to villous, up to 12 mm long. *Corolla* whitish, pale blue or mauve, 15−25 (−30) mm long; tube 8−12 mm long; upper lip falcate, 7−15 mm long; lower lip often shorter than the upper.

Occurs in the north-central to southern districts from south-western Transvaal and northern Cape Province through the western Orange Free State and Karoo to Namaqualand and southwards to Uniondale and Swellendam, with a single record from central S.W.A./Namibia; found on sandy soil in water-courses, on limestone formations and rocky hillsides, tending to spread as a weed along roadsides, on waste places and on overgrazed veld. Map 56.

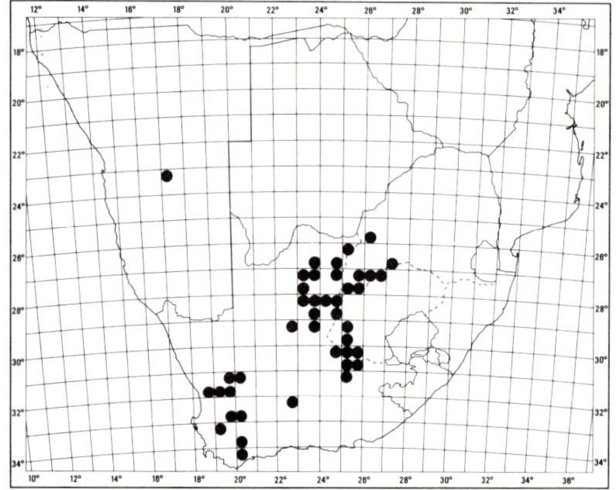

MAP 56. — **Salvia disermas**

Vouchers: *Acocks* 14408; *Flanagan* 1495; *Marloth* 14064; *Schlechter* 10926; *Verdoorn* 907.

Linnaeus records the origin of *S. disermas* as "Syria" but the type specimen (LINN 42/26), which agrees with the original description, matches the South African specimens cited above and is unlike anything from south-western Asia (see Hedge, l.c.).

Its relationship to *S. radula* (below) is discussed there.

23. **Salvia radula** *Benth.* in DC., Prodr. 12: 291 (1848); Skan in F.C. 5,1: 318 (1910); Hedge in Notes R. bot. Gdn Edinb. 33: 107 (1974). Type: Transvaal, Magaliesberg, *Burke* (K, holo.).

Perennial herb with one or more erect stems from a woody rootstock, 0,3−0,75 m tall; stems densely white-lanate below with some glandular hairs above. *Leaves* often crowded and larger near the base of the plant, petiolate; blade simple, broadly ovate to oblong-ovate, 55−130 × 30−70 mm, rugose and subglabrous above, densely white-lanate beneath, apex obtuse to rounded, base truncate to cordate, margin crenate to erose-dentate, occasionally obscurely lobed. *Inflorescence* of 15 or more verticils, spaced below, denser above; verticils usually 6-flowered. *Calyx* villous and gland-dotted, 12−15 mm long. *Corolla* white or pale mauve to blue, 18−25 mm long; tube 8−15 mm long; upper lip falcate,

FIG. 18. — 1, **Salvia radula,** habit, × 0,7; a, mature calyx, × 2; b, section through corolla, × 1,5; c, stamen, × 2,3; d, nutlet, × 6,7 (from Hedge, Notes R. bot. Gdn Edinb. 33: t.24, 1974, with his permission and that of Her Majesty's Stationery Office).

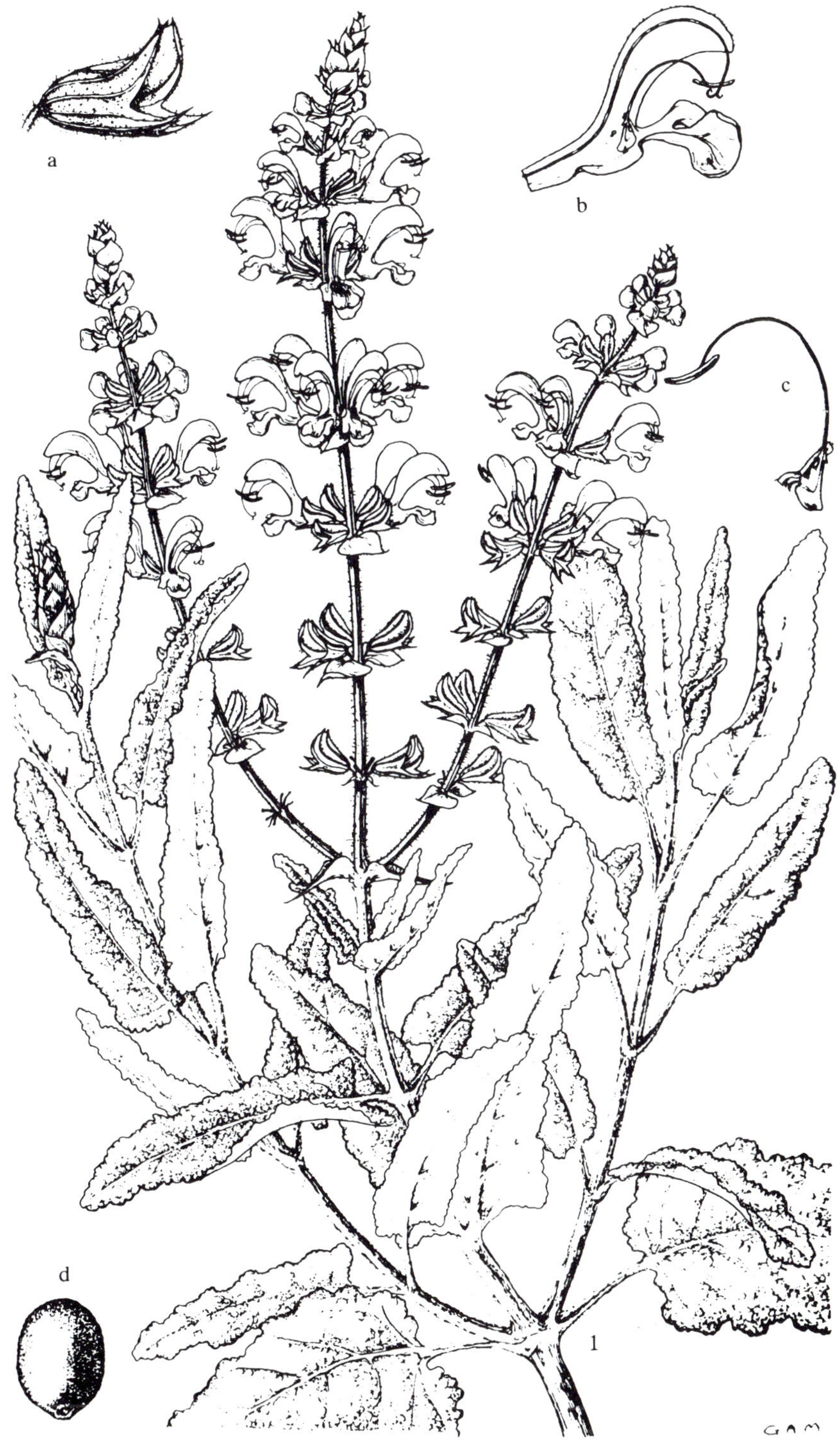

8–10 mm long; lower lip slightly shorter than the upper. Fig. 18.

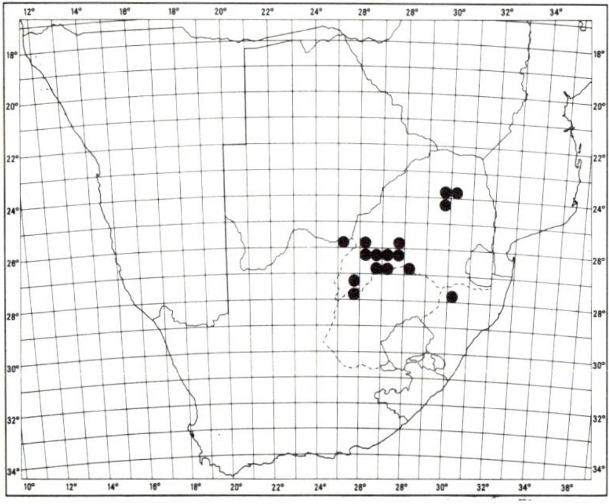

MAP 57. — **Salvia radula**

Distributed from south-western through central Transvaal to the eastern escarpment near Haenertsburg; on river banks, surface limestone and dolomitic wooded slopes, tending to spread along roads and into overgrazed veld. Map 57.

Vouchers: *Acocks* 12378; *Codd* 2127; *Prosser* 1912.

Closely related to *S. disermas* (above) but can usually be distinguished by its denser lanate indumentum on lower stems and undersides of leaves. The two species have met in the south-western Transvaal, aided by man's activities in road building and overgrazing of veld, and occasional intermediates are found in this area.

24. **Salvia tiliifolia** *Vahl*, Symb. Bot. 3: 7 (1794), as *tiliaefolia;* Correll & Johnston, Man. Vasc. Pl. Texas 1370 (1970); Standley & Williams in Fieldiana Bot. 24,9: 298 (1973). Type: from Mexico.

Annual herb 0,4–0,6 m tall; stems sparingly pilose, simple or branched. *Leaves* petiolate; blade soft, broadly ovate, 40–60 × 35–50 mm, green, sparingly pubescent, apex acuminate, base truncate, finely and regularly crenulate. *Inflorescence* simple or paniculate, of many fairly closely placed verticils; verticils 6–14-flowered. *Calyx* hispidulous, ribbed, enlarging to 7 mm long in fruit; upper lip entire. *Corolla* blue, 5–7 mm long.

A weed of gardens and waste places, recorded from several parts of Pretoria since 1943. Indigenous in Central America and introduced into the United States and Canada.

Vouchers: *Codd* 10737; *Repton* 1429.

25. **Salvia reflexa** *Hornem.*, Enum. Pl. Hort. Hafn. 1: 34 (1807); Steyermark, Fl. Missouri 1288 (1963); Correll & Johnson, Man. Vasc. Pl. Texas 1369 (1970); Hedge in Notes R. bot. Gdn Edinb. 33: 115 (1974). Type: from the U.S.A.

Annual erect free-flowering herb 0,5–0,75 m tall; stems usually solitary, branched above, greyish tomentose, glabrescent. *Leaves* shortly petiolate; blade lanceolate to linear-lanceolate, 30–60 × 5–8 mm, grey-green, soft, subglabrous above, tomentulose and gland-dotted beneath, apex obtuse, base attenuate, margin subentire to remotely and shallowly toothed. *Inflorescence* lax, of several to many 2(–3)-flowered verticils. *Calyx* deeply bilabiate, minutely hispidulous, ribbed, enlarging to 6 mm long in fruit; upper lip entire. *Corolla* blue, 5–6 mm long.

A weed of waste places first recorded from Pretoria and Krugersdorp in 1971, since collected from several localities in the Witwatersrand and Rustenburg areas, suggesting that it is already widely distributed; also common around Roma, Lesotho, where it has been seen for several years. Indigenous in the United States and Mexico and also introduced into tropical Africa.

Vouchers: *Hanekom* 1718; *Mogg* 35749.

26. **Salvia coccinea** *Etlinger*, Comm. Bot.-med. Salvia 23 (1777); Juss. ex Murr. in Comm. Gotting. 1: 86 (1778); Bailey, Cyclop. Hort. edn 21,3: 3064 (1963); Correll & Johnson, Man. Vasc. Pl. Texas 1369 (1970); Ross, Fl. Natal 304 (1972); Standley & Williams in Fieldiana Bot. 24,9: 280 (1973); Hedge in Notes R. bot. Gdn Edinb. 33: 114 (1974). Type: a cultivated plant.

S. pseudococcinea Jacq., Coll. 2: 302 (1788); Hook. in Curtis's bot. Mag. t.1864 (1828). *S. coccinea* var. *pseudococcinea* (Jacq.) Gray, Syn. Fl. 2,1: 368 (1878). Type: a cultivated plant.

Soft shrub 0,6–1,5 m tall; stems usually branched, herbaceous above, softly woody below, usually hispid. *Leaves* petiolate; blade ovate-deltoid, 35–50 × 23–35 mm, subglabrous and dark green above, tomentose and paler beneath, apex acute, base truncate to cordate, margin finely crenate. *Inflorescence* of up to 12 verticils, lax below, denser above; verticils (4−) 6−10-flowered. *Calyx* hispidulous, about 10 mm long; upper lip entire. *Corolla* scarlet to pink, 20−25 mm long; tube 15−20 mm long; upper lip straight, 5 mm long; lower lip slightly longer.

A garden escape which has become semi-naturalized in warm, moist parts of the eastern Transvaal and Swaziland. Indigenous in tropical America and Mexico.

Vouchers: *Bos* 1220; *Kluge* 810; *Scheepers* 41.

7305

15. SATUREJA

Satureja *L.*, Sp. Pl. 567 (1753); Gen. Pl. edn 5: 247 (1754); Benth., Lab. 351 (1834); in DC., Prodr. 12: 208 (1848); Benth. & Hook. f., Gen. Pl. 2,2: 1187 (1876); Briq. in Natürl. PflFam. 4,3a: 296 (1896); Hedberg, Afroalpine Vascular Plants 160−64, 317−18 (1957); Killick in Bothalia 7: 435 (1961); R. A. Dyer, Gen. 1: 529 (1975); often spelt *Satureia.* Type species: *S. hortensis* L.

Micromeria Benth. in Bot. Reg. sub. t.1282 (1829); Lab. 368 (1834); in DC., Prodr. 12: 212 (1848); Benth. & Hook. f., Gen. Pl. 2,2: 1188 (1876); Bak. in F.T.A. 5: 452 (1900); Cooke in F.C. 5,1: 306 (1910). Type species: *M. juliana* (L.) Benth.

Perennial herbs or soft shrubs. *Leaves* small, entire or obscurely toothed. *Flowers* solitary or in few−several-flowered cymes in the axils of undifferentiated leaves along the upper half of the stems (or sometimes lower). *Calyx* tubular or tubular-campanulate, 13−15-nerved, subequally 5-toothed, scarcely enlarging in fruiting stage. *Corolla* scarcely or well exserted from the calyx, dimorphic in some species; tube straight, cylindric below, campanulate above; upper lip short, broad, emarginate; lower lip longer, 3-lobed, the middle lobe the largest. *Stamens* 4, didynamous, included, attached near the throat, curved upwards, the two lower the longer. *Style* included, 2-lobed. *Nutlets* ovoid to oblong, smooth.

Non-Southern African species may be annual or perennial and in some the flowers are borne in the axils of reduced leaves towards the ends of the stems which then take the form of inflorescences; the calyx may be 5−10-ribbed and obscurely 2-lipped and the style may be unequally 2-lobed.

Species over 100, cosmopolitan; 4 indigenous in Southern Africa. Two European species, *S. hortensis* L. (Summer Savory) and *S. montana* L. (Winter Savory) are grown as pot-herbs, of which the former, an annual, is the more commonly grown in Southern Africa.

Bentham separated the genus *Micromeria* from *Satureja* on the basis of the 13−15-nerved, subequally 5-toothed calyx as against the 10-nerved, sometimes obscurely 2-lipped calyx of *Satureja*. Briquet, l.c., placed the genera *Micromeria* Benth. and *Calamintha* Mill. (calyx obscurely 2-lipped, flowers tend to be in terminal inflorescences) as synonyms of *Satureja* and this treatment has generally been followed, though with certain reservations, for the African species. Although such a grouping is heterogeneous it is felt that a world-wide revision of the whole group is required before maintaining separate genera in Southern Africa. Ongoing studies of the complex in Edinburgh indicate that true *Satureja* does not occur in Southern Africa. On this basis, *S. biflora* would be placed in *Micromeria* and the oldest name for the species may prove to be *M. imbricata* (Forssk.) Christen., while the remaining three species may need a new generic name (I.C. Hedge, personal communication).

1 Stems wiry, erect to spreading; leaves with the apex acute to obtuse, margin thickened; flowers in several−many-flowered, pedunculate cymes, rarely solitary; calyx tubular............................ 1. *S. biflora*

1 Stems herbaceous, prostrate or decumbent; leaves with apex obtuse to rounded, margin not thickened; flowers solitary or in up to 3 (rarely 5)-flowered cymes; calyx campanulate:

  2 Bracteoles small, linear; corolla less than 18 mm long:

    3 Leaves 5−11 × 4−10 mm; corolla 6−7 mm long .... 2. *S. compacta*

    3 Leaves 16−24 × 12−20 mm; corolla 10−15 mm long .... 3. *S. reptans*

  2 Bracteoles foliose; corolla 18−20 mm long.. 4. *S. grandibracteata*

1. **Satureja biflora** *(Buch.-Ham. ex D. Don) Briq.* in Natürl. PflFam. 4,3a: 299 (1896); Brenan in Mem. N.Y. bot. Gdn 9: 45 (1954); Hedberg, Afroalp. Vasc. Pl. 161 (1957); Killick in Bothalia 7: 435 (1961); Cufod. in Bull. Jard. bot. État Brux. 32, Suppl.: 821 (1962); Jacot Guill., Fl. Lesotho 238 (1971). Type: India, Upper Nepal, *Buchanan-Hamilton* s.n.

Thymus biflorus Buch.-Ham. ex D. Don, Prodr. Fl. Nepal. 112 (1825). *Micromeria biflora* (Buch.-Ham. ex D. Don) Benth., Lab. 378 (1834); in DC., Prodr. 12: 220 (1848); Hook. f., Fl. Brit. India 4: 650 (1885); Engler, Hochgebirgsfl. Trop. Afr. 365 (1892); Bak. in F.T.A. 5: 452 (1900); Cooke in F.C. 5,1: 306 (1910). Type: as above.

Micromeria ovata Benth., Lab. 377 (1834); in DC. Prodr. 12: 219 (1848); Engler, l.c. 364 (1892). *Satureja ovata* R. Br. in Salt, Abyss. App. 64 (1814), nom. nud. Type: Ethiopia, *Salt* s.n. (BM, holo.).

M. punctata Benth., Lab. 377 (1848); in DC., Prodr. 12: 220 (1848); Engler, l.c. 364 (1892). *Satureja punctata* (Benth.) Briq., l.c. 299 (1896); Brenan, l.c. 4 (1954); Hedberg, l.c. 161 (1957); E. & K. Walther i

c
b
1
a
d
f
e
R. Holcroft

Mitt. thüring. bot. Ges. 1: 7 (1957); Cufod., l.c. 823 (1962). *S. punctata* R. Br., l.c. 64 (1814), nom. nud. *S. biflora* var. *punctata* (Benth.) Fiori in Nuovo G. bot. ital. n.s. 20: 371 (1913). Type: Ethiopia, *Salt* s.n. (BM, holo.).

M. purtschelleri Gürke in Engl., l.c. 365 (1892). Type: Tanzania, Mt Kilimandjaro, *Meyer* 244 (B†).

Satureja biflora var. *rhodesica* E. & K. Walther, l.c. 7 (1957). Type: Malawi, Mt Mlanje, *G. Adamson* 368 (K, holo.).

— var. *villosa* E. & K. Walther, l.c. 7 (1957). Type: Tanzania?, Klinangop, *Dale* 2965 (K, holo.).

Perennial herb 0,2—0,6 m tall with several stems arising, often annually, from a woody base; stems slender, softly woody, usually erect, simple or sparingly branched, tomentose, bearded below the nodes. *Leaves* sessile or subsessile; blade elliptic to ovate, 5—12 × 3—8 mm, glabrous to tomentulose, gland-dotted below, apex acute to obtuse, base truncate, margin entire, thickened. *Flowers* in few—several-flowered, usually pedunculate cymes, in the axils of the leaves for almost the entire length of the stem, but mainly in the upper third; bracteoles small, linear. *Calyx* 15-ribbed, hispidulous, tubular, up to 4 mm long, subequally 5-toothed; teeth 1—1,5 mm long. *Corolla* white or pale mauve, 5—7 mm long; upper lip 1 mm long; lower lip 1,5—2 mm long. Fig. 19.

Distributed from India along the mountains of east tropical Africa to Southern Africa, where it is found from the Soutpansberg along the higher parts of eastern and central Transvaal to Lesotho, Transkei and eastern Cape, usually on rock slopes or moist places with grass and scattered bush. Map 58.

Vouchers: *Codd* 9787; *Dieterlen* 1346; *Galpin* 10101; *Meeuse* 9165; *Schlechter* 4534.

The leaves are pleasantly lemon-scented.

2. **Satureja compacta** *Killick* in Bothalia 7: 437 (1961); Ross, Fl. Natal 304 (1972). Type: Natal, Cathedral Peak Forest Station, *Killick* 1866 (PRE, holo.!).

Prostrate, mat-forming perennial herb; stems glandular-villous, 0,15—0,3 m long. *Leaves* shortly petiolate; blade broadly ovate to round, 5—11 × 4—10 mm, sparingly glandular-villous, rounded at apex and base, margin obscurely few-toothed. *Flowers* solitary in the axils of the upper leaves; pedicels 3—10 mm long with a pair of minute bracteoles. *Calyx* glandular-hispid, 4 mm long, deeply toothed. *Corolla* mauve, or white to yellowish with a deep purple throat, 6—7 mm long; tube 4—5 mm long; lobes 2 mm long.

Found in the Natal Drakensberg in mountain grassland at about 2 300 m altitude. Map 59.

Voucher: *Hilliard & Burtt* 9287.

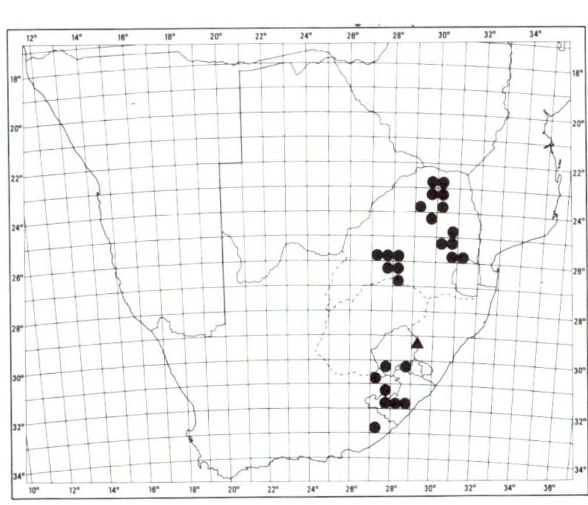

MAP 58. — ● **Satureja biflora**
▲ **S. grandibracteata**

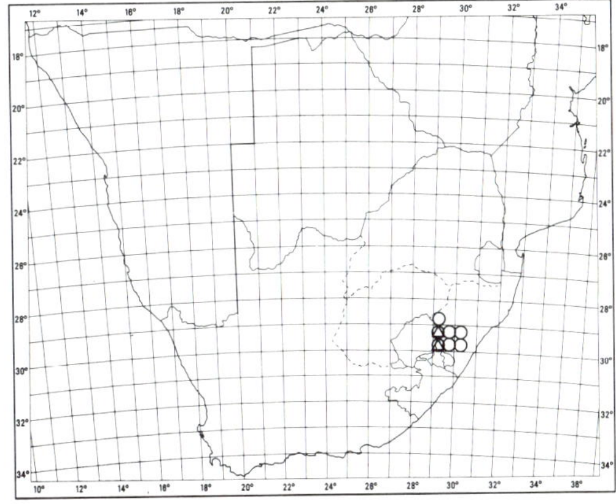

MAP 59. — △ **Satureja compacta**
○ **S. reptans**

FIG. 19. — 1, **Satureja biflora,** upper part of flowering stem, × 1; a, base of plant, × 1; b, flower, × 12; c, mature calyx, × 12; d, section through corolla, × 12; e, front of corolla, × 12; f, stigma and style, × 12 (*Mrs Jenkins*, living plant from Pilgrims Rest).

Closely related to *S. kilimandschari* (Gürke) Hedb. from East Africa which has rusty pubescence on stems and leaves and shorter pedicels.

3. **Satureja reptans** *Killick* in Bothalia 7: 436 (1961); Ross, Fl. Natal 304 (1972). Syntypes: Transkei, *Sutherland* (K); Natal, *Medley Wood* 3712 (K).

Micromeria pilosa Benth. in Hooker's Icon. Pl. 15: t.1522 (1866); Skan in F.C. 5,1: 307 (1910); non *S. pilosa* Velen. (1899). Type: as above.

Perennial herb; stems prostrate, slender, glandular-villous, 0,25—0,6 m long. *Leaves* sessile to shortly petiolate; blade ovate to broadly ovate, 16—24 × 12—20 mm, glandular-pubescent, apex obtuse to rounded, base truncate to subcordate, margin obscurely few-toothed to subentire. *Flowers* in 1—3 (—5)-flowered cymes; pedicels (strictly peduncle plus pedicel) slender, 10—25 mm long with a pair of minute bracteoles about the middle. *Calyx* glandular-hispid, up to 6 mm long, shortly toothed. *Corolla* white to pale blue with a median yellow stripe, 10—15 mm long; tube campanulate, 6—11 mm long; upper lip 2,5 mm long; lower lip 4 mm long (corolla minute, yellowish, up to 6 mm long in cleistogamous plants).

Found in the Natal Midlands and Drakensberg in mountain grassland at altitudes of 1 500 to 2 500 m. Map 59.

Vouchers: *Galpin* 11745; *Killick* 1272; 1429; *Medley Wood* 10894.

See *S. grandibracteata* (below) for the main differences between the two species.

4. **Satureja grandibracteata** *Killick* in Bothalia 7: 435 (1961); Ross, Fl. Natal 304 (1972). Type: Natal, Cathedral Peak Forest Station, *Killick* 1684 (PRE, holo.!).

Micromeria grandiflora Killick in Bothalia 6: 439 (1954), non Scheele (1843). Type: as above.

Perennial herb; stems decumbent, glandular-villous, 0,2—0,35 m long. *Leaves* sessile to subsessile; blade broadly ovate to subrotund, 15—20 × 10—18 mm, glandular-pubescent, apex rounded, base subcordate, margin obscurely few-toothed. *Flowers* solitary in the axils of the upper leaves; pedicels 10—20 mm long with a pair of leaf-like bracts about the middle. *Calyx* glandular-hispid, up to 10 mm long, deeply toothed. *Corolla* mauve, 18—20 mm long; tube campanulate, 14—16 mm long; upper lip 2,5 mm long; lower lip 4 mm long.

Known only from a small area on the Cathedral Peak Forest Station in the Natal Drakensberg, in mountain grassland at about 2 000 m altitude. Map 58.

Vouchers: *Codd & Dyer* 6241; *Killick* 1102.

Closely related to *S. reptans* (above) but it is a more robust plant with shorter internodes and larger bracteoles and corolla, shorter and stouter pedicels and longer calyx teeth. The leaves have a sharp, mentha-like scent.

7328 **16. MENTHA**

Mentha *L.*, Sp. Pl. 576 (1753); Gen. Pl. edn 5: 250 (1754); Smith in Trans. Linn. Soc. Lond. 5: 171 (1800); Benth., Lab. 168 (1833); in DC., Prodr. 12: 164 (1848); Benth. & Hook.f., Gen. Pl. 2,2: 1182 (1876); Briq. in Natürl. PflFam. 4,3a: 317 (1896); Bak. in F.T.A. 5: 451 (1900); Cooke in F.C. 5,1: 303 (1910); Harley in Fl. Europ. 3: 183 (1972); R. A. Dyer, Gen. 530 (1975); Codd in Bothalia 14: 169 (1983). Type species: *M. spicata* L.

Aromatic herbs, glabrous or pubescent. *Leaves* usually toothed. *Inflorescence* a terminal spike-like raceme of many-flowered verticils; flowers small; bracts leaf-like to smaller than the leaves; bracteoles linear. *Calyx* 10-nerved, subequally 5-toothed, scarcely accrescent. *Corolla* slightly longer than the calyx, obscurely bilabiate, 4-lobed; tube funnel-shaped. *Stamens* 4, attached at the middle of the corolla tube, subequal, spreading, normally exserted (shorter when abortive); filaments linear; anthers 2-thecous. *Disc* shallowly lobed. *Style* linear, exserted, shortly 2-fid. *Nutlets* ovoid, smooth or reticulate.

A cosmopolitan genus of about 20—30 species, occurring mainly in temperate regions; 2 species are indigenous one of which, *M. longifolia,* is divided into 3 subspecies. The widely cultivated *M. spicata,* used in the preparation of mint sauce, has been found as a garden escape and is included in the key.

Several species are grown for their essential oils or as culinary herbs, the best known being the Peppermint (*M. x piperita* L.) and the Spearmint (*M. spicata* L.), both of which have been grown commercially in South Africa. A variety of *M. arvensis* L., known as Japanese Mint, is also grown commercially, but has not been successful in South Africa because of susceptibility to rust. *M. pulegium* L., the Penny-Royal, is also grown as a culinary herb. Most species are used medicinally and this applies also to the indigenous species.

Although it is a very natural genus, the delimitation of species has been found difficult, especially with regard to the distinction between *M. longifolia* and *M. spicata.* The nomenclature of these two species has also been complicated by Linnaeus adopting a different treatment in edn 2 of his Species Plantarum (1763) from that adopted in the 1753 edition.

1 Leaves sessile or subsessile; inflorescence cylindrical, usually tapering towards the apex, 10—14 mm in diameter:

  2 Rhachis, pedicels and calyx pubescent:

    3 Leaves linear, 2—4 mm broad .... 1(a). *M. longifolia* subsp. *wissii*

    3 Leaves linear-lanceolate to lanceolate or oblong-lanceolate, more than 5 mm broad:

      4 Leaves fairly densely to densely pubescent on one or both surfaces
      .... 1(b). *M. longifolia* subsp. *capensis*

      4 Leaves glabrous or with a few scattered hairs beneath.................. 1(c). *M. longifolia* subsp. *polyadena*

  2 Rhachis, pedicels and calyx glabrous, though calyx teeth may be ciliate:

    5 Leaves lanceolate to elliptic-lanceolate, acuminate, margin shortly and often obscurely toothed
    .... 1(c). *M. longifolia* subsp. *polyadena*

    5 Leaves oblong to ovate-elliptic, acute to obtuse; calyx teeth often ciliate.......................... 2. *M. spicata*

1 Leaves usually petiolate; inflorescence of oblong or globose clusters, 14—20 mm in diameter
.... 3. *M. aquatica*

    1. **Mentha longifolia** *(L.) Huds.,* Fl. Angl. edn 1: 221 (1762); Briq. in Natürl. PflFam. 4,3a: 321 (1896); Cooke in F.C. 5,1: 304 (1910); Codd in Bothalia 14: 170 (1983). Type: from Europe.

M. spicata var. *longifolia* L., Sp. Pl. 576 (1753). *M. sylvestris* L., Sp. Pl. edn 2,2: 804 (1763); Smith in Trans. Linn. Soc. Lond. 5: 179 (1800); Bak. in F.T.A. 5: 451 (1900). Type: as above.

Perennial rhizomatous herb; stems erect to straggling, up to 1,5 m long, usually retrorse-tomentose, rarely (southern Cape) glabrous or subglabrous. *Leaves* sessile or subsessile; blade linear to linear-lanceolate, lanceolate or lanceolate-oblong, glabrous to variously pubescent (see subspecies), freely gland-dotted, apex acuminate, base truncate to obtuse, margin entire to shortly and distantly toothed. *Inflorescence* cylindrical, tapering at the apex, 30—100 × 10—12 (—14) mm, usually of many verticils, often somewhat lax below, dense above; rhachis usually densely retrorse-tomentose, rarely subglabrous or glabrous (southern Cape);

bracts much smaller than the leaves; bracteoles linear; pedicels usually hispid. *Calyx* tubular-campanulate, 2—2,5 (—3) mm long, densely to sparingly glandular-hispid, rarely glabrous (southern Cape). *Corolla* white to mauve, 3—5 mm long. *Stamens* exserted or occasionally abortive.

An extremely polymorphic species, widespread in Europe and the Mediterranean region to eastern Asia and the Canary Islands, extending to Ethiopia from where there is a gap to Zimbabwe and Southern Africa; on river banks and in moist places. Known in England as Horse Mint because the leaves are usually unpleasantly scented.

In Europe the leaves may be ovate to ovate-oblong or oblong-lanceolate but in Zimbabwe and Southern Africa the leaves are usually narrower in relation to their length, being lanceolate-oblong to linear-lanceolate or linear. In the latter region the material falls into three groups, each with a fairly distinct geographical distribution, to which the rank of subspecies is considered appropriate (see key).

(a) subsp. **wissii** *(Launert) Codd* in Bothalia 14: 170 (1983). Type: S.W.A./Namibia, Brandberg, *Wiss* 1418 (FR, holo.; M; PRE!).

M. wissii Launert in Mitt. bot. StSamml., Münch. 2: 311 (1957); Launert & Schreiber in F.S.W.A. 123: 19 (1969).

Leaves linear, grey-green, finely felted on both surfaces, 25—70 × 1,5—4 mm, margin entire or obscurely and distantly toothed.

Recorded from two localities in S.W.A./Namibia (the Brandberg and Naukluft) and from near Garies in Namaqualand; in watercourses and moist places. The leaves are said to be strongly and unpleasantly aromatic. Map 60.

Vouchers: *Dinter* 8288; *Merxmüller & Giess* 28155; *Pearson* 5641; *Strey* 2008.

(b) subsp. **capensis** *(Thunb.) Briq.* in Natürl. PflFam. 4,3a: 321 (1896); Cooke in F.C. 5,1: 304 (1910); Phillips in Ann. S. Afr. Mus. 16: 242 (1918); Wilman, Check List Griq. West 228 (1946); Jacot Guill., Fl. Lesotho 239 (1971); Ross, Fl. Natal 304 (1972); Codd in Bothalia 14: 170 (1983). Type: Cape, *Thunberg* s.n. (UPS, holo., microfiche 564/13446!).

M. capensis Thunb., Prodr. 95 (1800); Fl. Cap. edn Schult. 444 (1823); — subsp. *capensis*, Briq. in Bull.

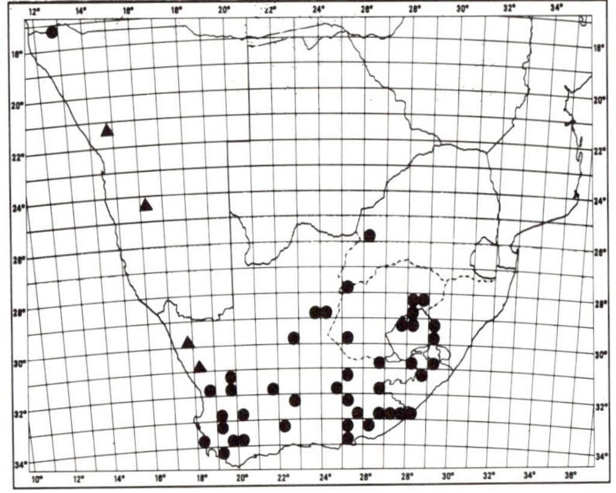

MAP 60. — ▲ **Mentha longifolia** subsp. **wissii**
● **M. longifolia** subsp. **capensis**

Soc. bot. Genève 5: 75 (1889). *M. longifolia* var. *capensis* (Thunb.) Briq. in Bull. Herb. Boissier 4: 687 (1896).

M. salicina Burch. ex Benth., Lab. 170 (1833); in DC., Prodr. 12: 168 (1848). *M. longifolia* subsp. *capensis* var. *salicina* (Burch. ex Benth.) Briq. in Natürl. PflFam. 4,3a: 321 (1896); Cooke, l.c. 304 (1910). Type: Cape, Roggeveld, Riet River, *Burchell* 1372 (K, holo.!).

M. lavandulacea sensu Benth. in E. Mey., Comm. 232 (1837); in DC., Prodr. 12: 165 (1848), partly.

— var. *latifolia* Benth. in DC., Prodr. 12: 165 (1848). Syntypes: Cape, Cafraria, Wittebergen, *Ecklon* s.n.; *Drège* s.n. (K!); Hay Div., Ongeluk, *Burchell* 2645 (K!).

M. capensis subsp. *bouvieri* Briq. in Bull. Soc. bot. Genève 5: 76 (1889). *M. longifolia* var. *bouvieri* (Briq.) Briq. in Bull. Herb. Boissier 4: 687 (1896). *M. longifolia* subsp. *bouvieri* (Briq.) Briq. in Natürl. PflFam. 4,3a: 321 (1896). Type: Cape, Uitenhage, Coega River, *Ecklon & Zeyher* 673 (G, holo.; SAM!).

M. longifolia var. *obscuriceps* Briq. in Bull. Herb. Boissier 2: 695 (1894); Cooke, l.c. 304 (1910). Type: Cape, a specimen in Herb. Delessert (G, holo.).

— var. *doratophylla* Briq., l.c. 695 (1894); Cooke, l.c. 305 (1910). Type: Cape, *Mund & Maire* s.n. (B, holo.).

— subsp. *capensis* var. *cooperi* Briq. ex Cooke, l.c. 304 (1910). Type: Fort Beaufort area, *Cooper* 555 (K, holo.!).

M. longifolia sensu Salter in Fl. Cape Penins. 695 (1950); sensu Jacot Guill., Fl. Lesotho 239 (1971).

FIG. 20. — 1, **Mentha longifolia** subsp. **capensis**, stem, × 1; 1a, inflorescence, ×1; 1b, flower, × 5 (living plant, BRI garden). 2, **M. longifolia** subsp. **polyadena**, stem and leaf, × 1 (*Pont* s.n.). 3, **M. spicata**, leaf, × 1 (garden plant). 4, **M. aquatica**, × 1 (*Breijer* sub TRV 19520).

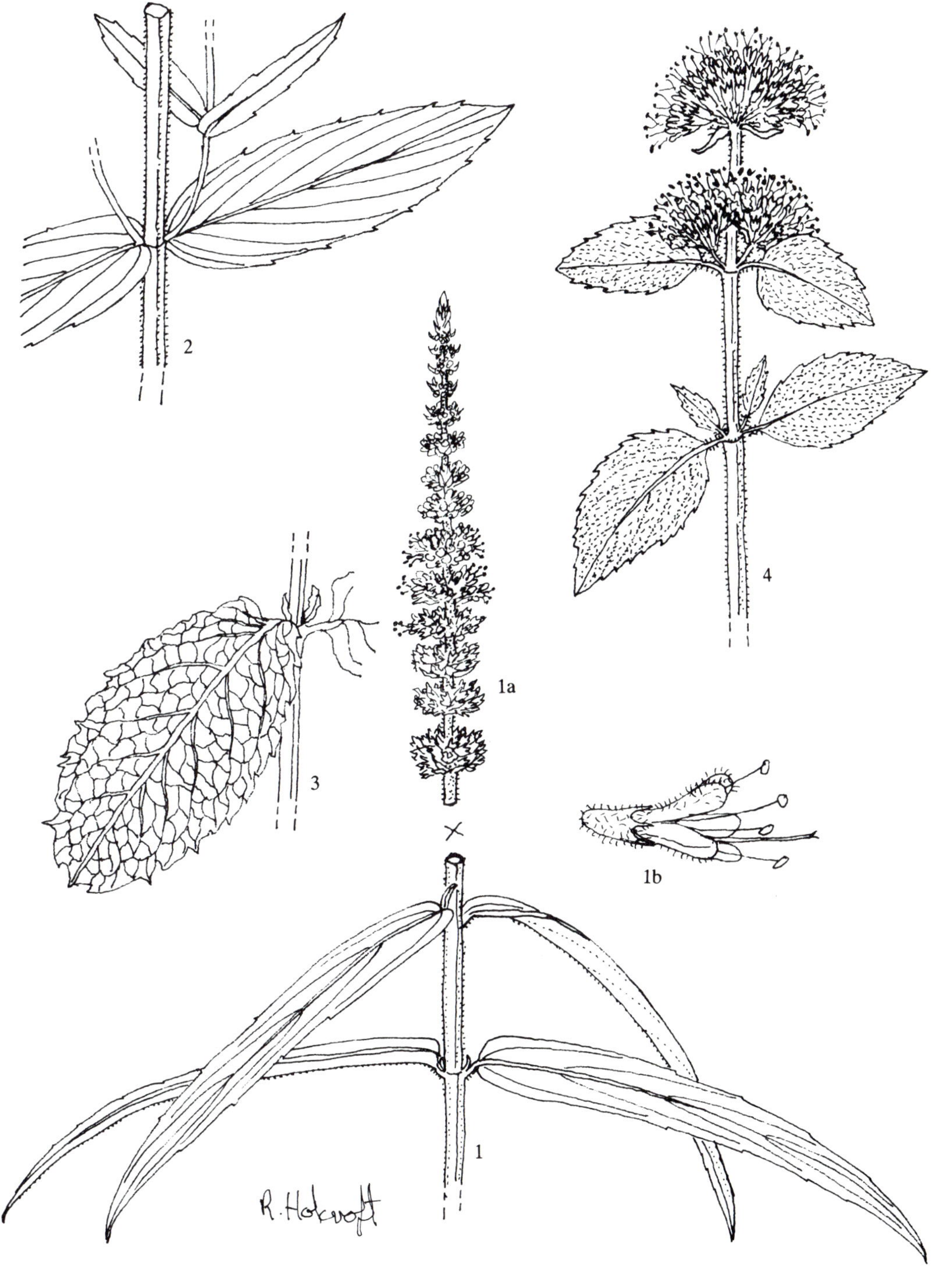

Leaves sparingly to finely pubescent and often dark coloured above, densely white-tomentose to finely or coarsely pubescent beneath, lanceolate to linear-lanceolate, (30−) 45−90 (−100) × (5−) 7−18 (−22) mm, apex acuminate, base obtuse to truncate, margin entire to shortly and distantly toothed. Fig. 20:1.

The typical form, with leaves often dark coloured above and densely white-tomentose beneath, occurs in the Orange Free State and adjacent Natal and south-western Transvaal, northern Cape and Lesotho, extending to the Transkei, to eastern Cape, the Cape Peninsula and along the south-western mountains to Calvinia district. To the north of this, in western Transvaal, northern S.W.A./Namibia and Zimbabwe, the leaves tend to be finely greyish-felted above and finely to coarsely grey-pubescent beneath. Occasional plants with this kind of pubescence occur also further south, even to the Peninsula. *M. capensis* subsp. *bouvieri* Briq. was based on such a plant from the Uitenhage area (*Ecklon & Zeyher* 673). Subsp. *capensis* appears to be absent from an area in the southern Cape between Humansdorp and Riversdale and, inland, to Montagu and Swartberg. In this area it is replaced by a form of subsp. *polyadena* and along the margins of the area occasional specimens are found that are somewhat intermediate between the two. On the other hand, the two subspecies overlap in Lesotho without any intermediates having been seen from this area. Map 60.

Vouchers: *Acocks* 9816; *Galpin* 2016; *Scheepers* 1858; *C. A. Smith* 5240.

The plant is described as 'peppermint-scented' or 'with a strong odour reminiscent of mint'. The leaves are boiled, sometimes with sugar, until a syrup is formed, which is used in the treatment of colds and bronchial complaints. Known in the Karoo and Namaqualand as Balderjan or variations of it, such as Ballerja, Balterja etc. It is also referred to as Wild Mint or Kruisement (Kruistemunt). In Lesotho the vernacular name for both subsp. *capensis* and subsp. *polyadena* is 'Koena', a crocodile, reputedly because the plants inhabit wet places.

(c) subsp. **polyadena** (*Briq.*) *Briq.* in Natürl. PflFam. 4,3a: 321 (1896); Cooke in F.C. 5,1: 303 (1910); Phillips in Ann. S. Afr. Mus. 16: 241 (1918); Jacot Guill., Fl. Lesotho 239 (1971); Codd in Bothalia 14: 172 (1983). Type: Transvaal, *Lincke* 97 (G).

M. viridis sensu Benth. in DC., Prodr. 12: 168 (1848), partly, as to *Burchell* 4798 (sphalm 4718), 7196; sensu Cooke, l.c. 305 (1910).

M. sylvestris L. subsp. *polyadena* Briq. in Bull. Soc. bot. Genève 5: 84 (1889). Type: as above.

M. longifolia sensu Compton, Fl. Swaziland 498 (1976).

Leaves glabrous on both surfaces or with a few scattered hairs, lanceolate to elliptic-lanceolate or oblong-lanceolate, 35−80 × 8−18 mm, apex acuminate, base obtuse to truncate, margin usually shortly toothed, occasionally subentire. *Rhachis* usually retrorse-tomentose, occasionally glabrous (see note below). Fig. 20:2.

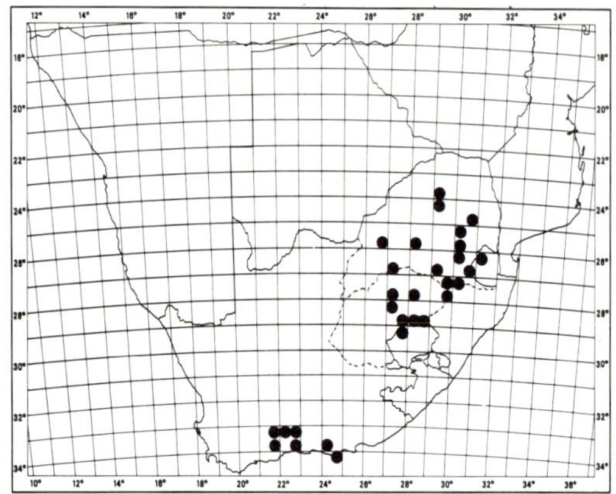

MAP 61. — **Mentha longifolia** subsp. **polyadena**

Recorded from two disjunct areas: (a) from the Transvaal, Swaziland, northern Natal, eastern Orange Free State and northern Lesotho; and (b) from the southern Cape Province between the Humansdorp and Riversdale districts and inland to the Swartberg. Found along water-courses, on river banks and in moist places. Map 61.

Vouchers: *Codd* 8262; *Galpin* 13078; *Leistner* 3019; *C. A. Smith* 1464.

The specimens from area (a) are relatively uniform with the rhachis retrorse-tomentose and the calyx densely glandular-pubescent. In area (b), on the other hand, there is a good deal of variation in the degree of pubescence. Some specimens are typical (*Oliver* 5661), some are sparsely pubescent (*Muir* 1973, 2004, *Dahlstrand* 3506), while others are completely glabrous (*Acocks* 18289, *Fourcade* 3235). All the specimens from area (b) agree well with subsp. *polyadena* so it would be illogical to place the glabrous specimens in a different species (*M. spicata*) as was done by Bentham, l.c. and Cooke, l.c., on this one character alone.

In *M. spicata* (= *M. viridis*), the commonly cultivated garden mint, the rhachis and calyx are glabrous and the leaves tend to be ovate-oblong with serrate margins. They can usually be readily distinguished from the glabrous specimens of *M. longifolia* subsp. *polyadena*.

2. **Mentha spicata** L., Sp. Pl. 576 (1753); Huds., Fl. Angl. 221 (1762); Bailey, Cycl. Hort. edn 21,2: 2035 (1963); Harley in Fl. Europ. 3: 186 (1972); Codd in Bothalia 14: 173 (1983). Type: from Europe, in Hb. Hort. Cliff. (BM).

M. spicata var. *viridis* L., Sp. Pl. 576 (1753). *M. viridis* (L.) L., Sp. Pl. edn 2,2: 804 (1763); Smith in Trans. Linn. Soc. Lond. 5: 185 (1800); Benth., Lab. 173 (1833); in DC., Prodr. 12: 168 (1848); Cooke in F.C. 5,1: 305 (1910); Wilman, Check List Griq. West 228 (1946); Salter in Fl. Cape Penins. 695 (1950). Type: as above.

Perennial rhizomatous herb; stems ascending, up to 0,6 m tall, glabrous to sparingly pubescent. *Leaves* sessile to shortly petiolate; blade lanceolate-oblong to ovate-oblong or ovate, 30—50 (−60) × 13—20 mm, glabrous or nearly so, freely gland-dotted on both surfaces, apex acute, base obtuse to truncate, margin serrate. *Inflorescence* cylindrical, 30—60 × 10—14 mm; rhachis and pedicels glabrous. *Calyx* tubular-campanulate, 2—2,5 mm long, glabrous; teeth sometimes ciliate. *Corolla* mauve to whitish, 4 mm long. Fig. 20:3.

Its origin is lost in antiquity, having probably arisen in cultivation in Europe in ancient times, possibly as a hybrid between *M. suaveolens* Ehrh. and *M. longifolia* (Harley, l.c.). It exists in a wide range of forms, the more desirable ones being propagated vegetatively. It is now widely naturalized throughout the world and has been recorded as a garden escape in South Africa, in moist places.

Vouchers: *Brink 62; Marloth 7328.*

Known as Spearmint, it is widely grown as a culinary herb e.g. for mint sauce and, commercially, for its essential oil which is used medicinally and in confectionery. Trials carried out in South Africa are reported by Baarschers, Horn & Rehm in S. Afr. J. Agr. Sci. 5: 66—77 (1962).

3. Mentha aquatica L., Sp. Pl. 576 (1753); Thunb., Fl. Cap. edn Schult. 444 (1823); Benth., Lab. 176 (1833); in E. Mey., Comm. 232 (1837); in DC., Prodr. 12: 170 (1848); Briq. in Natürl. PflFam. 4,3a: 320 (1896); Cooke in F.C. 5,1: 305 (1910); Phillips in Ann. S. Afr. Mus. 16: 242 (1918); Wilman, Check List Griq. West 227 (1946); Salter in Fl. Cape Penins. 696 (1950); Jacot Guill., Fl. Lesotho 239 (1971); Ross, Fl. Natal 304 (1972); Compton, Fl. Swaziland 498 (1976); Codd in Bothalia 14: 174 (1983). Type: from Europe.

M. dumetorum Schult. var. *natalensis* Briq. in Bull. Herb. Boissier 2: 702 (1894). Type: Natal, *Medley Wood 402* (B, holo.).

Perennial rhizomatous herb; stems ascending to 0,8 m tall or trailing in water to 1,5 m long, subglabrous or sparingly to densely pubescent. *Leaves* petiolate or rarely subsessile; blade lanceolate to broad-ly ovate, 20—55 × 5—26 mm, glabrous to fairly densely pubescent, freely gland-dotted, apex acute to acuminate, base cuneate to rounded, margin obscurely to distinctly toothed. *Inflorescence* terminal, of 1—3 (−4) spaced flower clusters up to 20 mm in diameter, the uppermost globose to oblong-capitate, subtended by reduced leaves (bracts), the lower clusters somewhat distant, globose, subtended by normal leaves; bracteoles linear; rhachis and pedicels subglabrous to densely pubescent. *Calyx* tubular, sparsely to densely pubescent, 3—4 mm long. *Corolla* pale to deep mauve, pinkish or purple, 5 mm long. Fig. 20:4.

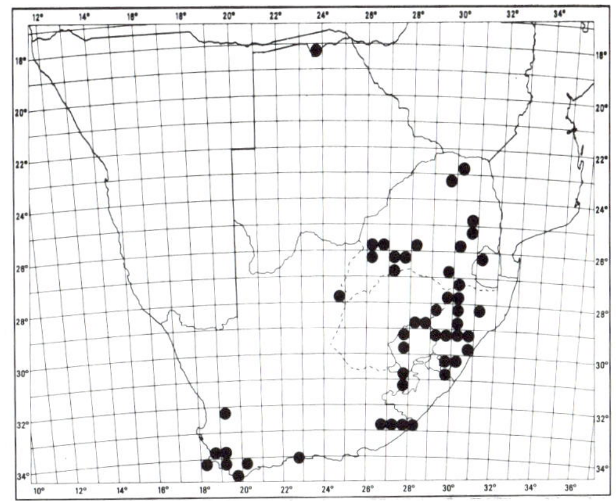

MAP 62. — **Mentha aquatica**

Widely distributed in Europe and around the Mediterranean, extending eastwards to Siberia; found in tropical Africa from Kenya to Malawi with a record from the swamps of northern Botswana; locally common in marshes and wet places in the higher rainfall areas of the Transvaal and adjacent parts of northern Cape and Swaziland, widespread in Natal, eastern Orange Free State and Lesotho, extending through the Transkei and eastern Cape, along the coast to the Peninsula and northwards along the mountains to Ceres. Map 62.

Vouchers: *Acocks 9707; 11354; 20145; Galpin 2680; Rodin 3938; Rogers 850.*

There is considerable variation in pubescence, size and shape of leaves but there is a complete range of intermediate forms.

Known as Water Mint or Kruisement, the leaves are said to have a strong minty scent. An infusion of the leaves is taken for colds and as a tonic, being highly regarded (*Hanekom 1379*) for promoting the flow of milk in nursing mothers (see also *Stachys linearis*).

7339   17. **TETRADENIA**

Tetradenia *Benth.* in Bot. Reg. sub t.1300 (1830); Lab. 164 (1833); in DC., Prodr. 12: 159 (1848); Benth. & Hook. f., Gen. Pl. 2,2: 1180 (1876); Benth. in Hooker's Icon. Pl. t.1282 (1879); Briq. in Natürl. PflFam. 4,3a: 331 (1897); Codd in Bothalia 14: 177 (1983). Type species: *T. fruticosa* Benth. (Malagasy Republic).

Iboza N.E. Br. in F.C. 5,1: 298 (1910); R.A. Dyer, Gen. 1: 533 (1975). Type: *I. riparia* (Hochst.) N.E. Br.

Moschosma auct., non Reichb.

Perennial shrublets or soft shrubs, occasionally reaching the stature of a small tree, usually leafless or nearly so at flowering stage; stems brittle or semisucculent, at first somewhat quadrangular and softly glandular-pubescent, terete and glabrescent with age. *Leaves* small or large, those subtending inflorescence branches smaller towards apex of inflorescence, often semisucculent, variously pubescent, crenate-dentate, aromatic. *Inflorescence* terminal, paniculate, often diffusely branched, the ultimate branchlets spike-like (referred to as "flower-spikes"); bracts small, ovate-deltoid to broadly ovate, imbricate in the bud stage, caducous or semipersistent. *Flowers* small, functionally unisexual or occasionally bisexual *(T. barberae)*, in whorls of 4−10, sessile or shortly pedicellate, dense or lax, mauve or whitish. *Calyx* minute, campanulate, 3-lobed, divided nearly to the base below, the upper lobe ovate, lateral lobes oblong, bifid or emarginate often giving the calyx a 5-toothed appearance; in female flowers the calyx enlarges slightly at maturity and the upper lobe becomes erect. *Corolla* small, tubular or funnel-shaped, limb spreading, asymmetrical, 4-lobed, the upper lobe emarginate or bifid so that the corolla may appear 5-lobed; lobes oblong, rounded, the lowest lobe usually the longest. *Stamens* 4, free, erect or spreading, absent in female flowers. *Disc* 1- or 2-lobed. *Ovary* present but usually infertile in male flowers; style exserted, deeply bilobed. *Nutlets* oblong-triquetrous.

Species probably 6; 3 in Southern Africa, one of which, *T. riparia*, is very variable and extends to Angola and through east tropical Africa to Ethiopia; 3 in Malagasy Republic of which one is closely allied to *T. riparia*.

The floral characters are very similar in all species and are not of much diagnostic value.

Although confused in the past with the genus *Basilicum* Moench (= *Moschosma* Reichb.), *Tetradenia* is related to *Mentha*, but differs in the plants being dioecious and more shrubby in habit.

1 Leaves small, ovate, 8−15 × 5−10 mm, bullate-rugose above, veins very prominent beneath; bracts
   ovate-deltoid, as long as broad .... 1. *T. barberae*

1 Leaves small or large, if less than 20 × 10 mm then not bullate-rugose above; bracts broader than long,
   rounded or abruptly apiculate at the apex:

  2 Leaves ovate-rotund, 12−30 × 10−30 mm (occasionally larger, under-surface finely velvety with
    sessile glands and no multicellular hairs; male flower-spikes 10−20 mm long .... 2. *T. brevispicata*

  2 Leaves variously shaped, usually larger than above, under-surface sparsely to densely pubescent with
    stalked glands and/or multicellular hairs; male flower-spikes 20−80 mm long .... 3. *T. riparia*

1. **Tetradenia barberae** *(N.E. Br.)* *Codd* in Bothalia 14: 178 (1983). Type: "Orange River Colony", *Mrs Barber* 7 (K, holo.!).

Iboza barberae N.E. Br. in F.C. 5,1: 302 (1910).

Twiggy shrublet 0,6−1 m tall; stems woody, terete, grey-brown, at first minutely tomentellous. *Leaves* shortly petiolate; blade small, ovate, 8−15 × 5−10 mm, bullate-rugose and finely glandular-scabrous above, conspicuously veined and densely glandular-tomentellous beneath, apex ob-tuse, base truncate, margin crenate, thick-ened below. *Inflorescence* evidently coeta-neous with the leaves, occasionally simple, usually with 1−3 pairs of branches near the base; terminal male flower-spikes dense, 30−95 mm long, lateral 15−55 mm long; bracts ovate-deltoid, acute, 3−3,5 × 2,5−3 mm, dotted with red sessile glands. *Calyx* 1,5 mm long, lateral lobes deeply toothed giving the impression of a 5-toothed calyx. *Corolla* 3 mm long. *Disc* with 1 lobe developed beyond the infertile ovary. *Female flowers* not seen, but occasional seeds are formed in the functionally male flowers.

A xerophytic shrublet of the lower Fish River valley, eastern Cape; in karroid scrub. Map 63.

Vouchers: *Bayliss* 3248; *Tsuane* A1126.

A few modern specimens, collected at and near Kaffir Drift on the Fish River, are the only specimens known, apart from the type, which was recorded from the "Orange River Colony", but this locality may be wrong.

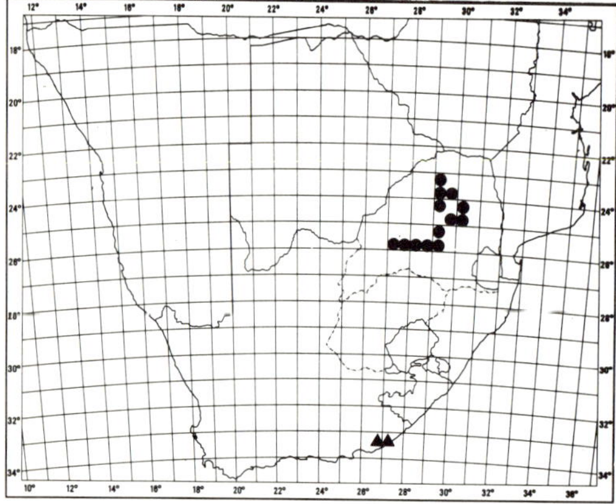

MAP 63. — ▲ **Tetradenia barberae**
● **T. brevispicata**

2. **Tetradenia brevispicata** *(N.E. Br.) Codd* in Bothalia 14: 179 (1983). Type: Transvaal, Wonderboom farm near Pretoria, *Burtt Davy* 1844 (K, holo.).

Iboza brevispicata N.E. Br. in F.C. 5,1: 302 (1910).

Twiggy shrub or small tree 0,6−2 (−3) m tall; stems slender, terete, greyish black with age, at first finely glandular-tomentellous, lacking stipitate glands or long multicellular hairs. *Leaves* rather small, petiolate; blade ovate-rotund to rotund, 12−30 (−55) × 10−30 (−50) mm, finely glandular-scabrous above, densely glandular-tomentellous beneath, the under-surface being obscured by a short cobwebby tomentum lacking stipitate glands, the nerves often fairly prominent, apex rounded, base truncate to subcordate,

margin crenate to deeply crenate-dentate; petiole 4−15 mm long. *Inflorescence* appearing after most of the leaves are shed, consisting of spikes or small panicles borne terminally and in the axils of the upper leaves; flower spikes dense, the male 10−20 (−25) mm long, the female shorter; bracts broadly ovate, acute, 1,5−1,75 × 1,5−2 mm. *Calyx* c. 1 mm long. *Corolla* white to mauve, c. 2 mm long. *Disc* 1 (−2)-lobed.

Found in central and northern Transvaal, on dry, wooded, rocky slopes; also in Zimbabwe. Map 63.

Vouchers: *Codd* 8778; *Pegler* 924; *Strey & Schlieben* 8616.

Characterized by the slender, twiggy stems with greyish black bark, the relatively small roundish, deeply crenate-dentate leaves which are finely tomentose below, and the short dense male flower-spikes.

3. **Tetradenia riparia** *(Hochst.) Codd* in Bothalia 14: 181 (1983). Type: Natal, *Krauss* 331 (MO!).

Moschosma riparium Hochst. in Flora 28: 67 (1845); Benth. in DC., Prodr. 12: 49 (1848); Briq. in Natürl. PflFam. 4,3a: 368 (1897); Wood & Evans, Natal Pl. 1: tt.1,2 (1898); Bak. in F.T.A. 5: 354 (1900). *Basilicum riparium* (Hochst.) Kuntze, Rev. Gen. Pl. 2: 512 (1891). *Iboza riparia* (Hochst.) N.E. Br. in F.C. 5,1: 300 (1910); Phillips in Flow. Pl. S. Afr. 20: t.767 (1940); Martineau, Rhod. Wild Flow. 69 (1953); Brenan in Mem. N.Y. bot. Gdn 9: 39 (1954); Andrews, Flow. Pl. Sudan 3: 212 (1956); Letty, Wild Flow. Transv. 289, t.144,3 (1962); Launert & Schreiber in F.S.W.A. 123: 14 (1969); Ross, Fl. Natal 306 (1972); Compton, Fl. Swaziland 507 (1976).

M. multiflorum Benth. in DC., Prodr. 12: 49 (1848); Bak. in F.T.A. 5: 354 (1900). *Basilicum multiflorum* (Benth.) Kuntze, Rev. Gen. Pl. 2: 512 (1891). *Iboza multiflora* (Benth.) E.A. Bruce in Kew Bull. 1940: 66 (1940); Agnew, Upland Kenya Wild Flow. 642 (1974). Syntypes: Ethiopia, *Schimper* 766 (K!); 1688 (K!).

M. myriostachyum Benth. in Benth. & Hook. f., Gen. Pl. 2,2: 1173 (1876). *Basilicum myriostachyum* (Benth.) Kuntze, Rev. Gen. Pl. 2: 512 (1891); Hiern, Cat. Afr. Pl. Welw. 1,4: 858 (1900). Type: Zambezi Region, no specimen cited.

M. urticifolium Bak. in F.T.A. 5: 353 (1900). *Iboza urticifolia* (Bak.) E.A. Bruce in Kew Bull. 1940: 66 (1940). Type: Tanzania, *Johnson* s.n. (K, holo.!).

Iboza galpinii N.E. Br. in F.C. 5,1: 300 (1910); Compton, Fl. Swaziland 507 (1976). Type: Transvaal, near Barberton, *Galpin* 972 (K, holo.; PRE!).

I. bainesii N.E. Br., l.c. 5,1: 301 (1900). Type: "South African Gold Fields," *Baines* s.n. (K, holo.!).

FIG. 21. — 1, **Tetradenia brevispicata,** male flower, × 6; 1a, leaf, × 1 *(Keytel* 744). 2. **T. riparia,** portion of inflorescence, × 1; 2a, male flower, × 6; 2b, calyx, × 6; 2c, bract, × 6; 2d, non-functional gynoecium, × 9 *(Codd* 8398); 2e, leaf, × 1 *(De Winter* 3597); 2f, leaf, × 1 *(Junod* 538); 2g, leaf, × 1 *(Medley Wood* 5760).

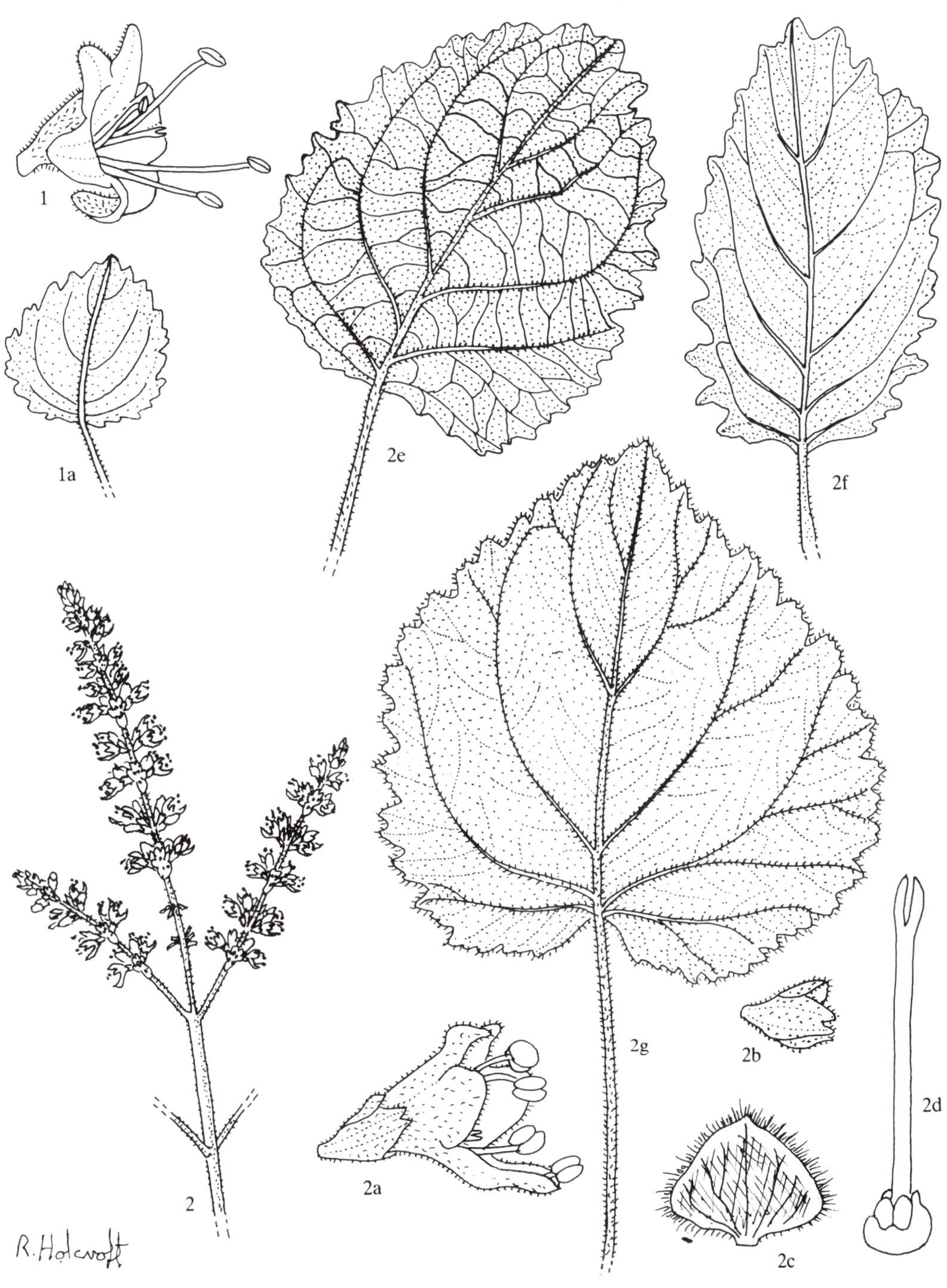

Soft shrub or small tree 1−3 (−5) m tall, freely branched; stems semisucculent, brittle, rather stout, at first 4-angled and glandular-pubescent, becoming terete and glabrous with age; bark pale brown. *Leaves* petiolate; blade ovate-oblong to rotund 35−80 × 35−70 mm, sparsely to densely glandular-pubescent on both surfaces, the under-surface varying from thinly pubescent on the veins to densely white tomentose over the whole surface, apex rounded, base rounded to truncate or cordate, margin coarsely crenate to crenate-dentate. *Inflorescence* a terminal, usually large panicle, diffusely branched and up to 300 × 200 mm in male specimens, smaller and more compact in the female, appearing usually after the leaves are shed; male flower-spikes dense to lax, 20−80 mm long, female flower-spikes dense, 10−25 mm long; bracts broadly ovate, 1,5−2 × 2−2,5 mm. *Calyx* 1 mm long, increasing to 2,5 mm in ripe female flowers. *Corolla* white to mauve, the male 3−3,5 mm long, slightly longer and more funnel-shaped than the female. *Disc* 1-lobed.

Occurs in Southern Africa from coastal Natal to Swaziland, Transvaal, south-eastern Botswana and the northern half of S.W.A./Namibia; extends to Angola

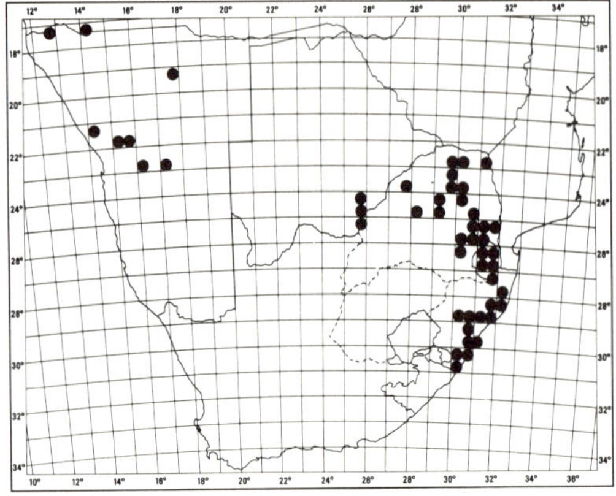

MAP 64. — **Tetradenia riparia**

and through east tropical Africa to Ethiopia. Found on wooded hillsides and stream-banks in relatively frost-free areas. Map 64.

Vouchers: *Burtt Davy* 390; *Dinter* 4673; *Galpin* 9724; *Medley Wood* 1001; 5760.

The above concept includes a good deal of variation in leaf size, shape and pubescence but no pattern emerges and so infraspecific taxa are not upheld.

7342   **18. HYPTIS**

Hyptis *Jacq.*, Collecteana 1: 101 (1787); Benth., Lab. 64 (1833); in DC., Prodr. 12: 85 (1848); Benth. & Hook. f., Gen. Pl. 2,2: 1178 (1876); Briq. in Natürl. PflFam. 4,3a: 333 (1897); Bak. in F.T.A. 5: 447 (1900); Cooke in F.C. 5,1: 297 (1910); Epling in Feddes Reprium 34: 73 (1933); R. A. Dyer, Gen. 530 (1975); nom. cons. Type species: *H. capitata* Jacq.

Mesosphaerum P.Br., Hist. Jamaic. 257 (1756); Kuntze, Rev. Gen. Pl. 2: 525 (1891). Type species: *M. suaveolens* (L.) Kuntze.

Bystropogon L'Hérit., Sert. Angl. 19 (1789). Type species: not designated.

Annual or perennial herbs or soft shrubs. *Flowers* in opposite, several—many-flowered cymes arranged laxly or densely in spike-like racemes or panicles; bracts leafly below, becoming smaller towards the apex; bracteoles linear to ovate. *Calyx* tubular-campanulate, ribbed, subequally 5-toothed, the tube somewhat accrescent in fruit; teeth subulate. *Corolla* small, 5-lobed, more or less bilabiate; tube slightly exceeding the calyx; lowest lobe saccate. *Stamens* 4, didynamous, declinate, inserted in the corolla throat; filaments linear, free; anthers 1-celled, scarcely exserted. *Style* shortly 2-fid or entire. *Nutlets* smooth or punctate-rugulose.

About 300 species, in the warmer parts of America; several species naturalized in the Old World, of which 3 have been recorded in Southern Africa.

1 Bracteoles linear-setose; cymes many-flowered usually forming dense, compact racemes or spikes:

  2 Leaves densely pubescent to whitish tomentose below; cymes secund, pectinate .... 1. *H. pectinata*

  2 Leaves subglabrous below; cymes glomerate forming a dense spike-like inflorescence .... 2. *H. spicigera*

1 Bracteoles ovate to oblong; cymes relatively few-flowered, forming lax racemes .... 3. *H. mutabilis*

1. **Hyptis pectinata** *(L.) Poit.* in Ann. Mus. natn. Hist. nat. 7: 474, t.30 (1806); Benth., Lab. 127 (1833); in DC., Prodr. 12: 127 (1848); A. Rich., Tent. Fl. Abyss. 2: 186 (1850); Bak. in F.T.A. 5: 488 (1900); Cooke in F.C. 5,1: 297 (1910); Nowicke & Epling in Ann. Mo. bot. Gdn 56: 84 (1969); Ross, Fl. Natal 304 (1972); Standley & Williams in Fieldiana Bot. 24,9: 255 (1973); Compton, Fl. Swaziland 498 (1976). Type: from Jamaica.

Nepeta pectinata L., Sp. Pl. edn 2,2: 799 (1763). *Bystropogon pectinatum* (L.) L'Hérit., Sert. Angl. 19 (1789). *Mesosphaerum pectinatum* (L.) Kuntze, Rev. Gen. Pl. 2: 525 (1891); Hiern, Cat. Afr. Pl. Welw. 1,4: 873 (1900). Type: as above.

Annual or short-lived perennial herb 0,6—2,3 m tall, softly woody at the base. *Leaves* petiolate; blade ovate to narrowly ovate, 15—45 × 10—30 mm, sparingly pubescent above, paler and usually softly white-tomentose beneath, apex obtuse, base truncate, margin finely and irregularly crenate-dentate; petiole 15—40 mm long. *Inflorescence* terminal, often branched, usually of horizontal, densely placed flower-clusters (scorpioid cymes), or sometimes lax; cymes usually in pairs from a common peduncle, many-flowered, secund, pectinate (comb-like); bracts leaf-like below, becoming smaller towards the apex; bracteoles linear-filiform, setose, 3 mm long. *Calyx* 2,5 mm long at flowering, increasing to 5 mm long, densely pubescent in the throat. *Corolla* whitish to mauve, 3,5 mm long. Fig. 22.

Indigenous in tropical America; widespread and apparently indigenous in tropical Africa from the Sudan to Senegal and southwards to Botswana, eastern Transvaal, Swaziland and Natal; usually on river banks and in moist places. Map 65.

Vouchers: *Codd* 7924; *Compton* 27751; *Medley Wood* 524; *Schlechter* 3080; *Thorncroft* 413.

2. **Hyptis spicigera** *Lam.*, Encycl. 3: 185 (1789); Benth., Lab. 78 (1833); in DC., Prodr. 12: 87 (1848); Bak. in F.T.A. 5: 448 (1900); Epling in Feddes Reprium 34: 96 (1933); Compton in Fl. Swaziland 499 (1976). Type: in Hb. Lam (P, fide Epling).

Annual erect herb 0,5—2 m tall, strongly aromatic; stems 4-angled with small recurved prickles. *Leaves* petiolate; blade

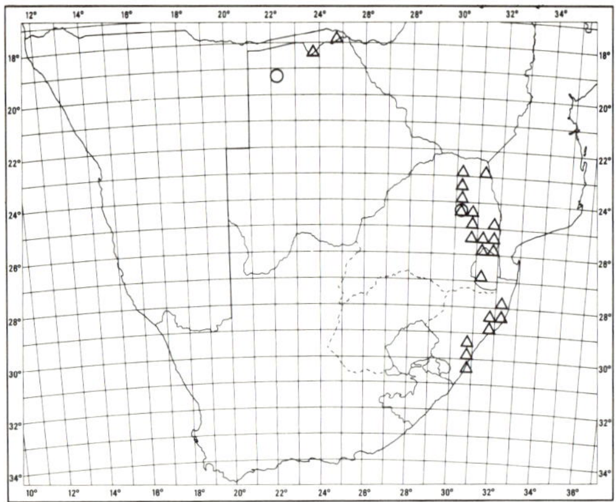

MAP 65. — △ **Hyptis pectinata**
○ **H. spicigera**

ovate to ovate-lanceolate, 30−80 × 10−30 mm, glabrous to sparingly pubescent, apex acute, base obtuse, margin serrate; petiole 10−35 mm long. *Inflorescence* terminal, spike-like, of densely placed many-flowered overlapping cymes; bracts leaf-like below, becoming smaller upwards; bracteoles numerous, linear, setose, 4 mm long, erect, giving the inflorescence a bristly appearance. *Calyx* densely glandular-hispid, enlarging to 6 mm long, mouth hirsute. *Corolla* whitish, 4−5 mm long.

Indigenous in tropical America; widespread as a weed in tropical Africa from the Nile Land to Senegal and southwards to Mozambique and Madagascar. Recorded from northern Botswana, eastern Transvaal and Swaziland. Map 65.

Vouchers: *Junod* sub TRV 10215; *Smith* 1693; *Wild & Drummond* 7119.

· **3. Hyptis mutabilis** *(A. Rich.) Briq.* in Bull. Herb. Boissier 4: 788 (1896); Epling in Feddes Reprium 34: 103 (1933); Standley & Williams in Fieldiana Bot. 24,9: 254 (1973). Type: from tropical America, specimen not found (fide Epling).

Nepeta mutabilis A. Rich. in Act. Soc. Hist. nat. Paris 1: 110 (1792). *Mesosphaerum mutabile* (A. Rich.) Kuntze, Rev. Gen. Pl. 2: 525 (1891). Type: as above.

H. spicata Poit. in Annu. Mus. natn. Hist. nat. 7: 474, t.28, f.2 (1806); Benth., Lab. 120 (1833); in DC., Prodr. 12: 121 (1848). *H. mutabilis* var. *spicata* (Poit.) Briq., l.c. 788 (1896); Epling in Feddes Reprium 34: 105 (1933); Nowicke & Epling in Ann. Mo. bot. Gdn 56: 8 (1969); Hilliard & Burtt in Notes R. bot. Gdn Edinb. 34: 285 (1976). Type: tropical America, *Richard* (P).

Annual herb up to 1,5 m tall, often much branched. *Leaves* petiolate; blade ovate or rhombic-ovate, 30−60 × 20−28 mm, variously pubescent, usually paler and softly tomentose beneath, apex acute, base cuneate to truncate, margin crenate-serrate; petiole 15−25 mm long. *Inflorescence* of lax racemose panicles; flower clusters (paired cymes) relatively small, glomerate, spaced up to 20 mm apart; bracts much reduced upwards; bracteoles ovate to oblong. *Calyx* 4−5 mm long in fruit, hispidulous. *Corolla* pale mauve to purple, 5−6 mm long.

Indigenous in tropical America. Recorded as a weed at Cedara Agricultural College, Natal, by Hilliard & Burtt, l.c.

Voucher: *Rhind* s.n. (NU).

FIG. 22. — 1, **Hyptis pectinata,** stem and leaves, × 1; a, part of inflorescence, × 1; b, detail of inflorescence, × 4; c, flower, × 9; d, section through corolla, × 9; e, mature calyx, × 9; f, nutlet, × 25 (*Liebenberg* 2958).

7345

19. AEOLLANTHUS

Aeollanthus *Mart. ex K. Spreng.*, Syst. Veg. 2: 678 (1825); Hedge in Notes R. bot. Gdn Edinb. 32: 47 (1972); R. A. Dyer, Gen. 530 (1975); Ryding in Nord. J. Bot. 1: 154 (1981); ibid. 2: 219 (1982). Type species: *A. suaveolens* Mart. ex K. Spreng.

Aeolanthus Mart., Amoen. Bot. Monac. 4: t.2 (1831); Benth., Lab. 61 (1833); in E. Mey., Comm. 230 (1837), as *Orollanthus;* in DC., Prodr. 12: 80 (1848); Briq. in Natürl. PflFam. 4, 3a: 349 (1897); Bak. in F.T.A. 5: 388 (1900); Cooke in F.C. 5,1: 294 (1910); Launert & Schreiber in F.S.W.A. 123: 7 (1969). Type species: *A. suavis* Mart.

Annual or perennial herbs or subshrubs; stems and leaves often fleshy. *Inflorescence* usually terminal, paniculate; flowers small, placed singly or in pairs in lax or dense spikes or racemes; bracts small. *Calyx* small and shortly 5-toothed at flowering, elongate and often becoming truncate at maturity, eventually circumscissile near the base. *Corolla* bilabiate; tube narrowly cylindrical at the base, widening upwards, straight or curved; upper lip obscurely 4-lobed, lower lip larger, concave, entire or toothed near the apex. *Stamens* didynamous, attached in the corolla mouth, declinate, usually lying in the lower corolla lip; filaments free; anthers confluent, 1-celled. *Style* shortly 2-fid, exserted beyond the stamens. *Nutlets* orbicular or ovoid, flattened, smooth.

An African genus of about 40 species found mainly south of the Sahara and in Ethiopia; 6 species in Southern Africa.

1 Leaves entire or toothed:
  2 Leaves lanceolate-elliptic to elliptic or ovate-lanceolate, subsessile or shortly petiolate, subentire; plants annual, erect .... 1. *A. suaveolens*
  2 Leaves ovate to obovate, usually long petiolate, toothed or subentire; plants annual or perennial, usually spreading:
    3 Bracts broadly ovate, obtuse, usually imbricate; bracts and stems usually hirsute to canescent .... 2. *A. buchnerianus*
    3 Bracts lanceolate to ovate-elliptic, acute or acuminate, overlapping only in the bud stage; bracts and stems glabrous to hispidulous:
      4 Corolla 7—11 mm long; flower spikes relatively compact, up to 40 mm long:
        5 Stems and leaves glabrous to minutely puberulous; leaf margin subentire to sparingly dentate .... 3. *A. parvifolius*
        5 Stems and leaves pubescent; leaf margin crenulate usually reddish purple .... 4. *A. rehmannii*
      4 Corolla 4—5 mm long; inflorescence usually freely branched; flower spikes elongate, slender, lax, 50—80 mm long .... 5. *A. neglectus*
1 Leaves pinnatifid .... 6. *A. namibiensis*

1. Aeollanthus suaveolens *Mart. ex K. Spreng.*, Syst. Veg. 2: 750 (1825). Type: ex hort. Munich, seed originally from S. America.

A. suavis Mart., Amoen. Bot. Monac. 4: t.2 (1831); Benth., Lab. 61 (1833); in DC., Prodr. 12: 80 (1848). Type: same as above.

A. heliotropioides Oliv. in Trans. Linn. Soc. Lond. 29: 138 (1875); Bak. in F.T.A. 5: 393 (1900); Morton in F.W.T.A. edn 2,2: 457 (1963). Type: Uganda, Umyoro, *Speke & Grant* s.n. (K, holo.).

Annual herb, erect, branched, 0,2—0,5 m tall; stems glabrous to hispidulous. *Leaves* sessile or shortly petiolate; blade lanceolate to elliptic or ovate-lanceolate, 30—40 × 8—15 mm, glabrous to hispidulous, apex obtuse to rounded, base cuneate, margin subentire. *Inflorescence* a fairly dense panicle; flowers secund, single, closely placed, subsessile; bracts ciliate, elliptic, acute, 2,5—3 mm long, overlapping towards the apex. *Calyx* hispidulous, 1 mm long at flowering, enlarging to 2,5 mm long. *Corolla* blue to purple, 4—5 mm long. *Nutlets* ovoid or oblong, smooth.

Distributed from Ghana to Tanzania and southwards to Zaire and Zimbabwe, with a single record from the Woodbush in Transvaal. Map 66.

Voucher: *Obermeyer* sub TRV 31861.

Hedge in Notes R. bot. Gdn Edinb. 32: 47 (1972) expressed the opinion that *A. heliotropioides* is probably a synonym of *A. suaveolens* and this view is supported by Ryding (in litt. Feb. 1982). It was apparently introduced into Brazil prior to the early nineteenth century and was cultivated in Munich Botanic Garden in 1825 from seed collected in "the gardens of Chinese in Santa Cruz, Brazil", where it was grown for its aromatic foliage.

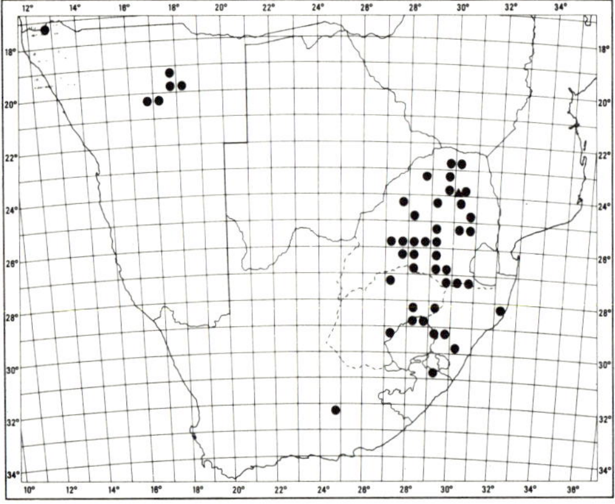

MAP 66. — ▲ **Aeollanthus suaveolens**
      ● **A. buchnerianus**

2. **Aeollanthus buchnerianus** *Briq.* in Bot. Jb. 19: 187 (1894); Bak. in F.T.A. 5: 392 (1900); Ryding in Nord. J. Bot. 1: 156 (1981). Type: Angola, Malanje, Bango, *Buchner* 571 (B, holo.†; K, lecto., fide Ryding).

A. njassae Gürke in Engl., Pflanzenw. Ost-Afr. C: 346 (1895); Bak., l.c. 393 (1900); Hedge in Notes R. bot. Gdn Edinb. 32: 45 (1972). Type: Malawi, Shire highlands, *Buchanan* 529 (B, syn.†; K, lecto., fide Ryding).

A. canescens Gürke in Bot. Jb. 22: 147 (1895); Cooke in F.C. 5,1: 294 (1910); Phillips in Ann. S. Afr. Mus. 16: 241 (1917); Launert & Schreiber in F.S.W.A. 123: 8 (1969); Ross, Fl. Natal 304 (1972); Jacot Guill., Fl. Lesotho 239 (1972). Type: Cape, Graaff-Reinet, *Bolus* sub Herb. Norm. Austr. Afr. 1345 (B, syn.†; K, lecto., fide Ryding).

A. nyikensis Bak. in Kew Bull. 1898: 160 (1898); in F.T.A. 5: 392 (1900). Type: Malawi, Nyika Plateau, *Whyte* 119 (K, holo.).

Plectranthus volkmannae Dinter in Feddes Reprium Beih. 53: 124 (1928), nom. nud. *P. rupicola* Dinter ex

Goossens in Trans. Roy. Soc. S. Afr. 21: 252 (1933), nom. nud. Specimen cited *Dinter* 5514.

Perennial semisucculent herb or soft shrub 0,15−0,5 m tall; stems erect or spreading, subglabrous to greyish velvety, often with scattered longish hairs. *Leaves* petiolate; blade ovate to broadly ovate, 20−45 (−60) × 18−40 (−50) mm, subglabrous to shortly pubescent, apex obtuse to rounded, base truncate, often decurrent on the petiole, margin crenate-dentate; petiole 15−40 mm long. *Inflorescence* fairly dense; flowers alternately single and in pairs; bracts broadly ovate, overlapping, 3−4 × 2,5−3,5 mm, obtuse to apiculate. *Calyx* 1 mm long at flowering enlarging to 3 mm. *Corolla* pale mauve to rosy pink, 4−5 mm long; lower lip with a deltoid tooth or projection at the base and somewhat hooded at the apex. Fig. 23: 2.

Recorded from northern S.W.A./Namibia, common at higher elevations in the Transvaal, extending to eastern Orange Free State and the adjoining parts of northern Natal, Lesotho and eastern Cape; although recorded from Swaziland by Compton, Fl. Swaziland 66 (1966), the specimens so named are *A. rehmannii*. Also found in Zimbabwe, Angola and Mozambique, northwards to Tanzania. Grows in shallow soil among rocks, in semi-shady places. Map 66.

Vouchers: *Codd* 2760; *Galpin* 9064; 11822; *Medley Wood* 7187.

Distinguished from other species in Southern Africa by the broadly ovate, overlapping bracts. There is a good deal of variation in the degree of pubescence; in S.W.A./Namibia the stems and leaves tend to be glabrous while in the eastern part of the distribution range they are usually canescent, often with longer hairs present as well. As pointed out by Ryding, l.c., *A. buchnerianus* belongs to a group of six tropical African species in which the lower lip of the corolla has a conspicuous projection at the base and is somewhat hooded at the apex resulting in an explosive pollination mechanism. The other five species do not extend into Southern Africa.

3. **Aeollanthus parvifolius** *Benth.* in DC., Prodr. 12: 80 (1848); Cooke in F.C. 5,1: 294 (1910); Ross, Fl. Natal 304 (1972); Compton, Fl. Swaziland 499 (1976). Type: Transkei, between St Johns and Umsikaba rivers, *Drège* (K, holo.).

FIG. 23. — 1, **Aeollanthus parvifolius,** flowering branch, × 1; 1a, flower, × 3; 1b, section through corolla, × 3; 1c, flowering calyx, × 8; 1d, mature dehisced calyx, × 8; 1e, persistent torus after dehiscence of calyx and nutlets, × 8 (*Van Jaarsveld* 180/75, cult.). 2, **A. buchnerianus,** portion of inflorescence, × 1; 2a, leaf, × 1; 2b, section through corolla, × 3; 2c, flowering calyx, × 8; 2d, mature dehisced calyx, × 8 (*Mennim* 14, cult.).

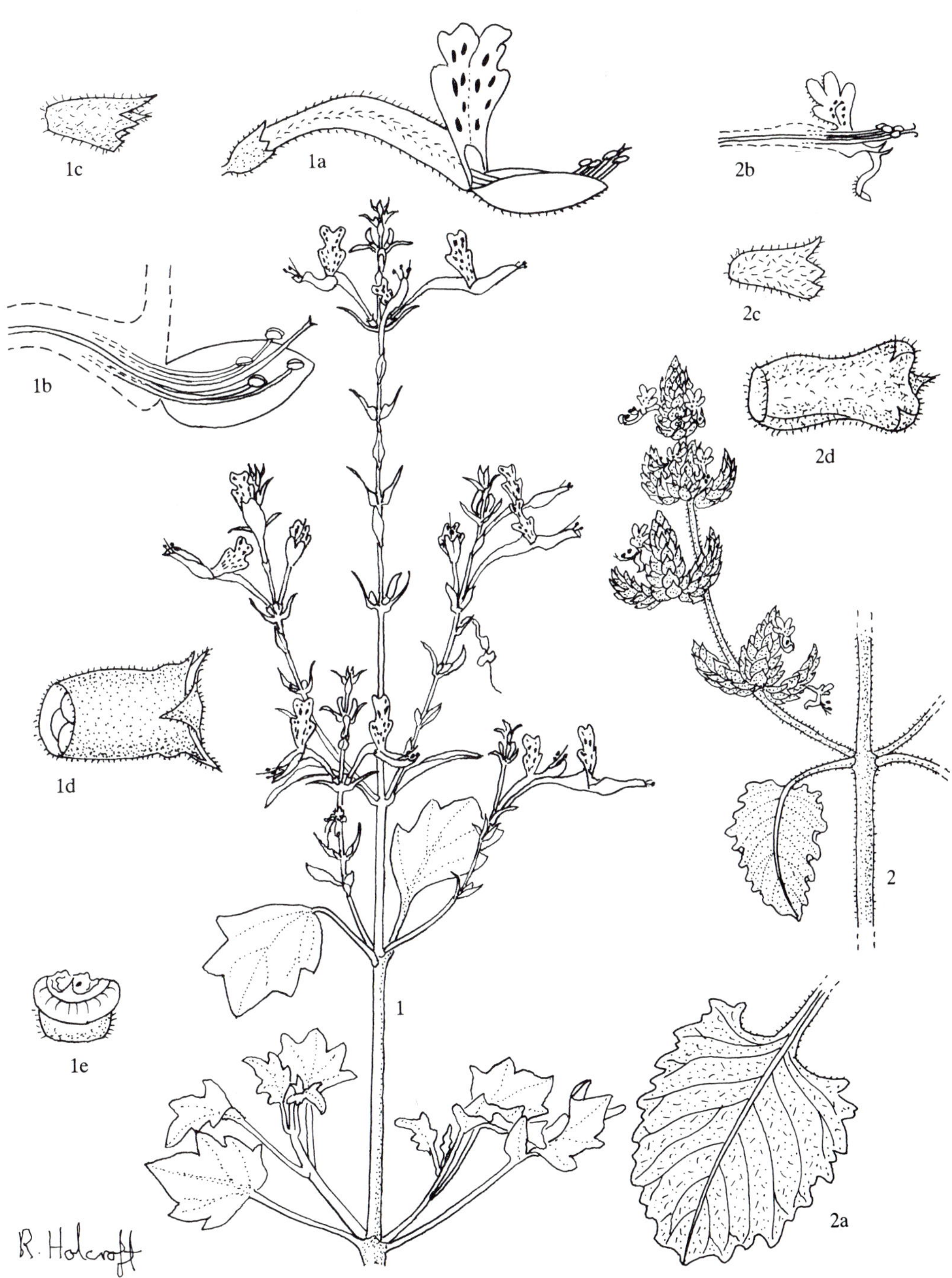

1c
1a
2b
2c
1b
2d
1d
2
1e
1
2a
R. Holcroft

A. suavis sensu Benth. in E. Mey., Comm. 230 (1838), as *Orollanthus suavis.*

Perennial semisucculent herb or subshrub, often woody at the base, branching, 0,2−0,5 m tall; stems spreading-ascending, glabrous to puberulous. *Leaves* softly fleshy, petiolate; blade ovate to subrotund, 12−28 × 8−25 mm, glabrous to puberulous, apex obtuse to rounded, base obtuse to truncate, margin sparingly and often obscurely toothed or subentire. *Inflorescence* often much branched; bracts lanceolate to elliptic, 2−2,5 × 0,5−1 mm, acute to acuminate, not overlapping except in the bud stage. *Calyx* 1 mm long at flowering, increasing to 3 mm long. *Corolla* white to pinkish with reddish purple markings on the upper lip, 7−12 mm long; tube curved near the base. Fig. 23: 1.

Found in the Transvaal at fairly high altitudes in the western Waterberg and on the Drakensberg escarpment, extending to Swaziland, coastal Natal and Transkei; usually among rocks. Map 67.

Vouchers: *Codd* 3734; 9311; *Galpin* 3494; *Pegler* 1560.

Characterised by the small, sparingly toothed, glabrous to puberulous leaves, and the relatively long, deflexed corolla tube.

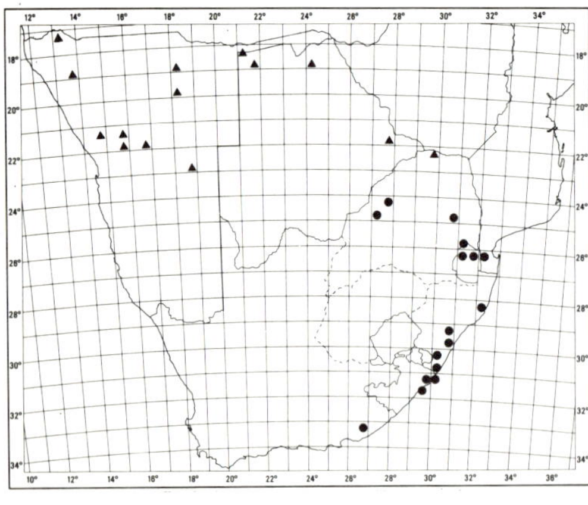

MAP 67. — ● **Aeollanthus parvifolius**
▲ **A. neglectus**

4. **Aeollanthus rehmannii** *Gürke* in Bull. Herb. Boissier 4: 819 (1896); Cooke in F.C. 5,1: 295 (1910); Ross, Fl. Natal 304 (1972); Compton, Fl. Swaziland 500 (1976). Syntypes: Transvaal, Houtbosch, *Rehmann* 6163; 6164.

A. crenatus S. Moore in J. Bot., Lond. 45: 94 (1907). Type: Zimbabwe, Matopo Hills, *Eyles* 1013 (BM, holo.).

Perennial semisucculent herb or subshrub, 0,2−0,5 m tall; branches spreading-ascending, minutely pubescent. *Leaves* softly fleshy, petiolate; blade ovate to broadly ovate, 15−35 (−40) × 10−22 (−28) mm, shortly and stiffly pubescent beneath, sparingly pubescent above, apex obtuse, base obtuse to truncate, margin crenulate-sinuate, often tinged with reddish purple. *Inflorescence* usually branched; bracts lanceolate to elliptic, 2−2,5 × 0,5−1 mm, acute, not overlapping except in the bud stage. *Calyx* 1 mm long at flowering, increasing to 3 mm long. *Corolla* white to pinkish mauve with reddish purple markings on the upper lip, 7−11 mm long; tube straight or nearly so.

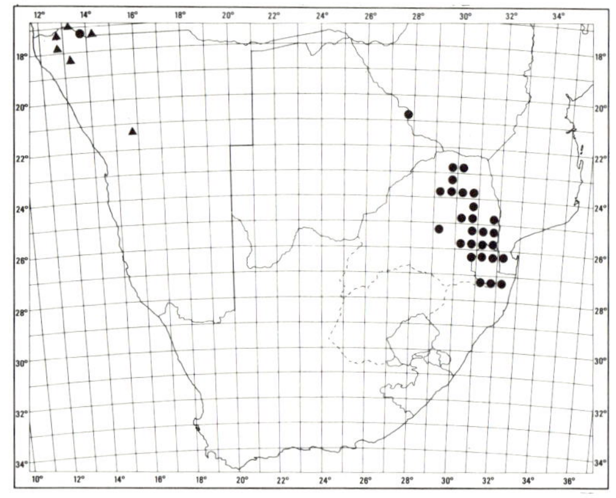

MAP 68. — ● **Aeollanthus rehmannii**
▲ **A. namibiensis**

Occurs in northern S.W.A./Namibia, Botswana, Transvaal, Swaziland and northern Natal, extending into tropical Africa; in rocky, wooded places. Map 68.

Vouchers: *Codd* 5244; 9518; *Compton* 30424; *Schlieben* 9411.

5. **Aeollanthus neglectus** *(Dinter) Launert* in Mitt. bot. StSamml., Münch. 2: 310 (1957); Launert & Schreiber in F.S.W.A. 123: 9 (1969). Syntypes: S.W.A./Namibia, Grossbarmen, *Dinter* 508; Wilhelmsberg, *Dinter* 573; Okahandja, *Dinter* 2590.

Plectranthus neglectus Dinter in Feddes Reprium 22: 380 (1926).

Annual semisucculent herb, branching from the base, 0,15−0,3 m tall; stems ascending, minutely puberulous. *Leaves* softly fleshy, drying membranous, long petiolate; blade broadly ovate to subrotund, 15−30 × 12−25 mm, glabrous to puberulous, apex obtuse to rounded, base truncate or abruptly cuneate, margin subentire or obscurely toothed. *Inflorescence* usually with many slender branches; flowers subsessile, widely spaced; bracts broadly elliptic or obovate-elliptic, acute to acuminate, 2−2,25 × 0,75−1,25 mm, not overlapping except in the bud stage. *Calyx* 1 mm long at flowering, elongating to 3 mm, curved at the apex. *Corolla* whitish to violet or pinkish, 3−5 mm long.

Found in the northern half of S.W.A./Namibia, Botswana and northern Transvaal, in rock crevices in semi-shady places. Also in Zimbabwe and southern Angola. Map 67.

Vouchers: *Codd* 4126; *Dinter* 7089; *Giess* 8449.

May be distinguished from *A. buchnerianus* (no. 2) by the freely branched inflorescence with longer, more slender and laxer floral branches. There has been a tendency to confuse this species with the two tropical species *A. pubescens* Benth. and *A. cameronii* Burkill. In *A. cameronii* the leaves are narrower and more distinctly crenate, while *A. pubescens* has longer leaves which are more pubescent than in *A. neglectus*.

6. Aeollanthus namibiensis *Ryding* in Nord. J. Bot. 2: 224 (1982). Type: S.W.A./Namibia, 30 km N.W. of Omaruru, *Wanntorp* 801 (S, holo.).

A. lobatus sensu Launert & Schreiber in F.S.W.A. 123: 8 (1969).

Annual branched herb 0,15−0,65 m tall, stems spreading-ascending, minutely puberulous. *Leaves* petiolate, pinnatifid; blade 15−40 × 10−20 mm, subglabrous, lobes linear to linear-spathulate, 4−10 × 1−2 mm, obtuse. *Inflorescence* sparingly branched, lax; flowers spaced, flower-spikes slender, 40−80 mm long; bracts lanceolate to ovate-lanceolate, 3−5 mm long, acute to acuminate. *Calyx* glandular-puberulous, 1−2 mm long at flowering, enlarging to 5 mm long with a bulbous base, narrowed above the middle, equally 5-toothed at maturity. *Corolla* mauve to lilac with purple spots, 6−7 mm long.

In northern S.W.A./Namibia; in pockets of soil among rocks. Map 68.

Vouchers: *De Winter & Leistner* 5414; *Merxmüller & Giess* 30551; *Vahrmeijer & Du Preez* 2624.

Closely related to *A. lobatus* N.E. Br. of southern Angola, differing mainly in the smaller flowers.

7345a   **20. ENDOSTEMON**

Endostemon *N.E. Br.* in F.C. 5,1: 295 (1910); Ashby in J. Bot., Lond. 74: 121 (1936); Launert & Schreiber in F.S.W.A. 123: 10 (1969); R.A. Dyer, Gen. 531 (1975). Type species: *E. obtusifolius* (E. Mey. ex Benth.) N.E. Br.

Orthosiphon sect. *Diffusi* Briq. in Natürl. PflFam. 4, 3a: 372 (1897). Type species: *O. diffusus* Benth.

Pseudocimum Brem. in Ann. Transv. Mus. 15: 251 (1933). Type species: *P. trichocalyx* Brem.

Perennial herbs or soft shrublets. *Leaves* subentire or toothed, aromatic. *Flowers* in 2−6 (−12)-flowered verticils in axillary or terminal racemes or panicles; bracts persistent, small or large. *Calyx* 5-toothed, bilabiate, accrescent; tube campanulate to tubular, gibbous, usually conspicuously ribbed; upper tooth the largest, ovate, erect, conspicuously veined, margin slightly decurrent; 4 lower teeth horizontal, lanceolate-deltoid to subulate or the lateral teeth occasionally oblong. *Corolla* subequally 4-lobed; tube cylindrical, slightly wider at the throat; lobes flat or nearly so with the uppermost and lowest lobe sometimes longer than the two lateral lobes. *Stamens* 4, included, inserted above the middle of the corolla tube; filaments very short, hairy, or absent; anthers 1-celled, reniform. *Style* included, simple or obscurely bifid. *Nutlets* suborbicular or oblong, sometimes mucilaginous on wetting.

Species 17, mainly African, extending into the southern Arabian Peninsula; 3 species in Southern Africa.

The genus is allied to *Orthosiphon* but differs in the 4-lobed corolla and the sessile to subsessile stamens inserted about the middle of the corolla tube.

1 Leaves broadly ovate, 10−30 mm broad; soft shrublet or herb up to 1,5 m tall .... 1. *E. obtusifolius*

1 Leaves linear or lanceolate to oblanceolate, 1−10 mm broad; dwarf shrublet or herb 0,15−0,4 m tall:

  2 Verticils 2-flowered; bracts small, 3−5 mm long; calyx throat villous; lateral calyx teeth subulate
  .... 2. *E. tenuiflorus*

  2 Verticils 4−6-flowered; bracts leaf-like, 8−15 mm long; calyx throat not villous; lateral calyx teeth
  oblong .... 3. *E. tereticaulis*

1. **Endostemon obtusifolius** *(E. Mey. ex Benth.) N.E. Br.* in F.C. 5,1: 296 (1910); Ashby in J. Bot., Lond. 74: 131 (1936); Ross, Fl. Natal 304 (1972); Compton, Fl. Swaziland 500 (1976). Type: Transkei, between Umtentu and Umzimkulu Rivers, *Drège* (K, holo.).

Ocimum obtusifolium E. Mey. ex Benth. in E. Mey., Comm. 227 (1838); in DC., Prodr. 12: 38 (1848); Briq. in Natürl. PflFam. 4, 3a: 371 (1897).

O. rariflorum Hochst. in Flora 28: 67 (1845). Type: Natal, Umlaas River, *Krauss 8*.

O. laxiflorum Bak. in F.T.A. 5: 348 (June 1900); in Hiern, Cat. Afr. Pl. Welw. 1,4: 850 (Aug. 1900). Syntypes: Angola, *Welwitsch 5552; 5554* (BM).

Straggling to erect herb or soft shrub 0,5−1,5 m tall, much branched; stems hispid. *Leaves* petiolate; blade broadly ovate, 15−40 × 12−30 mm, upper surface thinly pilose, under-surface hispidulous and reticulate veined, apex obtuse to rounded, base obtuse to truncate, margin shallowly crenate-serrate; petiole 5−15 mm long.

Racemes lax, 150−300 mm long, of many spaced verticils; verticils (2−) 3−8 (−12)-flowered; bracts ovate, acuminate, 4−5 mm long. *Calyx* hispidulous, 3,5 mm long at flowering, enlarging to 6−7 mm long. *Corolla* white, 5 mm long. Fig. 24:1.

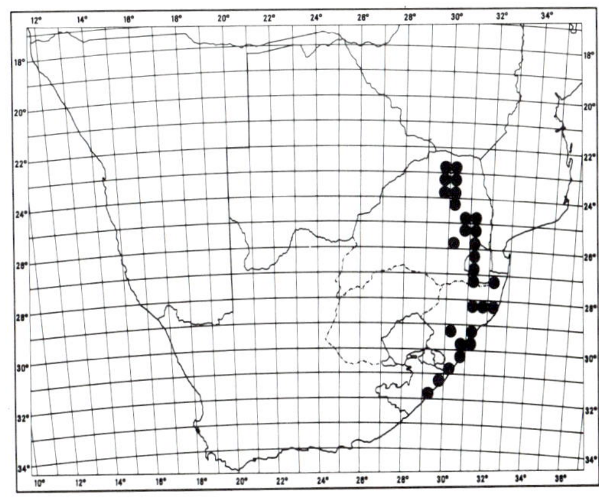

MAP 69. — **Endostemon obtusifolius**

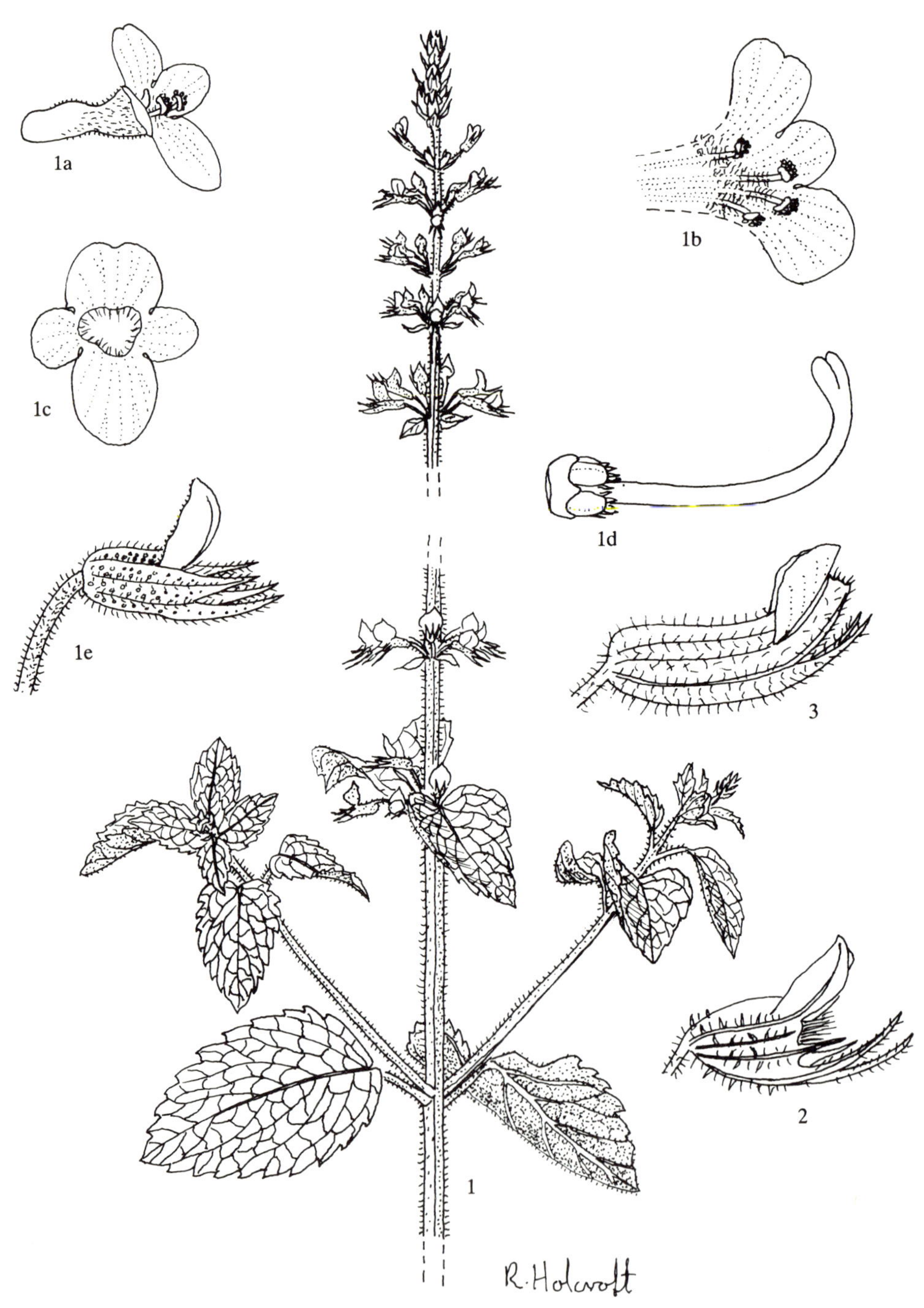
1a
1b
1c
1d
1e
2
3
1
R. Holcroft

Occurs in the Transvaal from the Soutpansberg, along the foothills of the eastern escarpment to Barberton, extending to Swaziland, semi-coastal and coastal Natal and Transkei; also in southern Angola, Zimbabwe and Malawi. Grows on wooded stream banks and at forest margins. Map 69.

Vouchers: *Acocks* 13368; *Codd* 8386; 9438; *Medley Wood* 12592; *Schlechter* 4526.

2. **Endostemon tenuiflorus** *(Benth.)* *Ashby* in J. Bot., Lond. 74: 125 (1936); Launert & Schreiber in F.S.W.A. 123: 10 (1969). Type: Arabia Felix, *Botta* s.n. (K, holo.).

Orthosiphon tenuiflorus Benth. in DC., Prodr. 12: 50 (1848); Briq. in Natürl. PflFam. 4, 3a: 373 (1897); Bak. in F.T.A. 5: 366 (1900).

Ocimum depauperatum Vatke in Linnaea 43: 84 (1880 − 1882). Type: Somalia, *Hildebrandt* 1561 (BM; K, fide Ashby).

Pseudocimum trichocalyx Brem. in Ann. Transv. Mus. 15: 252 (1933). Type: Transvaal, Soutpan, *Bremekamp* 251 (PRE, holo.!).

Perennial soft viscid dwarf shrublet 0,15−0,25 m tall, woody below and freely branched; stems ascending, glandular-hispidulous. *Leaves* subsessile, often coriaceous; blade linear to linear-oblanceolate, 12−40 × 2−5 mm, echinulate with numerous sunken glands, apex obtuse, base attenuate, margin remotely and obscurely toothed, often revolute. *Racemes* lax, 50−70 mm long, of several spaced verticils; verticils 2-flowered; bracts 1,5−5 mm long. *Calyx* aculeate, strongly veined, villous in the throat, enlarging to 3,5 mm long at maturity; tube campanulate; lateral teeth subulate, shorter than the lower pair. *Corolla* whitish to mauve or pink, 8−9 mm long; tube cylindric, 6−7 mm long; lobes 2 mm long. Fig. 24: 2.

Found in north-western S.W.A./Namibia, Botswana and in the northern and eastern Transvaal lowveld, in dry *Colophospermum-Commiphora-Acacia* woodland. Also in Zimbabwe and Malawi, with a gap in distribution, appearing again in Ethiopia, Somalia, the southern Arabian Peninsula and Socotra. Map 70.

Vouchers: *Brenan* 14166; *Codd & Dyer* 3831; 4670; *Van der Schijff* 3584.

Distinguished from *E. tereticaulis* (below) by the 2-flowered verticils, minute bracts, villous calyx throat, subulate lateral calyx teeth and longer corolla tube.

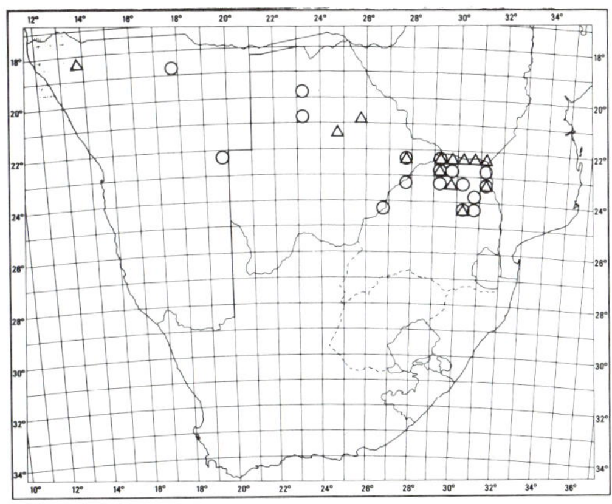

MAP 70. — △ **Endostemon tenuiflorus**
○ **E. tereticaulis**

3. **Endostemon tereticaulis** *(Poir.) Ashby* in J. Bot., Lond. 74: 129 (1936); F.W. Andr., Fl. Pl. Anglo-Egypt. Sudan 3: 209 (1956); Morton in F.W.T.A. edn 2,2: 452 (1963); Launert & Schreiber in F.S.W.A. 123: 11 (1969). Type: from W. tropical Africa, in Hb. Desfontaines (G).

Ocimum tereticaule Poir. in Lam., Encycl. Suppl. 1: 592 (1811); Benth., Lab. 14 (1832); in DC., Prodr. 12: 41 (1848); Briq. in Natürl. PflFam. 4, 3a: 372 (1897); Bak. in F.T.A. 5: 347 (1900).

O. thonningii Schumach. & Thonn. in Schumach., Beskr. Guin. Pl. 4: 269 (1827). Type: Senegal, *Thonning* 78 (C, holo.).

Orthosiphon cleistocalyx Vatke in Linnaea 37: 317 (1872). Type: Ethiopia, *Schimper* 385 (B†).

O. gofensis S. Moore in J. Bot., Lond. 39: 263 (1901). Type: Ethiopia, *Delamere* s.n. (BM).

O. kelleri Briq. in Bull. Herb. Boissier sér. 2, 3: 988 (1903). Type: Somalia, *Keller* 232 (K).

Endostemon ocimoides Brem. in Ann. Transv. Mus. 15: 250 (1933). Type: Transvaal, between Leipzig and Bochum, *Bremekamp* 153 (PRE, holo.!).

Perennial, dwarf soft shrublet 0,2−0,5 m tall, woody below and freely branched; stems ascending, villous. *Leaves* shortly petiolate; blade oblanceolate to obovate, 20−28 × 5−12 mm, upper surface thinly pubescent, under-surface more densely to

---

FIG. 24. — 1, **Endostemon obtusifolius**, flowering stem, × 1; 1a, flower, × 5; 1b, corolla, opened longitudinally, × 5; 1c, front of corolla, × 5; 1d, gynoecium, × 10; 1e, mature calyx, × 5 (*Holcroft* s.n.). 2, **E. tenuiflorus**, mature calyx, × 5 (*Kerfoot* 8019). 3, **E. tereticaulis**, mature calyx, × 5 (*Schlieben & Strey* 8342).

appressed grey-villous and freely gland-dotted, apex rounded, base cuneate, margin obscurely crenulate, occasionally revolute. *Racemes* semi-lax, 50—90 mm long, of several to many fairly closely placed verticils; verticils usually 6-flowered; bracts leaf-like, 8—15 mm long. *Calyx* subglabrous to hispid, 5—6 mm long at maturity; tube tubular; lateral teeth oblong, as long as the lower subulate pair. *Corolla* mauve to purple, 5 mm long. Fig. 24: 3.

Found in north-eastern S.W.A./Namibia, Botswana and at low altitudes in north-western, northern and north-eastern Transvaal, in dry open woodland in sandy and rocky places. Also from Senegal to Somalia and through tropical East Africa to Zimbabwe and Mozambique. Map 70.

Vouchers: *Acocks* 16774; *Codd* 6617; *De Winter* 2477; *Meeuse* 10623.

Characterized by the leaf-like bracts, the calyx with a tubular tube and winged lateral teeth, and the small mauve to purple corolla.

7347 21. PYCNOSTACHYS

Pycnostachys *Hook.*, Exot. Fl. 3: t.202 (1825); Benth., Lab. 61 (1833); in DC., Prodr. 12: 83 (1848); Benth. & Hook. f., Gen. Pl. 2,2: 1177 (1876); Briq. in Natürl. PflFam. 4,3a: 350 (1897); Bak. in F.T.A. 5: 378 (1900); Cooke in F.C. 5,1: 290 (1910); Perkins in Notizbl. bot. Gart. Mus. Berl. 8: 63 (1921); E. A. Bruce in Kew Bull. 1939: 563 (1939); R. A. Dyer, Gen. 531 (1975). Type species: *P. coerulea* Hook.

Echinostachys E. Mey., Comm. 243 (1837). Type species: *E. reticulata* E. Mey.

Perennial erect herbs or soft shrubs. *Leaves* opposite or whorled. *Flowers* blue or mauve, rarely whitish, in dense terminal spikes; bracts small, distinct from the leaves. *Calyx* subequally 5-toothed; teeth subulate, rigid, spinescent. *Corolla* bilabiate; tube cylindric below, enlarging near the throat, deflexed; upper lip 4-lobed, shorter than the lower; lower lip large, boat-shaped. *Stamens* 4, didynamous, declinate, inserted in the corolla throat and lying in the lower lip; filaments shortly united at the base. *Disc* produced in front. *Style* slender, shortly 2-fid at the apex, slightly exceeding the stamens in length. *Nutlets* ovoid, black or brown.

About 40 species, all African, one extends to Malagasy Republic; 3 species in Southern Africa.

1 Leaves sessile or subsessile (petioles up to 5 mm long); blade linear to lanceolate or narrowly elliptic:
  2 Corolla 4–6 mm long; calyx teeth 2–3 mm long.. 1. *P. coerulea*
  2 Corolla 8–18 mm long; calyx teeth 4–6 mm long .... 2. *P. reticulata*
1 Leaves petiolate (petioles of lower leaves 10–40 mm long); blade ovate-lanceolate to broadly ovate
.... 3. *P. urticifolia*

1. **Pycnostachys coerulea** *Hook.*, Exot. Fl. 3: t.202 (1825); Benth., Lab. 61 (1833); in DC., Prodr. 12: 83 (1848); Bak. in F.T.A. 5: 382 (1900) Bruce in Kew Bull. 1939: 582 (1939); Agnew, Upl. Kenya Wild Flow. 632 (1974). Type: Madagascar, ex hort. Kew (leg. *Bojer & Helsinger*) (K, holo.).

P. micrantha Gürke in Engl., Pflanzenw. Ost-Afr. C: 345 (1895); Bak., l.c. 381 (1900); Perkins in Notizbl. bot. Gart. Mus. Berl. 8: 69 (1921). Type: Tanzania, *Stuhlmann* 1630.

P. stenostachys Bak., l.c. 380 (1900). Type: Uganda, *Speke & Grant* s.n. (K, holo.).

P. brevipetiolata De Wild., Pl. Bequaert. 4: 394 (1928). Type: Zaire, *Bequaert* 5972 (BR, holo.).

Erect sparingly branched herb 0,6–1,2 m tall; stems puberulous. *Leaves* sessile; blade linear-lanceolate to narrowly lanceolate or narrowly elliptic, 70–100 × 10–15 mm, subglabrous, apex acuminate, base cuneate, margin distantly and shortly toothed. *Inflorescence* usually solitary or occasionally stem branched towards the apex, 25–50 × 8–10 mm; bracts oblong-linear, 2,5 mm long, ciliate. *Calyx* ciliate; teeth 2–3 mm long. *Corolla* blue, 4–5 mm long.

Found in northern Botswana, in marshy grassland and on floating reed-beds. Also in tropical Africa, northwards to Uganda and Kenya, and extending to Malagasy Republic. Map 71.

Vouchers: *Curson* 374; *Smith* 1440; 1530.

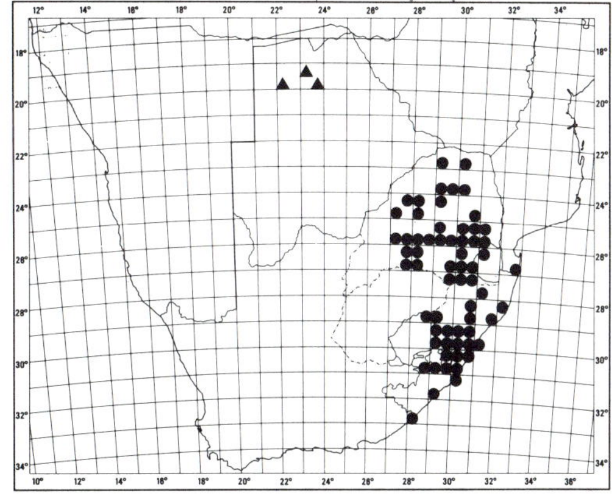

MAP 71. — ▲ **Pycnostachys coerulea**
   ● **P. reticulata**

2. **Pycnostachys reticulata** *(E. Mey.)* **Benth.** in DC., Prodr. 12: 83 (1848); Bak. in F.T.A. 5: 382 (1900); Cooke in F.C. 5,1: 291 (1910); Perkins in Notizbl. bot. Gart.

1a
1b
2a
2b
2c
1
2
R. Holcroft

Mus. Berl. 8: 71 (1921); Bruce in Kew Bull. 1939: 584 (1939); Ross, Fl. Natal 304 (1972); Compton, Fl. Swaziland 500 (1976). Type: Natal, near Durban, *Drège* s.n. (K, holo.).

Echinostachys reticulata E. Mey., Comm. 243 (1837); Hochst. in Flora 28: 68 (1845).

P. reticulata var. *angustifolia* Benth. in DC., Prodr. 12: 83 (1848). Syntypes: Natal, Durban, *Krauss* 329; Transvaal, Magaliesberg, *Burke* s.n.

P. kirkii Bak., l.c. 381 (1900). Syntypes: Malawi, *Kirk* s.n.; *Buchanan* 700.

P. uliginosa Gürke in Bot. Jb. 30: 396 (1901); Perkins, l.c. 72 (1921). Type: Malawi, *Goetze* 806.

P. purpurascens Briq. in Bull. Herb. Boissier sér. 2,3: 998 (1903); Cooke, l.c. 292 (1910); Phillips in Flower. Pl. S. Afr. 13: t.513 (1933). Type: Transvaal Witwatersrand, *Hutton* 878.

P. schlechteri Briq., l.c. 999 (1903). Type: Cape, Mount Frere, *Schlechter* 6406 (PRE!).

P. holophylla Briq., l.c. 1000 (1903). Type: Transvaal, Johannesburg, *CSCA Herbarium* No. 347.

Erect herb 0,2−2 m tall; stems solitary or few from the base, softly woody below, simple or sparingly branched. *Leaves* sessile or subsessile; blade linear-lanceolate to linear-elliptic, elliptic or oblong, 40−110 × 8−25 (−30) mm, subglabrous to puberulous or sometimes pubescent beneath, apex acute to acuminate, base cuneate, margin obscurely to sharply and regularly toothed; petiole up to 5 mm long. *Inflorescence* solitary or several borne on branches produced on the upper part of the stem, the central one the largest, 30−50 (−80) × 20 × 25 mm; bracts linear, ciliate, 4 mm long. *Calyx* puberulous, purple; teeth 4−6 mm long. *Corolla* pale blue or sky blue to pale mauve or pinkish or almost white, 8−18 mm long. Fig. 25: 1.

Common in the Transvaal from the Soutpansberg to the central highlands and Witwatersrand, westward to Rustenburg and along the eastern escarpment to south-eastern Transvaal, Swaziland and Natal from the Drakensberg to the coast, extending to East Griqualand and coastal Transkei to Kentani; in moist, grassy places. Also from Zimbabwe to Malawi and Tanzania. Map 71.

Vouchers: *Acocks* 23576; *Codd* 9557; *Galpin* 12041; *Schlechter* 2802.

The width and pubescence of the leaves vary a good deal. The type of *P. reticulata* has oblong-lanceolate, markedly reticulate and pubescent leaves whereas the type of *P. schlechteri* represents the other extreme with linear-lanceolate, subglabrous leaves. The extremes are linked by a range of intermediates.

3. Pycnostachys urticifolia *Hook.* in Curtis's bot. Mag. t.5365 (1863); Bak. in F.T.A. 5: 386 (1900); Cooke in F.C. 5,1: 291 (1910); Perkins in Notizbl. bot. Gart. Mus. Berl. 8: 74 (1921); R. A. Dyer in Flower. Pl. S. Afr. 14: t.560 (1934). Type: Malawi, ex Hort. Kew "from seed sent by Drs Kirk and Meller" (K, holo.).

P. pubescens Gürke in Engl., Pflanzenw. Ost-Afr. C: 345 (1895); Bak., l.c. 386 (1900). *P. urticifolia* var. *pubescens* (Gürke) Gürke in Bot. Jb. 22: 146 (1895). Syntypes: Malawi, *Buchanan;* Mozambique, *Carvalho*.

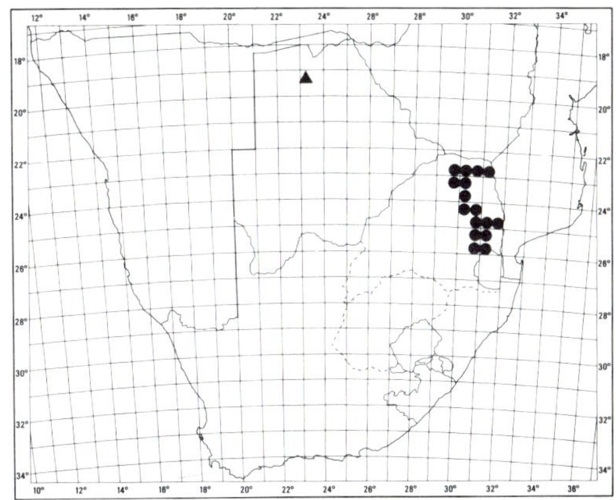

MAP 72. — ● **Pycnostachys urticifolia**
▲ **Neohyptis paniculata**

Erect herb or soft shrub 1−2,5 m tall, woody at the base, branched or sometimes several-stemmed from the base; stems usually branched especially towards the apex, occasionally simple. *Leaves* petiolate; blade narrowly to broadly ovate, (45−) 50−120 × (30−) 40−70 mm, subglabrous to densely pubescent on both sides, apex acute, base obtuse to truncate, margin regularly crenate; petiole 10−50 mm. *Inflorescence* borne on the ends of the ascending branches, the central one the

FIG. 25. — 1, **Pycnostachys reticulata,** flowering stem, × 1; 1a, section through corolla, × 2; 1b, flowering calyx, × 2 (*Bredell* s.n.). 2, **P. urticifolia,** leaf, × 1; 2a, corolla, × 2; 2b, section through corolla, × 2; 2c, bract, calyx and base of corolla, × 2 (living plant, BRI garden).

largest (50—) 70—100 × 25—30 mm; bracts linear to spathulate, ciliate, 4—5 mm long. *Calyx* sparingly pubescent, reddish purple; teeth 8—10 mm long. *Corolla* gentian blue or rarely whitish, 12—20 mm long. Fig. 25: 2.

Common in the north-eastern and eastern Transvaal as far south as Barberton; in moist places, grassy stream banks or at forest margins. Also from Zimbabwe and Mozambique to Malawi and Tanzania. Map 72.

Vouchers: *Codd & De Winter* 5551; *Galpin* 943; *Obermeyer* 490; *Scheepers* 273.

7347a

22. NEOHYPTIS

Neohyptis *J.K. Morton* in J. Linn. Soc., Bot. 58: 272 (1962); in F.W.T.A. edn 2,2: 466 (1963). Type species: *N. paniculata* (Bak.) J.K. Morton.

Stems erect or decumbent, quadrangular. *Inflorescence* of short, dense, spike-like racemes borne terminally and in the axils of the leaves of the upper half of the stem; bracts persistent. *Calyx* tubular-campanulate, slightly ventricose when mature, equally 5-toothed; teeth lanceolate, acute. *Corolla* bilabiate; tube straight; upper lip erect, 4-lobed; lower lip concave, spreading. *Stamens* 4, didynamous, declinate, not exceeding the lower corolla lip; filaments fused in pairs towards the base, attached in the throat. *Nutlets* small, glabrous.

1 species, found in west and east tropical Africa, extending to Angola, Zambia and northern Botswana.

Neohyptis paniculata *(Bak.) J.K. Morton* in J. Linn. Soc., Bot. 58: 273 (1962); in F.W.T.A. edn 2, 2: 466 (1963). Type: Angola, Pungo Andongo, *Welwitsch 5528* (K, lecto.).

Geniosporum paniculatum Bak. in F.T.A. 5: 351 (1900); Hiern, Cat. Afr. Pl. Welw. 1,4: 853 (1900).

— var. *debile* Hiern, l.c. 853 (1900). Type: Angola, Pungo Andongo, *Welwitsch 5527* (K, holo.).

Hyptis baumii Gürke in Baum, Kunene-Samb. Exped. 354 (1903). *Plectranthus guerkei* Briq. in Annu. Conserv. Jard. bot. Genève 7–8: 323 (1904), non *P. baumii* Gürke. Type: Angola, Onschingwe, *Baum* 789.

Annual herb, sparingly branched. *Leaves* sessile; blade ovate to ovate-lanceolate, 20–30 × 7–10 mm, glabrous above, hispidulous on the nerves beneath, apex subacute, base obtuse, margin obscurely crenate. *Flower-spikes* usually sessile, occupying the upper half of the stem, 1–3 in each leaf axil, 10–35 × 6–8 mm, many-flowered; bracts ovate, crowded, persistent, 3 mm long, each subtending 1–3 small subsessile flowers. *Calyx* pubescent, 2,5–3 mm long. *Corolla* whitish to lilac, 4–4,5 mm long.

Found on seasonally inundated flood plains in northern Botswana, extending through Angola to West Tropical Africa. Map 72.

Vouchers: *Smith* 615; 2770.

In general appearance this species resembles *Hyptis* (no. 18) but differs in the calyx not being ribbed and the teeth being lanceolate, not subulate, and the corolla is distinctly bilabiate. Its relationship is nearer to *Plectranthus* (no. 23) but if differs in the dense bracteate spike-like racemes, and the lower lip of the corolla being shallowly concave, not boat-shaped. It has a distinctive facies, unlike any species of *Plectranthus*.

7350   **23. PLECTRANTHUS**

Plectranthus *L'Hérit.,* Stirp. Nov. fasc. 4: t.41, 42 (March 1788); Benth., Lab. 29 (1832); in
DC., Prodr. 12: 62 (1848); Benth. & Hook. f., Gen. Pl. 2: 1175 (1876); Briq. in Natürl.
PflFam. 4, 3a: 352 (1897); Bak. in F.T.A. 5: 398 (1900); Cooke in F.C. 5,1: 266 (1910);
Morton in J. Linn. Soc., Bot. 58: 231 (1962); Launert & Schreiber in F.S.W.A. 123: 1−32
(1969); Blake in Contr. Queensl. Herb. 9: 1−120 (1971); Codd in Bothalia 11: 271 (1975);
R.A. Dyer, Gen. 522 (1975). Lectotype (Bullock & Killick in Taxon 6: 239, 1957): *P.
fruticosus* L'Hérit.

Germanea Lam., Encycl. 2: 690 (April 1788); Hiern, Cat. Afr. Pl. Welw. 1: 865 (1900). Type: based on two
species, *G. urticifolia* Lam. (which is a synonym of *P. fruticosus* L'Hérit.) and *G. maculosa* Lam.

Coleus Lour., Fl. Cochin. 372 (1790); emend. Benth., Lab. 47 (1832), partly; in DC., Prodr. 12: 70 (1848);
Benth. & Hook. f., l.c. 2: 1176 (1876); Briq., l.c. 4,3a: 359 (1897); Bak., l.c. 5: 422 (1900); Cooke, l.c. 5,1: 289
(1910); Phillips, Gen. edn 2: 649 (1951); all partly. Type species: *C. amboinicus* Lour.

Neomuellera Briq. in Bot. Jb. 19: 186 (1894). Type species: *N. welwitschii* Briq.

Burnatastrum Briq. in Natürl. PflFam. 4, 3a: 358 (1897). Lectotype (Codd in Bothalia 11: 374, 1975): *B.
spicatum* (E. Mey. ex Benth.) Briq.

Ascocarydion G. Tayl. in J. Bot., Lond. 69, Suppl. 2: 162 (1931). Type species: *A. mirabile* (Briq.) G. Tayl.

Annual or perennial herbs or subshrubs; stems and leaves herbaceous, semi-succulent
or succulent. *Inflorescence* paniculate, racemose or subspicate, usually terminal; flowers in
verticils, few-flowered cymes or dichasia, or occasionally solitary; bracts small, clearly
differentiated from the leaves. *Calyx* 2-lipped to subequally 5-toothed; when 2-lipped, the
upper lip consisting of a large single tooth, lower lip of 4 lanceolate-deltoid to subulate
teeth; tube glabrous or villous within, sometimes gibbous at the base. *Corolla* bilabiate;
tube usually bent and variously expanded near the base, occasionally expanding gradually,
rarely straight; upper lip usually 4-lobed, shorter than the lower boat-shaped lip. *Stamens* 4,
rarely 2 abortive (*P. zluensis,* no. 37), attached at the corolla mouth, free or united in a
sheath at the base, declinate in the lower lip of the corolla; anthers 1-thecous. *Style* lying
with the stamens in the lower lip of the corolla; stigma shortly 2-lobed. *Nutlets* ovoid or
oblong, smooth.

Species about 350, Africa to Asia and Australia; of the 44 species dealt with below, 42 are indigenous to
Southern Africa; the two semi-naturalized species are *P. ornatus* Codd (no. 14) and *P. barbatus* Andr. (no. 15).

The subject of generic delimitation in this and allied genera was discussed in Bothalia 11: 371 (1975) where a
rather broad circumscription of *Plectranthus* was adopted and the species were grouped into several subgenera.
Subgen. *Plectranthus* is represented in the present treatment by species nos. 16−44, in which the upper calyx tooth
is distinctly larger than the lower 4, the stamens are all free to the base, and the flowers are arranged in few- to
many-flowered sessile cymes or, occasionally, in few-flowered pedunculate cymes. This subgenus may again be
subdivided into 2 sections:

1. Species 16−21, which correspond more or less to Bentham's *Plectranthus* sect. *Coleoides,* characterized by
having flowers in dense clusters of 3−15 in the axil of each bract; the bracts are usually deciduous before the
flowers begin to open; and the calyx, which is attached to the pedicel at a sharp angle (declinate), is distinctly
gibbous at the base. The section is distributed from south to north-east Africa, India and East Indies and includes
most if not all the Australian species. Throughout the section, species limits are difficult to determine, being based
on characters such as habit, tomentum, size, shape and toothing of leaves and size of flowers, all of which tend to
grade from one extreme to the other without sharp demarcation. Most species which have been investigated
cytologically have 2n = 42 chromosomes.

2. Species 22−44, representing sect. *Plectranthus,* with flowers in few-flowered sessile or shortly branched
cymes in the axil of each bract; the bracts (often very small) persist beyond the flowering stage; and the calyx is not
markedly gibbous at the base. The section is concentrated largely in Africa south of the Sahara, with outliers in
West Tropical Africa and extending eastwards to southern Asia. In Southern Africa the species are relatively
clear-cut, being based on often striking differences in shape and size of the corolla, degree of exsertion of the
stamens, and supported by vegetative characters. Two main trends in corolla shape may be recognized:

(a) Species 22−39 in which the corolla tube expands abruptly at the base, where it may be saccate or even
spurred, and then may or may not narrow towards the throat.

(b) Species 40−44 in which the corolla does not expand abruptly at the base but may or may not gradually
expand towards the throat. In nos. 40 and 41 the tube is straight and in nos. 42−44 it is somewhat sigmoid. Species
which have been investigated cytologically mostly have 2n = 28 chromosomes.

1 Flowers yellow, in pseudoracemes borne terminally as well as from the upper nodes of the usually leafless stems:

  2 Plants annual; stems with conspicuous bristles; corolla 4−5 mm long .... 1. *P. tetragonus*

  2 Plants perennial with edible tuberous rootstock; stem without bristles; corolla 14−16 mm long .... 2. *P. esculentus*

1 Flowers white or shades of blue, violet or purple (rarely yellow), disposed in verticils, cymes or dichasia; inflorescence usually terminal, paniculate, racemose or subspicate, borne on leafy stems:

  3 Mature calyx subequally 5-toothed, often erect or finally circinnate and sometimes ventricose (in *P. cylindraceus* (no. 8) the uppermost calyx tooth is slightly larger than the other 4 but is difficult to see because of the dense covering of hairs):

    4 Flowers in 10−20-flowered sessile cymes; inflorescence branches slender, up to 350 mm long, peduncle up to 300 mm long; plants with long horizontal tuberous roots .... 3. *P. xerophilus*

    4 Flowers in pedunculate or sessile cincinni (often compact and glomerate in *P. cylindraceus* (no. 8) and *P. spicatus* (no. 7) or in 3-flowered cymes; roots not tuberous:

      5 Flowers in 3-flowered pedunculate cymes, forming a diffusely branched panicle 300−400 mm long .... 4. *P. candelabriformis*

      5 Flowers in pedunculate or sessile paired cincinni; inflorescence less than 300 mm long:

        6 Leaves broader than 30 mm, chartaceous or leathery; inflorescence a lax or dense panicle, flowers blue:

          7 Leaves thick-textured, under-surface densely grey velvety-tomentose; robust plants with erect, sparingly branched tomentose stems up to 2 m tall .... 5. *P. mirabilis*

          7 Leaves thin-textured, under-surface subglabrous to sparingly pubescent; herbaceous, branched plants usually less than 1 m tall .... 6. *P. hereroensis*

        6 Leaves less than 30 mm broad, semisucculent; inflorescence subspicate or sparingly branched, flowers in clusters, mauve, purple or, rarely, whitish:

          8 Corolla 7−8 mm long, purple, subglabrous; flowers in loose clusters .... 7. *P. spicatus*

          8 Corolla 4−5 mm long, mauve (rarely whitish or pale yellow), villous; flowers in densely glomerate clusters .... 8. *P. cylindraceus*

  3 Mature calyx with upper tooth distinctly broader than the rest, oblong to ovate or subrotund, remaining 4 teeth deltoid to subulate; calyx finally horizontal, teeth spreading:

    9 Upper tooth of calyx horizontal, oblong to ovate, usually rounded at the apex; flowers in glomerate, densely tomentose clusters:

      10 Leaves obovate, cuneate at the base; corolla 4−5 mm long; stamens free to the base .... 8. *P. cylindraceus*

      10 Leaves ovate to subrotund, broadly truncate to cordate at the base; corolla more than 5 mm long; stamens united at the base:

        11 Stems erect, woody at the base; corolla whitish, 10−12 mm long .... 9. *P. unguentarius*

        11 Stems decumbent, succulent; corolla mauve to whitish, 7−9 mm long .... 10. *P. amboinicus*

    9 Upper tooth of calyx erect, ovate-deltoid to broadly ovate or subrotund, apex acute to apiculate; inflorescence paniculate, racemose or subspicate:

      12 Mature calyx villous in the throat; stamens united at the base; inflorescence subspicate with pedicels erect, appressed to the rhachis:

        13 Bracts rounded at the apex, subpersistent; stems procumbent, slender, sparingly branched .... 11. *P. tetensis*

        13 Bracts acute to abruptly acuminate, early deciduous, forming a conspicuous 4-angled coma at the apex of the inflorescence; stems erect to procumbent, sometimes mat-forming:

          14 Erect or spreading semisucculent herbs up to 0,6 m tall; leaves ovate-lanceolate to obovate, 20−50 × 15−35 mm:

            15 Corolla less than 10 mm long; annual plants .... 12. *P. caninus*

            15 Corolla exceeding 10 mm long; perennial or weakly perennial plants:

              16 Corolla 10−20 mm long; inflorescence elongate, 70−150 mm long with 5−12 spaced fruiting verticils below the flowers; indigenous .... 13. *P. neochilus*

16 Corolla 20−25 mm long; inflorescence compact, 30−50 (−90) mm long with 1 or 2, rarely more, spaced fruiting verticils below the flowers; cultivated or semi-naturalized .... 14. *P. ornatus*

14 Erect bushy herb or soft shrub up to 2 m tall; leaves not succulent, ovate to broadly ovate-elliptical, tomentose, 50−90 × 30−50 mm; cultivated or semi-naturalized .. 15. *P. barbatus*

12 Mature calyx glabrous in the throat; stamens free to the base; inflorescence usually paniculate or racemose:

17 Bracts deciduous before the flowers open (occasionally persisting in abnormal cases); fruiting calyx gibbous ventrally; flowers in dense verticils, (3−) 4−12 to each bract scar:

18 Stems erect or decumbent; flowers mauve to purple (rarely white):

19 Leaves deeply dentate; rhachis coarsely glandular-hispid, pubescence often yellowish (S.W.A./Namibia) .... 16. *P. dinteri*

19 Leaves crenate-dentate; rhachis sparsely to fairly densely glandular-tomentose, pubescence greyish:

20 Leaves 40−100 mm long, if less, then sparingly to fairly densely strigose:

21 Leaves densely tomentose on both surfaces:

22 Stems 0,3−0,6 m tall; inflorescence 80−300 mm long, simple or with a pair of branches near the base .... 18(a). *P. hadiensis* var. *hadiensis*

22 Stems 0,5−1,5 m tall; inflorescence 200−600 mm long, usually with 1 or 2 pairs of branches near the base .... 18(b). *P. hadiensis* var. *tomentosus*

21 Leaves sparingly to fairly densely strigose .... 18(c). *P. hadiensis* var. *woodii*

20 Leaves 25−40 mm long; stems 1−several often from a burnt perennial base; small bushes up to 0,4 m tall .... 19(c). *P. madagascariensis* var. *ramosior*

18 Stems procumbent; flowers white, mauve or blue:

23 Corolla 7−18 mm long:

24 Leaves deeply dentate or deeply and coarsely crenate-scalloped:

25 Leaves deeply dentate, densely tomentose; flowers usually white ...... 17. *P. grandidentatus*

25 Leaves deeply and coarsely crenate-scalloped, medium to densely strigose; flowers purple-blue to lilac .... 20. *P. mutabilis*

24 Leaves crenate-dentate:

26 Leaf-blade 40−100 × 32−100 mm, densely tomentose; flowers usually mauve (rarely white) .... 18(b). *P. hadiensis* var. *tomentosus*

26 Leaf-blade 15−40 (−45) × 12−35 (−40) mm, sparingly to densely short tomentose; flowers usually white (rarely mauve) ..... 19(a). *P. madagascariensis* var. *madagascariensis*

23 Corolla 5−6 mm long:

27 Leaves coarsely crenate with 3−4 pairs of rounded teeth; corolla white (Transkei and southern Natal) .... 19(b). *P. madagascariensis* var. *aliciae*

27 Leaves obscurely crenate-dentate with 5−7 pairs of shallow teeth; corolla blue-mauve (KwaZulu, coastal) .... 21. *P. psammophilus*

17 Bracts (often very small) persisting beyond the flowering stage; fruiting calyx enlarged and often oblique but not conspicuously gibbous ventrally; flowers in lax verticils with 1−3 flowers to each bract or in pedunculate 3−8-flowered cymes:

28 Corolla tube expanding abruptly at or near the base and often saccate or spurred dorsally, usually declinate near the base: (second half of couplet on p. 4: 141)

29 Fertile stamens 2, staminodes 2; stems and under-surface of leaves softly velvety pubescent .... 37. *P. zuluensis*

29 Fertile stamens 4:

30 Corolla tube less than 10 mm long:

31 Leaf-blade less than 40 mm in length (occasionally longer in *P. oertendahlii* (no. 25) but then leaf-blade subrotund and noticeably lighter-veined):

32 Under-surface of leaf dotted with minute red gland-dots (also on calyx and corolla):

33 Corolla tube scarcely narrowed near the throat; corolla lips 5—7 mm long; stamens 4—6 mm long; young stems and petioles subglabrous to pubescent .... 22. *P. verticillatus*

33 Corolla tube narrowed near the throat; corolla lips 3—5 mm long; stamens 0,5—3 mm long; young stems and petioles greyish to rusty strigose or densely grey tomentose:

    34 Young stems, petioles and leaves greyish to rusty strigose; petioles up to 25 mm long .... 23. *P. strigosus*

    34 Young stems, petioles and leaves densely grey tomentose; petioles up to 10 mm long .... 24. *P. purpuratus*

32 Under-surface of leaf dotted with minute colourless or honey-coloured gland-dots:

  35 Corolla white or whitish, or with a few purple spots:

    36 Corolla tube 8 mm or longer, narrowing conspicuously towards the throat; leaves noticeably lighter-veined .... 25. *P. oertendahlii*

    36 Corolla tube 4—5 mm long, narrowing slightly towards the throat; leaves not lighter-veined .... 2. *P. elegantulus*

  35 Corolla blue or mauve:

    37 Corolla tube 4—7 mm long, upper lip 1,5—5 mm long:

      38 Corolla pale blue, narrowing towards the throat, upper lip 4—7 mm long; stems much swollen at the base .... 27. *P. ernstii*

      38 Corolla sky-blue, widening slightly towards the throat, upper lip 1,5—2 mm long; stems not swollen at the base .... 36. *P. dolichopodus*

    37 Corolla tube 8 mm or longer, upper lip about 10 mm long and equally broad .... 38(a). *P. saccatus* var. *saccatus*

31 Leaf blade normally more than 40 mm long (sometimes smaller in *P. ciliatus* but then leaves and calyx ciliate with multicellular purple-striped hairs):

  39 Flowers in 3—8-flowered, often pedunculate cymes; corolla villous; upper lip of corolla 2 mm long .... 34. *P. rehmannii*

  39 Flowers in 1—3-flowered sessile cymes; corolla glabrous or, if hairy, then upper lip 4—6 mm long:

    40 Leaf margin with few (6—14) pairs of large teeth 8—10 mm long which bear small secondary teeth; corolla whitish with a fringe of hairs on the lower lip .... 35. *P. swynnertonii*

    40 Leaf margin and corolla not as above:

      41 Under-surface of leaf dotted with minute colourless or honey-coloured gland-dots:

      42 Corolla white, mauve or pink often speckled with purple; upper lip 2,5—7 mm long:

        43 Leaves not tomentose; corolla subglabrous:

          44 Leaf margin and calyx ciliate with purplish multicellular hairs; corolla whitish freely speckled with purple .... 29. *P. ciliatus*

          44 Leaf margin not ciliate; calyx with occasional multicellular hairs; flowers mauve or pink with darker markings .... 30. *P. fruticosus*

        43 Leaves tomentose; corolla villous .... 31. *P. oribiensis*

      42 Corolla blue, with or without purple spots:

        45 Corolla sky-blue; tube 5 mm long, upper lip 2 mm long .... 36. *P. dolichopodus*

        45 Corolla mauve-blue; tube 8 mm or more long, upper lip 10 mm long .... 38(a). *P. saccatus* var. *saccatus*

      41 Under-surface of leaf dotted with minute red gland-dots:

        46 Leaves often thin-textured, apex acute, base abruptly cuneate, margin with coarse teeth often bearing small secondary teeth .... 32. *P. grallatus*

46 Leaves usually thick-textured, apex obtuse to rounded, base truncate, shortly attenuate or markedly decurrent on the petiole, margin regularly crenate .... 33. *P. rubropunctatus*

30 Corolla tube 10 mm long or longer:

47 Corolla distinctly narrowing towards the throat; stamens 3—5 mm long:

48 Corolla white or tinged with mauve; leaves lighter-veined, under-surface with minute honey-coloured gland-dots .... 25. *P. oertendahlii*

48 Corolla mauve to violet; leaves not lighter-veined, under-surface with minute red gland-dots .... 26. *P. praetermissus*

47 Corolla not or scarcely narrowing towards the throat; stamens 7—10 mm long:

49 Leaves broadly ovate to ovate-deltoid, subglabrous, base truncate; upper lip of corolla 10—16 mm long:

50 Corolla tube 8—18 mm long .... 38(a). *P. saccatus* var. *saccatus*

50 Corolla tube 20—26 mm long .... 38(b). *P. saccatus* var. *longitubus*

49 Leaves broadly elliptical to obovate-elliptical, sparingly strigose, base cuneate; upper lip of corolla 5—6 mm long, tube 23-27 mm long .... 39. *P. hilliardiae*

28 (from p. 4: 139) Corolla tube expanding gradually from the calyx mouth or nearly parallel-sided for entire length, straight or curved:

51 Corolla tube straight, more than 12 mm long:

52 Corolla tube 20—25 mm long, nearly parallel-sided .... 40. *P. ambiguus*

52 Corolla tube 12—18 mm long, widening slightly towards the throat .... 41. *P. ecklonii*

51 Corolla tube curved, 7—10 mm long, rather like a miniature "Dutchman's Pipe":

53 Stems decumbent, greyish-tomentulose; leaves semi-succulent, 20—30 × 18—30 mm .... 42. *P. dolomiticus*

53 Stems erect to spreading, sparingly to densely pubescent; leaves herbaceous, 40—140 × 35—110 mm:

54 Leaves coarsely dentate, truncate at the base; corolla purple .... 43. *P. petiolaris*

54 Leaves regularly crenate-dentate, cordate; corolla white with vertical mauve stripes on upper lip .... 44. *P. laxiflorus*

1. **Plectranthus tetragonus** *Gürke* in Bot. Jb. 19: 109 (1894); Bak. in F.T.A. 5: 401 (1900); Hutch. & Dandy in Kew Bull. 1926: 481 (1926); Launert & Schreiber in F.S.W.A. 123: 26 (1969); Codd in Bothalia 11: 376 (1975). Type: Tanzania, Usambara, Mashena, *Holst* 3573.

P. melanocarpus Gürke, l.c. 109 (1894); Bak., l.c. 402 (1900); Hutch. & Dandy, l.c. 481 (1926). *Englerastrum melanocarpus* (Gürke) Th. Fries jun., l.c. 71 (1924). *Coleus melanocarpus* (Gürke) Robyns & Lebrun, l.c. 106 (1929). Type: Tanzania, Massai steppe, *Fischer* 511.

P. biflorus Bak., l.c. 402 (1900). Type: Malawi, between Kondowe and Karonga, *Whyte* s.n. (K, holo.).

Englerastrum tetragonus (Gürke) Th. Fries jun. in Notizbl. bot. Gart. Mus. Berl. 9: 73 (1924). *Coleus tetragonus* (Gürke) Robyns & Lebrun in Annls Soc. scient. Brux. sér. B, 49: 106 (1929).

Erect, annual herb 0,3—0,6 m tall; stems solitary with conspicuous patent bristles along the upper part. *Leaves* petiolate; blade membranous, ovate, 35—70 × 20—35 mm, subglabrous, apex acute, base obtuse to cuneate or attenuate on the petiole, margin coarsely toothed. *Inflorescences* produced mainly after the leaves are shed, terminal and from the nodes often from near the base to the apex of the plant, racemose, 40—120 mm long, simple or branched; flowers numerous, solitary or occasionally in pairs on slender pedicels, opposite or alternate. *Calyx* 2 mm long at flowering, increasing to 10 mm in fruit, tubular and slightly curved; upper tooth ovate, erect; lower 4 teeth lanceolate-subulate, horizontal. *Corolla* yellow, 4—5 mm long, slightly geniculate and expanding near the base. *Stamens* free or shortly united at the base. Fig. 26: 1.

Recorded from northern S.W.A./Namibia, Botswana, north-western Transvaal and Swaziland, and extending to Mozambique and through Zimbabwe and Zambia to Tanzania; in sandy soil in dry woodland. Map 73.

Vouchers: *Ellis* 3076; *Giess* 15117; *Van Jaarsveld* 3326.

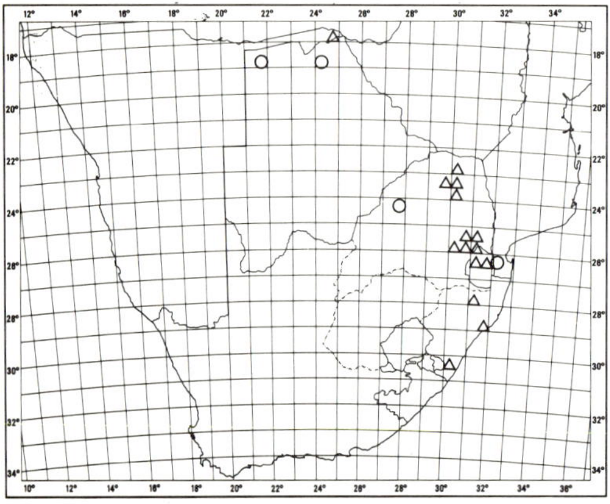

MAP 73. — ○ **Plectranthus tetragonus**
      △ **P. esculentus**

2. Plectranthus esculentus *N.E.Br.* in Kew Bull. 1894: 12 (1894); in Hooker's Icon. Pl. 25: t.2488 (1896); Cooke in F.C. 5,1: 285 (1910); Ross, Fl. Natal 305 (1972); Codd in Bothalia 11: 377 (1975). Type: Natal, cult. Botanic Garden, Durban, *Medley Wood* 3633 (K, holo.!).

P. floribundus N.E. Br., l.c. 12 (1894); in Hooker's Icon. Pl. 25: t.2489 (1896); Cooke, l.c. 273 (1910). *Englerastrum floribundum* (N.E. Br.) Th. Fries jun. in Notizbl. bot. Gart. Mus. Berlin 9: 73 (1924). *Coleus floribundus* (N.E. Br.) Robyns & Lebrun in Revue Zool. Bot. afr. 16: 359 (1928); in Annls Soc. scient. Brux. sér. B, 49: 96 (1929), nom. illegit. Lectotype (Robyns & Lebrun, 1928): Natal, Inanda, *Medley Wood* 646 (K, lecto.!; PRE!).

— var. *longipes* N.E. Br., l.c. 13 (1894); Bak. in F.T.A. 5: 403 (1900). *E. floribundum* var. *longipes* (N.E. Br.) Th. Fries jun., l.c. 77 (1924). *C. floribundus* var. *longipes* (N.E. Br.) Robyns & Lebrun, l.c. 360 (1928). Lectotype (Robyns & Lebrun, 1928): Zimbabwe, Umzingwani River, *Baines* s.n. (K, lecto.).

Coleus dazo A. Chev., Veg. Utiles de l'Afr. Trop. Franc. 1, 1: 106 (1905). Type: from West Africa.

Coleus esculentus (N.E. Br.) G. Tayl. in J. Bot., Lond. 69, Suppl. 2: 158 (1931).

Erect herb or suffrutex from a tuberous-rooted base; stems 1—several arising annually, sparingly branched, 0,6—1,2 m tall, usually leafless at flowering. *Leaves* fairly thick-textured, subsessile; blade oblong-elliptic to oblanceolate, 50—80 × 13—25 mm, scabrid, under-surface with brown gland-dots, apex obtuse to rounded, base obtuse to cuneate, margin obscurely denticulate. *Inflorescences* occupying the upper 0,2—0,6 m of the stem, consisting of 2—4 short racemes (occasionally branched) from each node; flowers solitary in the axil of each bract; bracts persistent; pedicels 3—5 mm long. *Calyx* 9—10 mm long in fruit, glandular-hispidulous, 5-toothed, the upper tooth the largest. *Corolla* yellow, 14—16 mm long; tube geniculate, expanding above the middle; upper lip 2 mm long; lower lip deeply boat-shaped, 7—8 mm long. *Stamens* usually united at the base, 6—7 mm long.

Distributed from Equatorial Africa southwards to Angola, the eastern Transvaal, Swaziland and coastal Natal, in dry wooded country; often spread by cultivation because the tubers are edible. Map 73.

Vouchers: *Galpin* 591; *Gerstner* 5821; *Rudatis* 1105.

Characterized by the yellow flowers borne in short pseudo-racemes after the leaves have been shed, and the thickened, edible roots.

3. Plectranthus xerophilus *Codd* in Bothalia 11: 282 (1974); ibid. 11: 378 (1975); in Flower. Pl. Afr. 44: t.1728 (1977). Type: Transvaal, Lydenburg District, near Marone, *Codd & Dyer* 7729 (PRE, holo.!).

Perennial shrub 1—1,7 m tall with thick horizontal tuberous roots; stems slender, sparingly branched, terete to obscurely 4-angled, grey-tomentose. *Leaves* subsessile to shortly petiolate; blade subcoriaceous, ovate to elliptic, 35—90 × 25—70 mm, upper surface bullate, grey-green, under-surface reticulate, grey-tomentose with multicellular hairs, short gland-tipped hairs and reddish brown gland-dots; apex obtuse to rounded, base subtruncate, margin coarsely crenate. *Inflorescence* terminal, on slender peduncles up to 300 mm long, simple or with 1—3 pairs of basal branches, racemes slender, up to 350 mm long; flowers densely clustered in 12—20-flowered verticils spaced

---

FIG. 26. — 1, **Plectranthus tetragonus**, flowering stem, × 1; 1a, flower, × 5; 1b, section through corolla, × 5; 1c, mature calyx, × 5 (*Hardy* 5630). 2, **P. xerophilus**, part of inflorescence, × 1; 2a, leaf, × 1; 2b, mature calyx, × 5; 2c, flower, × 5 (after Flower. Pl. Afr. 44: t.1728, 1974).

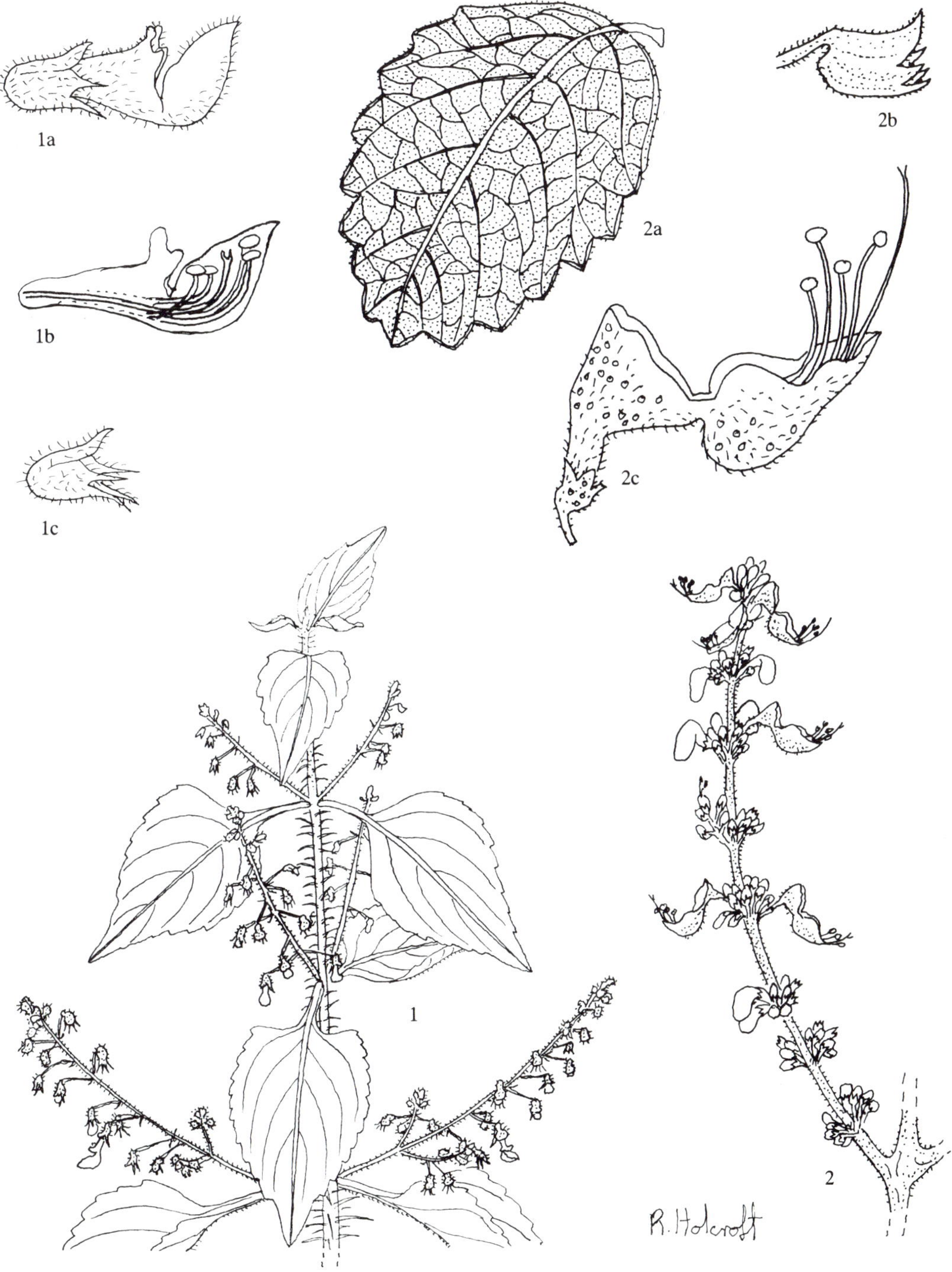

1a
1b
1c
2a
2b
2c
1
2
R. Holcroft

3—25 mm apart; bracts early deciduous. *Calyx* 4 mm long in fruit, subequally 5-toothed, crisped tomentose and gland-dotted, the uppermost tooth slightly larger than the rest. *Corolla* violet to mauve-purple, crisped-tomentose without, c. 10 mm long; tube curved upwards and expanding towards the throat; upper lip short, hooded; lower lip boat-shaped, 4—6 mm long. *Stamens* shortly united at the base, 7—8 mm long. Fig. 26: 2.

Found in the eastern and northern Transvaal at medium to low altitudes on hot, dry, rocky slopes. Map 74.

Vouchers: *Codd* 10010; 10489; *De Winter* 7725; *Meeuse* 10199; 10351.

A very distinct species with long slender inflorescences, flowers in crowded verticils, subequally 5-toothed calyx, a wide-mouthed corolla with a somewhat hooded upper lip, and stamens shortly united at the base.

4. **Plectranthus candelabriformis** Launert in Mitt. bot. StSamml., Münch. 7: 300 (1968); Launert & Schreiber in F.S.W.A. 123: 24 (1969); Codd in Bothalia 11: 380 (1975). Type: S.W.A./Namibia, 16 km E. of Runtu, *Merxmüller & Giess* 1912 (M, holo.!).

Erect perennial branched herb or suffrutex up to 1 m tall; branches ascending, sparingly pubescent with longish multicellular hairs. *Leaves* petiolate; blade thin-textured, ovate, 60—150 × 35—80 (—110) mm, sparingly pubescent, under-surface with orange gland-dots, apex acute, base rounded to subcordate, margin regularly crenate-dentate; petiole 20—60 mm long. *Inflorescence* a diffusely branched terminal panicle with 1 or 2 main branches near the base; flowers in 3-flowered pedunculate cymes in the axils of persistent bracts; peduncles c. 20 mm long, pedicels 5—10 mm long. *Calyx* up to 9 mm long in fruit, subequally 5-toothed, ventricose, freely orange gland-dotted. *Corolla* violet, 6,5—7,5 mm long; tube geniculate near the base and expanding slightly towards the throat; upper lip 4 mm long, lower lip boat-shaped, 3,5 mm long. *Stamens* free to the base, 3 mm long.

Extends from Tanzania to Zambia and the extreme north of S.W.A./Namibia; on sandy soil in thickets, grassy depressions and disturbed areas. Map 74.

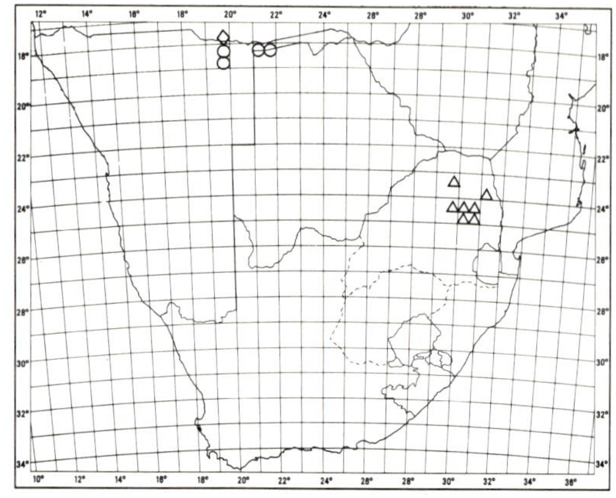

MAP 74. — △ **Plectranthus xerophilus**
◊ **P. candelabriformis**
○ **P. mirabilis**

Vouchers: Only the type specimen seen.

A distinct species with subequally 5-toothed, somewhat ventricose calyx and flowers in 3-flowered pedunculate cymes.

5. **Plectranthus mirabilis** *(Briq.)* Launert in Mitt. bot. StSamml., Münch. 7: 299 (1968); Launert & Schreiber in F.S.W.A. 123: 25 (1969); Codd in Mitt. bot. StSamml., Münch. 10: 248 (1971); in Bothalia 11: 381 (1975). Lectotype: Angola, Malange, *Mechow* 489 (Z).

Coleus mirabilis Briq. in Bot. Jb. 19: 183 (1894); Bak. in F.T.A. 5: 440 (1900); Codd in Flower. Pl. Afr. 36: t.1417 (1963). — var. *hypisodontus* Briq., l.c. (1894). *Ascocarydion mirabile* (Briq.) G. Tayl. in J. Bot., Lond. 69, Suppl. 2: 162 (1931).

— var. *mechowianus* Briq., l.c. 19: 183 (1894). Type: Angola, between Malange and Cuango Rivers, *Mechow* s.n.

— var. *poggeanus* Briq., l.c. 19: 183 (1894). Type: Upper Congo, Lulua River, *Pogge* 350.

— var. *buchnerianus* Briq., l.c. 19: 183 (1894). Syntypes: Angola, Moma, near Malange, *Buchner* 81, 82, 83, 84, 85.

C. leucophyllus Bak. in Kew Bull. 1895: 292 (1895); in F.T.A. 5: 442 (1900). Type: Malawi, Mivero, *Carson* 26.

Erect perennial woody herb or suffrutex 1—3,5 m tall; stems 1—several arising annually from the base, unbranched or sparingly branched, grey-tomentose. *Leaves* petiolate; blade fairly thick-textured, ovate to ovate-lanceolate, 60—120 × 30—60 mm, upper surface dull green, under-surface densely grey-tomentose, orange gland-dotted on both surfaces; apex acute, base obtuse to cuneate, margin regularly and finely crenate-dentate except in the lower third; petiole 10—20 mm long. *Inflorescence* a compact terminal panicle with usually 1 or 2 pairs of main branches near the base; flowers densely placed in opposite and decussate, pedunculate dichasia; bracts early deciduous and present only as an apical coma in the bud stage. *Calyx* 7—8 mm long in fruit, becoming erect and ventricose, subequally 5-toothed, glandular-hispid. *Corolla* deep blue, 13—15 mm long; tube at first narrow and ascending then sharply recurved about the middle and expanding to the throat; upper lip 4 mm long, lower lip boat-shaped, 8—9 mm long. *Stamens* united at the base, 7—9 mm long.

Found in Zaire, Malawi, Zambia, Angola and northern S.W.A./Namibia; in peaty soil in moist grassy depressions and on river banks. Map 74.

Vouchers: *Maguire* 1700; *Merxmüller & Giess* 2155; *Schoenfelder* 1049.

With its tall erect stems, grey-white foliage and deep blue flowers, this is one of the most striking members of the genus. In the inflorescence and floral characters it is allied to *P. hereroensis* (below).

6. Plectranthus hereroensis *Engl.* in Bot. Jb. 10: 267 (1888); Dinter in Feddes Reprium 22: 380 (1926); Taylor in J. Bot., Lond. 69, Suppl. 2: 160 (1931); Launert & Schreiber in F.S.W.A. 123: 25 (1969). Type: S.W.A./Namibia, Hereroland, Kaiser Wilhelmsberg near Okahandja, *Marloth* 1350 (B, holo. †; G!, GRA!; K, lecto.!; M!; PRE!; SAM!).

P. matabelensis Bak. in F.T.A. 5: 417 (1900). Syntypes: Zimbabwe, Matabeleland, Shasha River, *Holub* 1403—1406 (K).

Neomuellera damarensis S. Moore in J. Bot., London 39: 265 (1901). Type: S.W.A./Namibia, Damaraland, *Een* s.n. (BM, holo.).

P. myrianthus Briq. in Bull. Herb. Boissier sér. 2, 3: 1001 (1903); Cooke in F.C. 5, 1: 271 (1910); Codd in Mitt. bot. StSamml., Münch. 10: 248 (1971). *Coleus myrianthus* (Briq.) Brenan in Mem. N.Y. bot. Gdn 9: 43 (1954). Type: Transvaal, Witwatersrand, *Hutton* 877 (Z, holo.!; GRA!; K!; NH!).

P. otaviensis Dinter in Feddes Reprium (Beih.) 53: 116, 117 (1928), nomen subnudum based on *Dinter* 5699 from Otavi (B!; PRE!; SAM!).

P. aurifer Dinter ex Launert in Mitt. bot. StSamml., Münch. 2: 312 (1975); Dinter in Feddes Reprium (Beih.) 53: 117 (1928), nomen subnudum. Type: S.W.A./Namibia, Nossib, *Dinter* 7367 (M, holo.!; K!).

Erect annual or weakly perennial herb up to 1 m tall; stem usually solitary, branching above. *Leaves* petiolate; blade thin to medium-textured, ovate to ovate-triangular, 40—70 (—90) × 35—60 (—70) mm, subglabrous to finely pubescent, under-surface with reddish to brownish gland-dots, apex acute, base truncate, margin finely to coarsely crenate-dentate; petiole 20—70 mm long. *Inflorescence* terminal, paniculate, lax or dense; flowers arranged in opposite and decussate, pedunculate, lax or dense dichasia; peduncle 8—20 mm long; bracts persistent. *Calyx* 5—7 mm long in fruit, becoming erect and ventricose, subequally 5-toothed. *Corolla* pale to deep blue (rarely white), 6—7 mm long; tube at first narrow and ascending then sharply decurved about the middle and expanding to the throat; upper lip 2 mm long, lower lip deeply boat-shaped, 4—6 mm long. *Stamens* free or shortly united at the base, 4—5 mm long.

Found in northern S.W.A./Namibia, Botswana and central and northern Transvaal, extending to Angola, Zambia and Zimbabwe; on south-facing and wooded hillsides at medium to high altitudes. Map 75.

Vouchers: *Codd* 4193; 6555; 8629; *Giess* 3686; 10370; 15151.

There is a good deal of variation in size and pubescence of the leaves and the size and number of leaf-margin teeth. An interesting feature is that the stamens are sometimes free and sometimes united at the base, a distinction which, in the past, was used to separate the genus *Coleus* from *Plectranthus*. It is probable that certain species described from Zimbabwe may prove to be synonyms.

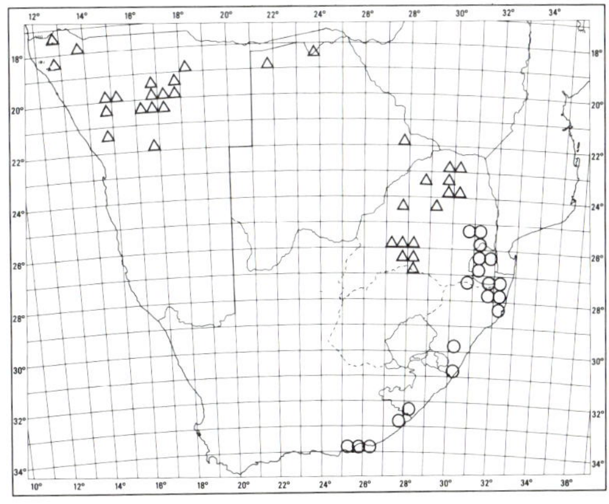

MAP 75. — △ **Plectranthus hereroensis**
   ○ **P. spicatus**

7. **Plectranthus spicatus** *E. Mey. ex Benth.* in E. Mey., Comm. 230 (1837); *Drège,* Zwei Pfl. Doc. 133, 141 (1843); Benth. in DC., Prodr. 12: 60 (1848); Cooke in F.C. 5,1: 270 (1910); Codd in Mitt. bot. StSamml., Münch. 10: 248 (1971); Ross, Fl. Natal 305 (1972); Codd in Bothalia 11: 383 (1975); Compton, Fl. Swaziland 504 (1976). Lectotype: Cape, "Glen-filling", *Drège* 4731b in Herb. Benth. (K, lecto.!; MO!; P!; S!).

P. subspicatus Hochst. in Flora 28: 67 (1845). Type: Cape. Uitenhage, *Krauss* 1112.

Burnatastrum spicatum (E. Mey. ex Benth.) Briq. in Natürl. PflFam. 4,3a: 358 (1897).

Perennial succulent plant; stems several from the base, decumbent with inflorescences ascending up to 0,6 m, finely tomentose to subglabrous. *Leaves* subsessile to shortly petiolate; blade fleshy, drying fairly thick-textured, obovate, 25—50 × 8—25 mm, subglabrous to finely pubescent, veins indistinct, under-surface with red gland-dots; apex obtuse, base cuneate, margin with a few irregular teeth mainly in the upper half. *Inflorescence* subspicate, simple or occasionally with a pair of branches near the base, 90—300 mm long; flowers in opposite and decussate, compact, several- to many-flowered dichasia spaced 5—20 mm apart; bracts shed at early flowering stage. *Calyx* 5 mm long in fruit, slightly ventricose, circinnate with the mouth erect, subequally 5-toothed. *Corolla* purple, 7—8 mm long; tube at first narrow and ascending then sharply decurved about the middle and expanding towards the throat; upper lip 2,5 mm long, lower lip boat-shaped, 2,5—3 mm long. *Stamens* free to the base, 2,5—3 mm long.

Found in the eastern Transvaal lowveld, eastern Swaziland and the coastal areas of Natal and eastern Cape as far south as Humansdorp; in dry woodland often associated with other succulent plants, in rocky places or brackish flats. Map 75.

Vouchers: *Codd* 6500; 9630; *Hilliard & Burtt* 10315; *Pegler* 2026.

The calyx changes in shape as it matures. In the flowering stage it is horizontal with the uppermost tooth larger than the rest. As it becomes older the tube curves upwards, becoming swollen at the base, and all five teeth are erect with the uppermost only slightly larger than the rest.

8. **Plectranthus cylindraceus** *Hochst. ex Benth.* in DC., Prodr. 12: 60 (1848); A. Rich., Tent. Fl. Abyss. 2: 182 (1851); Briq. in Natürl. PflFam. 4, 3a: 354 (1897); Bak. in F.T.A. 5: 414 (1900); Andrews, Flow. Pl. Sudan 3: 223 (1956); Launert & Schreiber in F.S.W.A. 123: 24 (1969); Codd in Mitt. bot. StSamml., Münch. 10: 248 (1971); Ross, Fl. Natal 305 (1972); Agnew, Upland Kenya Wild Flow. 635 (1974); Codd in Bothalia 11: 385 (1975); Compton, Fl. Swaziland 502 (1976). Type: Ethiopia, Samen, near Gapdia, *Schimper* 113 (K, holo.!; BM!; G!; P!).

P. marrubioides Hochst. ex Benth. in DC., Prodr. 12: 60 (1848); A. Rich., l.c. 181 (1851); Briq., l.c. 354 (1897); Bak., l.c. 414 (1900). Type: Ethiopia, Samen, near Jaja, *Schimper* 1925 (K, holo.!; BM!; G!; P!).

P. moschosmoides Bak., l.c. 414 (1900). Type: Angola, Huila, *Welwitsch* 5489 (K, holo.!; BM!).

Germanea cylindracea (Hochst. ex Benth.) Hiern, Cat. Afr. Pl. Welw. 1,4: 861 (1900).

P. villosus T. Cooke in Kew Bull. 1909: 378 (1900); in F.C. 5,1: 275 (1910). *P. glomeratus* R. A. Dyer in Flower. Pl. S. Afr. 24: sub. t.946 (1944), nom. superfl. Type: Natal, Entumeni, *Medley Wood* 3955 (K, holo.!; NH!).

P. densiflorus T. Cooke, l.c., 378 (1909); in F.C. 5,1: 276 (1910). Type: Natal, near the Mooi River, *Medley Wood* 4475 (K, holo.!; GRA!; NH!; SAM!.).

P. spiciformis R. A. Dyer in Flower. Pl. S. Afr. 24: t.946 (1944). Type: Transvaal Hammanskraal, *Mogg* sub PRE 27138 (PRE, holo.!).

Perennial succulent plant forming a dense cluster of basal leaves from which arise annually several to many stems; stems erect or decumbent, 0,6—1,5 m long, finely pubescent. *Leaves* sessile or shortly petiolate; blade fleshy, drying thin or thick-textured, broadly obovate to oblong-obovate, 25—50 × 15—40 mm, tomentulose on both surfaces, under-surface with pale to yellow gland-dots; apex obtuse, base cuneate, margin with few, irregular teeth. *Inflorescence* terminal, subspicate, dense or interrupted, 80—350 mm long, usually with 1—several pairs of branches; flowers in dense, villous, many-flowered, opposite and decussate, subsessile dichasia. *Calyx* 3 mm long in fruit, slightly curved upwards, villous, 5-toothed, the uppermost tooth distinctly larger than the rest. *Corolla* pale mauve and white, whitish or occasionally yellowish, 4—5 mm long; tube straight, expanding slightly towards the throat; upper lip 1 mm long, lower lip concave, 2 mm long, both lips pubescent. *Stamens* free to the base, 2,5—4 mm long.

Widespread in Africa from Ethiopia southwards to Angola, northern S.W.A./Namibia, Botswana, Transvaal (excluding the Highveld), eastern Swaziland, Natal midlands and coast to the Umtamvuna River; often growing communally in dry woodland under thorn trees and on brackish soil, as well as in crevices on rocky slopes. Map 76.

Vouchers: *Codd* 6037; 8365; 8679; *De Winter* 2882; *Galpin* M288.

The flowers are among the smallest in the genus. The structure of the glomerate flower clusters is difficult to discern because of the dense covering of hairs but is essentially the same as that of *P. spicatus* (above); however, the calyx is not markedly circinnate so that this species is intermediate between *P. spicatus* (i.e. the genus *Burnatastrum* Briq.) and the more conventional *Plectranthus* spp.

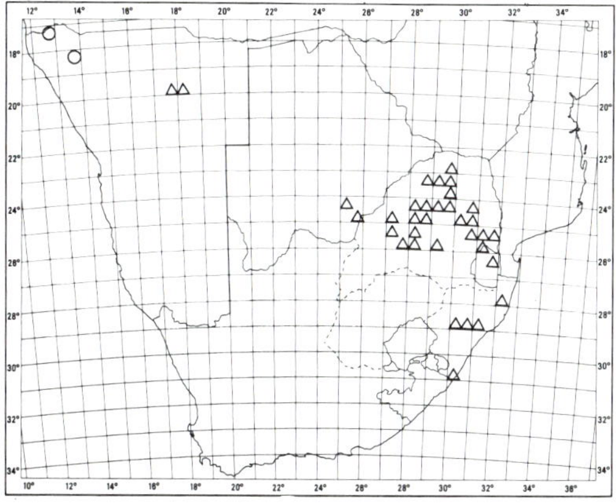

Mᴀᴘ 76. — △ **Plectanthus cylindraceus**
　　　　　　○ **P. unguentarius**

9. **Plectranthus unguentarius** *Codd* in Bothalia 11: 387 (1975). Type: S.W.A./ Namibia, Kaokoveld, 17 km S. of Kaoko Otavi, *De Winter & Leistner* 5595 (PRE, holo.!).

P. amboinicus sensu Launert & Schreiber in F.S.W.A. 123:24 (1969).

Perennial erect semi-succulent suffrutex 1−1,5 m tall; stems woody at the base, sparingly branched, densely canotomentose. *Leaves* petiolate; blade semifleshy, drying fairly thick-textured, broadly obovate to subrotund, 40−60 × 40−70 mm, densely pubescent, under-surface with reddish brown gland-dots, apex rounded, base cuneate to abruptly attenuate, margin crenate. *Inflorescence* terminal, spike-like, up to 350 mm long, simple or with a pair of branches near the base; flowers in very dense ± 20-flowered cymes, densely tomentose, producing ± 40-flowered verticils, crowded towards the apex and spaced 10−30 mm apart lower down. *Calyx* 5−6 mm long in fruit, glandular-tomentose; uppermost tooth much larger than the remaining 4 teeth, ovate-oblong, abruptly acute at the apex. *Corolla* white, 10−12 mm long; tube slightly bent about the middle and expanding to the throat; upper lip 1,5 mm long. *Stamens* united at the base for 1−2 mm, 5−7 mm long.

Known only from the Kaokoveld in northern S.W.A./Namibia, in dry Mopane woodland on high rocky situations. Map 76.

Vouchers: In addition to the type, only one specimen seen, *Davies, Thompson & Miller* 88.

The roots are pleasantly aromatic and are used in the preparation of a pomade by the local inhabitants.

The species is related to *P. amboinicus* (below) but is more robust, with erect stems, more densely-flowered verticils and larger flowers.

10. **Plectranthus amboinicus** *(Lour.) Spreng.*, Syst. Veg. 2: 690 (1825), as 'Amboinensis'; Launert in Mitt. bot. StSamml., Münch. 7: 298 (1969); Ross, Fl. Natal 305 (1972); Codd in Bothalia 11: 388 (1975). Type: from Amboina, Moluccas (only a few scarcely recognizable fragments exist in BM, but the description is detailed and adequate; Launert, l.c., has designated a specimen from Siam, *Kerr* s.n. in BM, as being representative of the species).

Coleus amboinicus Lour., Fl. Cochin. 372 (1790); Briq. in Natürl. PflFam. 4,3a: 359 (1897); Merrill in Addisonia 20: 11 (1937); Codd in Mitt. bot. StSamml., Münch. 10: 248 (1971); Compton, Fl. Swaziland 505 (1976). *Majana amboinica* (Lour.) Kuntze, Rev. Gen. 2: 524 (1891).

C. aromaticus Benth. in Wall., Pl. As. Rar. 2: 15 (1831); Lindl. in Bot. Reg. 18: t.1520 (1832); Benth., Lab. 51 (1832); in DC., Prodr. 12: 72 (1848); Hook. f., Fl. Brit. Ind. 4: 625 (1885); Trimen, Handb. Fl. Ceylon 3: 374 (1895). *Plectranthus aromaticus* Roxb., Hort. Beng. 45 (1814), nom. nud. *P. aromaticus* (Benth.) Roxb., Fl. Ind. edn 2,3: 22 (1832). Type: India, Patna, *Buchanan-Hamilton* (in Herb. Wallich, K, holo.!).

C. crassifolius Benth. in Wall., Pl. As. Rar. 2: 15 (1831); Lab. 52 (1832). Type: India, *Wight* (in Herb. Wallich, K, holo.!).

C. amboinicus var. *violaceus* Gürke in Bot. Jb. 19: 210 (1894); Bak. in F.T.A. 5: 434 (1900); Hiern, Cat. Afr. Pl. Wclw. 1,4: 865 (1900). Type: Tanzania, Lake Chala, *Volkens* 321 (BM!; K!).

Perennial, succulent, many-stemmed herb; stems decumbent, up to 1,5 m long, pubescent, with ascending inflorescences. *Leaves* petiolate; blade fleshy, drying thick-textured, broadly ovate to ovate-deltoid, 25−45 × 25−40 mm, densely pubescent, both surfaces with pale to brownish gland-dots, apex obtuse to rounded, base truncate to abruptly attenuate, margin finely crenate; petiole 4−10 mm long. *Inflorescence* slender, spike-like, 100−300 mm long; flowers in densely glomerate verticils spaced 10−30 mm apart; bracts persistent to flowering stage. *Calyx* 5−6 mm long in fruit, glandular-villous; uppermost tooth much larger than the rest, oblong to broadly oblong, abruptly apiculate. *Corolla* lilac, mauve or whitish, 7−9 mm long; tube slightly bent about the middle and expanding to the throat; upper lip 1,5−2 mm long, lower lip boat-shaped, 4 mm long. *Stamens* united at the base for 1−2 mm, 4−5 mm long.

Occurs naturally from Kenya southwards to Angola in the west and, in the east, to Mozambique, Swaziland and northern Natal; at low altitudes in woodland or coastal bush, on rocky slopes and loamy or sandy flats. Map 77.

Vouchers: *Compton* 28621; 29081; *Ward* 3983.

This was one of the plants taken by the early voyagers from Africa to the Far East and is now widely cultivated in the tropics of both hemispheres. The leaves are strongly and pleasantly aromatic and are used medicinally and for flavouring food, being known as Soup Mint, French Thyme, Spanish Thyme, Country Borage and Indian Mint. According to Trimen, l.c., it is employed as a medicine, especially for cattle, and a plant was always found growing in a little box suspended from the ox-carts.

11. **Plectranthus tetensis** *(Bak.) Agnew,* Upland Kenya Wild Flow. 635 (1974); Codd in Bothalia 11: 390 (1975). Type: Mozambique, near Tete, *Kirk.* s.n. (K, holo.!).

C. decumbens Gürke in Bot. Jb. 19: 211 (1894); Bak., l.c. 431 (1900); Compton, Fl. Swaziland 505 (1976); non *Plectranthus decumbens* Hook. f. (1864). Syntypes: Kenya, Duruma district, *Hildebrandt* 230; Tanzania, Kilimanjaro, *Volkens* 237 (BR!).

Coleus tetensis Bak. in F.T.A. 5: 431 (1900).

C. vagatus E. A. Bruce in Bothalia 6: 227 (1951); Codd in Mitt. bot. StSamml., Münch. 10: 248 (1971).

Plectranthus vagatus (E. A. Bruce) Codd in Ross, Fl. Natal 305 (1971), non rite publ.

Perennial, semi-succulent, several-stemmed procumbent herb; stems pubescent, sparingly branched, up to 0,7 m long with ascending inflorescences. *Leaves* shortly petiolate; blade softly succulent, ovate to obovate, 15−25 × 12−20 mm, pubescent, under-surface with orange-red gland-dots, apex obtuse, base cuneate to obtuse, obscurely crenate-dentate. *Inflorescence* a terminal, dense spike-like raceme 50−80 mm long, not markedly 4-angled at the apex; bracts persistent, fleshy, rounded; flowers in 4−6-flowered sessile cymes, forming 8−12-flowered verticils; pedicels erect, appressed to the rhachis. *Calyx* 4−5 mm long in fruit, red gland-dotted without, densely villous inside, bilabiate, the upper lip consisting of a large broadly ovate tooth, the lower lip of 4 subequal deltoid-subulate teeth. *Corolla* mauve-purple, 15−18 mm long; tube narrow and ascending then geniculate and expanding about the middle; upper lip 3,5−4 mm long, lower lip deeply boat-shaped, 9−11 mm long. *Stamens* 9−11 mm long, united at the base for 2 mm. Fig. 27: 2.

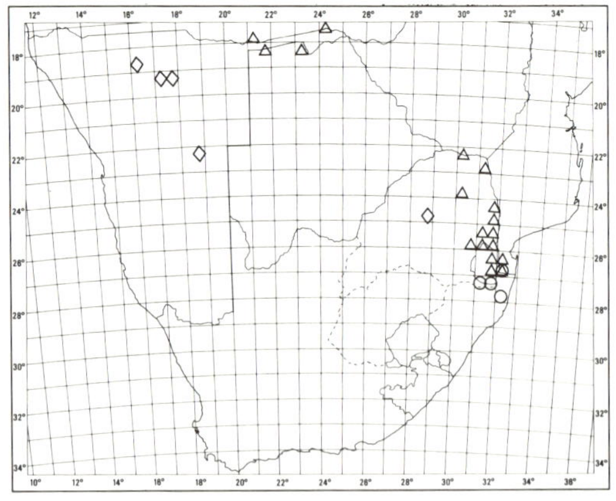

MAP 77. — ○ **Plectranthus amboinicus**
△ **P. tetensis**
◊ **P. caninus**

FIG. 27. — 1, **Plectranthus neochilus**, flowering stem, × 1; 1a, leaf, × 1; 1b, calyx, front view, × 3; 1c, section through corolla, × 2,5 (living plant, BRI garden). 2, **P. tetensis**, flowering stem, × 1; 2a, leaf, × 1; 2b, calyx, front view, × 3; 2c, section through corolla, × 3 (*Hardy* 5605).

1b
1c
2b
2a
1a
1
2
2c
R. Holcroft.

Distributed from Kenya and Tanzania through Mozambique and Zimbabwe to northern S.W.A./Namibia, Botswana, northern and eastern Transvaal, eastern Swaziland to coastal northern Natal; usually associated with dry thorn-scrub on brackish flats. Map 77.

Vouchers: *Codd* 6083; *Hardy* 5605; *Van der Schijff* 3020; 3536.

Obviously related to *P. caninus* (no. 12) and *P. neochilus* (no. 13) but distinguished by the trailing stems and fleshy persistent bracts. The leaves are not unpleasantly scented.

12. **Plectranthus caninus** *Roth*, Nov. Pl. Sp. 279 (1821); Launert & Schreiber in F.S.W.A. 123: 24 (1969); Agnew, Upland Kenya Wild Flow. 635 (1974); Codd in Bothalia 11: 390 (1975). Type: India, *Heyne* s.n. (Herb. Wallich, K!).

Coleus spicatus Benth. in Wall., Pl. As. Rar. 2: 15 (1831); Lab. 49 (1832); in DC., Prodr. 12: 71 (1848); Wight, Ic. t.1431 (1849); Hook. f., Fl. Br. India 4: 624 (1885). Type: India, *Wight* s.n. (K, holo.!).

C. caninus (Roth) Vatke in Linnaea 37: 318 (1871), excl. *Schimper* 622; Gürke in Bot. Jb. 19: 212 (1894); Briq. in Natürl. PflFam. 4: 3a: 359 (1897); Codd in Bothalia 7: 433 (1961).

C. flavovirens Gürke in Engl., Pflanzenw. Ost.-Afr. C 347 (1895). Type: East Africa, *Volkens* 1771 (BR!).

C. omahekense Dinter in Feddes Reprium (Beih.) 53: 123 (1928). Syntypes: S.W.A./Namibia, Grootfontein, Etemba, *Dinter* 3265; Otjikuara, *Dinter* 3265.

Annual or weak perennial, erect, branching, semi-succulent herb, 0,15−0,4 m tall; stems villous. *Leaves* petiolate; blade slightly fleshy, oblanceolate, obovate-oblanceolate, elliptic or ovate-lanceolate, 30−55 × 25−35 mm, sparingly pubescent, under-surface with reddish gland-dots and short gland-tipped hairs, apex acute to obtuse, base cuneate, margin subentire to obscurely few-toothed; petiole 4−20 mm long. *Inflorescence* a terminal, dense, spike-like raceme 25−90 mm long, simple or occasionally with a pair of branches near the base; bracts forming a 4-angled apical coma, early deciduous; flowers in 3−4-flowered sessile cymes, forming 6−8-flowered closely placed verticils; pedicels erect. *Calyx* 5 mm long in fruit, similar to *P. tetensis* (above). *Corolla* blue-purple, 8−10 mm long; tube slightly geniculate and expanding about the middle; upper lip 1,5 mm long, lower lip boat-shaped, 5−6 mm long. *Stamens* 5−6 mm long, united at the base for 1,5 mm.

Recorded from India and from Ethiopia through east tropical Africa to Zimbabwe, Zambia and into northern S.W.A./Namibia and northern Transvaal; often growing communally under trees in dry open woodland or on rocky outcrops. Map 77.

Vouchers: *Giess* 12554; *Mauve* 5284; *Tölken* 5420.

Closely related to *P. neochilus* (below) but has more compact (not interrupted) inflorescences and shorter corolla. The leaves are unpleasantly aromatic.

13. **Plectranthus neochilus** *Schltr.* in J. Bot., Lond. 34: 394 (1896); Cooke in F.C. 5, 1: 285 (1910); Launert & Schreiber in F.S.W.A. 123: 25 (1969); Ross, Fl. Natal 305 (1972); Codd in Bothalia 11: 392 (1975). Type: Transvaal, Barberton, Rimers Creek, *Galpin* 968 (K, holo.!; GRA!; NH!).

Coleus schinzii Gürke in Bull. Herb. Boissier 6: 555 (1898); Bak. in F.T.A. 5: 430 (1900). Type: S.W.A./Namibia, Ovamboland, Tsumeb, *Schinz* 56 (Z, holo.!).

C. pentheri Gürke in Annln naturh. Mus. Wien 20: 48 (1905); Cooke in F.C. 5,1: 289 (1910); Bruce in Hooker's Icon. Pl. 34: t.3375 (1938). Type: Cape Province, Albany District, Breakfast Vlei, *Krook* in Herb. Penther 1716 (W, holo.!; PRE!).

C. carnosus Dinter, ined.; Dinter ex Eliovson, S. Afr. Flow. for the Gdn 165 (1955), illustr. only.

C. neochilus (Schltr.) Codd in Bothalia 7: 432 (1961); Letty, Wild Flow. Transv. 288, t.143,2 (1962).

Perennial or sometimes annual, decumbent to erect, often much branched and bushy, succulent herb 0,12−0,5 m tall; stems sparingly to densely villous; roots sometimes tuberous. *Leaves* petiolate; blade succulent, often viscid, tending to fold along the midrib, obovate to elliptic-ovate, 20−50 × 15−35 mm, pubescent, under-surface with orange gland-dots, apex obtuse, base cuneate to attenuate, margin obscurely few-toothed; petiole 5−15 mm long. *Inflorescence* a terminal spike-like raceme 70−150 mm long; bracts forming a 4-angled apical coma, greenish white tipped with purple, early deciduous; flowers in 3-flowered sessile cymes, forming 6-flowered verticils; verticils dense above, laxer and 5−15 mm apart below; pedicels erect. *Calyx* 6 mm long in fruit, similar to *P. tetensis* (no. 11). *Corolla* mauve-purple, 12−20 mm long; tube slightly geniculate about the middle and expanding towards the throat; upper lip bluish white, 2 mm long; lower lip boat-shaped, 8−11 mm long. *Stamens* 8−11 mm long, united at the base for 2−3 mm. Fig 27: 1.

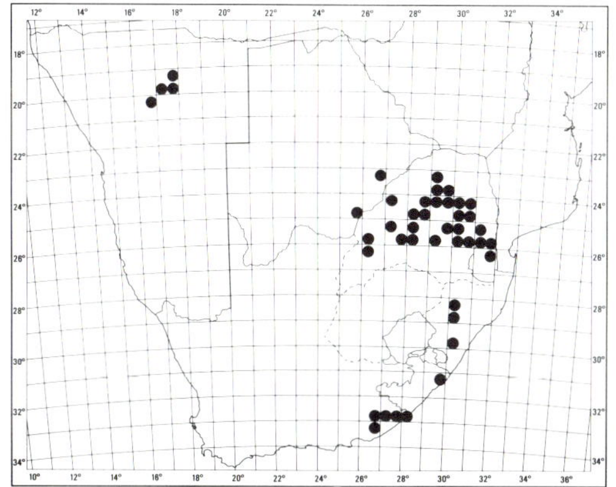

MAP 78. — **Plectranthus neochilus**

Recorded from Zambia, Zimbabwe, northern S.W.A./Namibia, Botswana, central and eastern Transvaal, Swaziland, Natal Midlands and eastern Cape as far south as Albany District; usually under trees in open woodland and among rocks (especially dolomite) in grassland. Map 78.

Vouchers: *Acocks* 9547; 11262; *Codd* 2570; 8602; 9531; *Sidey* 3670.

The species varies a good deal in vegetative characters. In S.W.A./Namibia the plants often behave as annuals with ascending stems, whereas in the eastern parts of its range the plants tend to be perennial and the stems decumbent; some plants occurring in grassy places have tuberous roots. In all forms the leaves are unpleasantly scented and the floral characters are relatively constant. The distinctions between *P. neochilus* and the closely related *P. caninus* (no. 12) and *P. ornatus* (below) are discussed in Bothalia 11: 393 (1975).

14. **Plectranthus ornatus** *Codd* in Bothalia 11: 393 (1975). Type: Ethiopia, *Schimper* II. 1328 (P!).

Coleus comosus Hochst. ex Gürke in Bot. Jb. 19: 212 (1894); Bak. in F.T.A. 5: 426 (1900); Bruce in Hooker's Icon. Pl. 34: t.3374 (1938); non *Plectranthus comosus* Sims (1822). *C. spicatus* sensu A. Rich., Tent. Fl. Abyss. 2: 183 (1851), as to syn. and spec. cited. *C. caninus* sensu Vatke in Linnaea 37: 318 (1871); sensu Engl., Hochgebirgsfl. Trop. Afr. 359 (1892).

Perennial decumbent to trailing succulent herb, branching freely at the base, up to 0,3 m tall. *Leaves* shortly petiolate; blade succulent, drying thick-textured, obovate to broadly obovate, 20−30 × 15−25 mm, sparingly to fairly densely pubescent, under-surface with orange gland-dots, strongly veined, apex rounded, base cuneate, margin finely crenate-dentate in the upper half; petiole 2−10 mm long. *Inflorescence* a terminal dense spike-like raceme

40−60 (−90) mm long; bracts forming a 4-angled apical coma, greenish white tipped with purple, early deciduous; flowers in 3-flowered sessile cymes, forming 6-flowered verticils; verticils crowded except for 1−3 shortly spaced below; pedicels erect. *Calyx* 6 mm long in fruit, similar to *P. tetensis* (no. 11). *Corolla* bluish mauve with purple mottling on the upper lip, 20−25 mm long; tube slightly geniculate and expanding towards the throat; upper lip 6 mm long, lower lip boat-shaped, 12−15 mm long, sometimes bifurcate at the apex. *Stamens* 12−14 mm long, united at the base for 3−4 mm.

Indigenous in Ethiopia to Tanzania at relatively high altitudes, growing among rocks in semi-shade. Cultivated and semi-naturalized in Southern Africa.

Voucher: *Codd* 8238.

Related to *P. neochilus* (above) but may be separated by the shorter, more compact inflorescence and the longer corolla, especially the longer upper lip of the corolla, while the lower lip is often split longitudinally at the apex. The leaves are unpleasantly scented.

15. **Plectranthus barbatus** *Andr.**, Bot. Rep. t.594 (1809); Agnew, Upland Kenya Wild Flow. 636 (1974); Codd in Bothalia 11: 394 (1975). Type: Bot. Rep. t.594, ex hort. England, seed from Ethiopia, no specimen preserved.

P. forskohlii sensu Ait. f., Hort. Kew. edn 2,3: 425 (1811); sensu Sims in Curtis's Bot. Mag. t.2036 (1819). *Coleus forskohlii* sensu Briq. in Natürl. PflFam. 4,3a: 359 (1897).

Coleus barbatus (Andr.) Benth. in Wall., Pl. As. Rar. 2: 15 (1831); Lab. 49 (1832); in DC., Prodr. 12: 71 (1848); Hook. f., Fl. Brit. Ind. 4: 625 (1885); Trimen, Handb. Fl. Ceylon 3: 373 (1895); Bak. in F.T.A. 5: 429 (1900); Bruce in Kew Bull. 1935: 322 (1935); Andrews Flow. Pl. Sudan 3: 208, t.53 (1935).

Erect, bushy, softly semi-succulent woody herb or soft shrub up to 3 m tall; stems densely woolly tomentose. *Leaves* petiolate; blade semi-succulent, ovate to broadly elliptical, 40−90 × 25−50 mm, densely woolly tomentose, under-surface copiously gland-dotted, apex obtuse to rounded, base obtuse to cuneate, margin regularly crenate-dentate; petiole 10−20 mm long. *Inflorescence* a terminal spike-like raceme 200−230 mm long, enclosed in large imbricate bracts in the bud stage, elongating

---

* Willemse in Kew Bull. 40: 96 (1985) considers that the correct name for this species is *P. comosus* Sims.

and becoming lax with age; bracts ovate, acuminate, early deciduous; flowers in 3—4-flowered sessile cymes forming 6—8-flowered verticils. *Calyx* 7 mm long in fruit, like *P. tetensis* (no. 11), glandular hispid. *Corolla* pale blue-mauve, 17—20 mm long; tube geniculate about the middle and expanding to the mouth; upper lip 3 mm long, lower lip boat-shaped 10—13 mm, long. *Stamens* 10—14 mm long, united at the base for 3 mm.

A native of India and probably introduced to East Africa at an early stage. Cultivated in various parts of the world, including Southern Africa, where it has become semi-naturalized.

Vouchers: *Codd* 6631; *Strey* 3872.

16. **Plectranthus dinteri** *Briq.* in Bull. Herb. Boissier sér. 2,3: 1070 (1903); Codd in Bothalia 11: 396 (1975). Type: S.W.A./Namibia, Hereroland, Waterberg, *Dinter* 336.

P. zatarhendi sensu Launert & Schreiber in F.S.W.A. 123: 26 (1969).

Erect to spreading, annual or perennial semi-succulent herb about 0,4 (rarely to 1 m) tall, sparingly branched; stems densely glandular-tomentose. *Leaves* petiolate; blade softly semi-succulent, ovate to broadly ovate, 30—90 × 25—70 mm, coarsely glandular-pubescent, under-surface with red gland-dots, apex acute to obtuse, base truncate, margin coarsely to deeply dentate; petiole 25—40 mm long. *Inflorescence* terminal, simple or with a pair of branches near the base, racemes 100—250 mm long; flowers in 3—6-flowered sessile cymes, forming 6—12-flowered verticils; verticils 10—25 mm apart; bracts early deciduous. *Calyx* 4 mm long in fruit, glandular-scabrid and red gland-dotted; tube declinate, gibbous at the base; uppermost tooth the largest, erect, ovate, acute, remaining 4 teeth subequal, lanceolate-subulate. *Corolla* mauve-purple, red gland-dotted, 8—10 mm long; tube geniculate about the middle and expanding to the throat; upper lip 2 mm long, lower lip boat-shaped, 4—5 mm long. *Stamens* free to the base, 4—5 mm long.

Found in the north-central part of S.W.A./Namibia, in sandy places and rock crevices, particularly on the dolomite formation. Map 79.

Vouchers: *Dinter* 2426; 5606; *Giess* 9600; 12556; *Hardy* 2130.

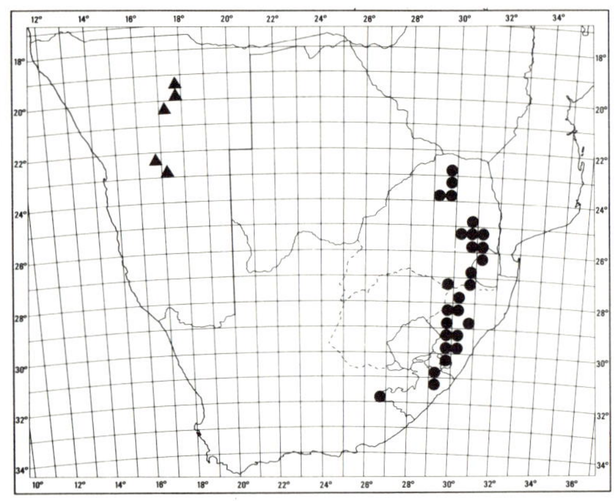

MAP 79. — ▲ **Plectranthus dinteri**
● **P. grandidentatus**

17. **Plectranthus grandidentatus** *Gürke* in Bull. Herb. Boissier 6: 554 (1898), partly; Cooke in F.C. 5,1: 278 (1910); Codd in Mitt. bot. StSamml., Münch. 10: (1971); Ross, Fl. Natal 305 (1972); Codd in Bothalia 11: 396 (1975); Compton, Fl. Swaziland 503 (1976). Lectotype: Cape, East Griqualand, Emyembe Mtn, *Tyson* sub Herb. Austr. Afr. 1517 (K, lecto.!).

Perennial semi-succulent procumbent herb; stems up to 2 m long, fairly densely to densely tomentose. *Leaves* petiolate; blade semi-succulent, ovate to broadly ovate, 20—70 × 18—75 mm usually densely pubescent, under-surface with red to brownish gland-dots, apex acute, base truncate, margin deeply dentate with 4—7 pairs of triangular teeth 3—7 mm long; petiole 15—45 mm long. *Inflorescence* terminal, often on short lateral shoots, usually simple or occasionally with 1—2 pairs of branches near the base; flowers in 3—6-flowered sessile cymes, forming 6—12-flowered verticils; verticils 5—10 mm apart. *Calyx* 4 mm long in fruit, glandular-scabrid, *Corolla* white (rarely purple), finely pubescent and gland-dotted, 7—13 mm long; tube slightly bent about the middle and expanding to the throat; lower lip boat-shaped, 4—8 mm long. *Stamens* free to the base, 5—8 mm long.

Distributed from the Soutpansberg to eastern Transvaal, Swaziland, Natal midlands and eastern

Cape to about Queenstown; in relatively dry, rocky places in open woodland. Map 79.

Vouchers: *Codd* 5933; 8621; *Dyer* 4870; *Story* 4209.

P. grandidentatus is a fairly clear-cut entity with trailing or straggly stems, floccose-tomentose stems and leaves, deeply dentate leaves, which are broadly truncate at the base, and usually white flowers. There is a good deal of variation in leaf size and in size of marginal teeth so that at one end of the scale, it approaches *P. madagascariensis* (no. 19) and, at the other end, it tends to grade into *P. hadiensis* (no. 18), possibly as a result of hybridization. The typification of the species is discussed in Bothalia 11: 397 (1975).

18. **Plectranthus hadiensis** *(Forssk.) Schweinf. ex Sprenger,* Wein. III. Gart. Zeitung 19: 2 (1894); C. Christensen in Dansk. bot. Ark. 4: 21 (1922); Wood in Kew Bull. 37: 599 (1983). Type: Yemen, Hadiyah, *Forsskål* (C, holo.!). Wood, l.c., has shown that this specimen, previously considered to be the type of *Ocimum zatarhendi* Forssk., does not agree with the description of that species, but agrees in every respect with the description of *O. hadiense* Forssk. There is circumstantial evidence that at some time early in the nineteenth century the specimen was wrongly annotated.

Perennial semi-succulent herb; stems erect to decumbent, 0,5–1,5 m tall, sparsely to densely tomentose. *Leaves* petiolate; blade medium- to thick-textured, ovate to subrotund, (35–)40–105 × (25–)30–100 mm, sparingly strigose to densely woolly-tomentose, gland-dotted, apex acute to rounded, base cuneate to subcordate, margin shallowly to fairly distinctly crenate-dentate; petiole 10–40 mm long. *Inflorescence* terminal, simple or with 1–2 pairs of branches near the base, racemes 80–500 mm long; bracts usually early deciduous but sometimes persisting to the flowering stage (in var. *tomentosus*); flowers in sessile 4–15-flowered cymes, forming 8–25-flowered verticils; verticils 10–30 mm apart; pedicels 2–4 mm long. *Calyx* 5 mm long in fruit, glandular-scabrid, shape as in *P. dinteri* (no. 16). *Corolla* usually shades of mauve to purple, rarely white, 7–13 mm long, finely pubescent and gland-dotted on the lips; tube expanding gradually from the base and bent about the middle; lower lip boat-shaped, 4–7 mm long. *Stamens* free to the base, 5–8 mm long.

According to the present concept, the species is found from the Transkei, through Natal, Swaziland, Transvaal and tropical east Africa to Somalia and the southern Arabian Peninsula, occurring at forest margins in dry woodland and among rocks in grassland.

A good deal of variation is included in the concept and 3 varieties are recognized in Southern Africa.

The varieties are keyed out in the key to species.

(a) var. **hadiensis.**

Ocimum hadiense Forssk., Fl. Aegypt.-Arab. 109 (1775). *Plectranthus forsskalaei* Vahl, Symb. Bot. 1: 44 (1790), nom. superfl. *P. hadiensis* (Forssk.) Schweinf. ex Sprenger, Wein III. Gart. Zeitung 19: 2 (1894); C. Christensen in Dansk. bot. Ark. 4: 21 (1922); Wood in Kew Bull. 37: 599 (1983). See note on typification above.

P. pachyphyllus Gürke ex T. Cooke in F.C. 5,1: 185 (1910). Type: Natal, Inchanga, *Rehmann* 7878 (Z, holo.!).

P. zatarhendi sensu E.A. Bruce in Kew Bull. 1935: 590 (1935). *P. zatarhendi* var. *zatarhendi* sensu Codd in Bothalia 11: 398 (1975).

Stems 1 – few from a perennial base, erect or decumbent, 0,3–0,6 m long, sparingly branched, densely glandular-tomentose; leaves broadly ovate, 35–80 × 20–55 mm, densely tomentose on both surfaces, shallowly crenate-dentate; inflorescence simple or occasionally with a pair of branches near the base, racemes 80–300 mm long; flowers 4–8 in the axil of each bract, bracts early deciduous.

Found in the midlands and semi-coastal Natal and mountainous parts of eastern and central Transvaal, among rocks in dry woodland or on exposed rocky places in grassland where it is subjected to periodic burning. It extends through east tropical Africa to the southern Arabian Peninsula. Map 80.

Vouchers: *Breyer* sub TRV 17783; *Galpin* 13300; *Medley Wood* 4775; *Strey* 5164.

The typical form of the species has relatively short, somewhat decumbent stems, large, densely tomentose, rather shallowly crenate-dentate leaves and a short, simple or rarely branched inflorescence with 4–8 flowers in the axil of each bract.

(b) var. **tomentosus** *(Benth.) Codd,* comb. nov.

P. tomentosus Benth. in E. Mey., Comm. 229 (1837); Drège, Zwei Pfl. Doc. 159, 160 (1843); Benth. in DC.. Prodr. 12: 67 (1848); Wood, Natal Pl. 4: t.316 (1906); Cooke in F.C. 5,1: 186 (1910), partly; Dyer in Flower Pl. S. Afr. 24: t.960 (1944); Compton, Checklist Fl Swaziland 67 (1966); Codd in Mitt. bot. StSamml. Münch. 10: 248 (1971); Ross, Fl. Natal 305 (1972). *P. zatarhendi* var. *tomentosus* (Benth.) Codd in Bothalia 11: 399 (1975); Compton, Fl. Swaziland 504 (1976) Type: Natal, Port Natal, *Drège* (K, holo.!; MO!; P!).

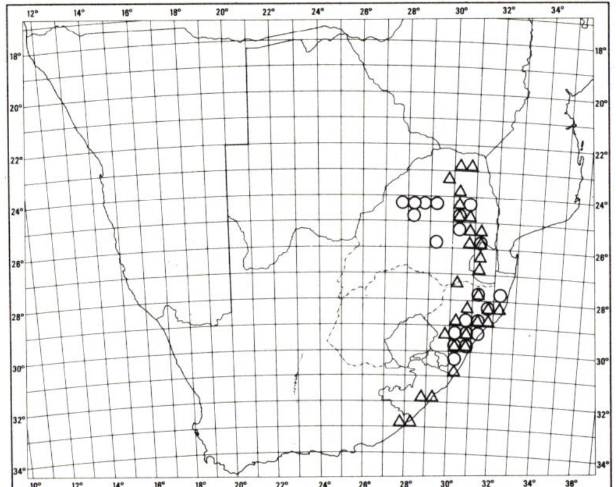

MAP 80. — ○ **Plectranthus hadiensis** var. **hadiensis**
△ **P. hadiensis** var. **tomentosus**

Stems 1 — few, densely to shaggily tomentose, when erect usually solitary, branching above, up to 1,5 m tall, or decumbent to 0,7 m long; leaves broadly ovate to subrotund, 40−100 × 32−100 mm, densely tomentose on both surfaces, shallowly to fairly distinctly crenate-dentate; inflorescence usually with 1 or 2 pairs of branches near the base, occasionally simple, racemes 150−500 mm long, flowers 5−15 in the axil of each bract, bracts sometimes persisting to the flowering stage.

Found mainly in semi-coastal areas from about the Kei River to coastal Natal, extending inland to Swaziland and the eastern Transvaal; in dry woodland and rocky grassland. Introduced into Ceylon and India, where it is cultivated to some extent. Map 80.

Vouchers: *Codd* 9589; 9613; *Compton* 27760; *Dyer* 4352; *Medley Wood* 5752.

In its typical form it is a robust erect branched plant with a large branched inflorescence of pale mauve flowers. However, plants with similar leaves and tomentum but with shorter, decumbent stems are found, particularly in Swaziland and eastern Transvaal, suggesting a gradation into var. *hadiensis*. Specimens are also seen in these areas with fairly deeply crenate-dentate leaves which cannot always be separated with certainty from *P. grandidentatus* (no. 17). It is possible that hybridization between the two occurs, which might account for occasional plants with mauve flowers being placed in *P. grandidentatus* (no. 17) and others with white flowers in *P. hadiensis* var. *tomentosus*.

(c) var. **woodii** *(Gürke) Codd*, comb. nov.

P. woodii Gürke in Bot. Jb. 26: 76 (1898) (sphalm. "Wodii"); Cooke in F.C. 5,1: 287 (1910); Codd in Mitt. bot. StSamml., Münch. 10: 248 (1971); Ross, Fl. Natal 305 (1972). *P. zatarhendi* var. *woodii* (Gürke) Codd in Bothalia 11: 401 (1975). Lectotype: Natal, Ipolweni, *Wood* s.n. (GRA, lecto.!).

P. draconis Briq. in Bull. Herb. Boissier sér. 2,3: 1071 (1903); Cooke, l.c. 288 (1910). Type: Natal, Biggarsberg, *Rehmann* 7092 (Z, holo.!).

Stems 1 − few arising annually from a perennial base, decumbent or suberect, 0,3−0,6 m long, sparingly branched, glandular-puberulous to sparsely or fairly densely strigose; leaves ovate-elliptical to broadly ovate, (30−)35−60 × 25−50 mm, shortly hispid to sparingly or fairly densely strigose, shallowly to fairly distinctly crenate-dentate; inflorescence simple or with a pair of branches near the base, racemes 100−350 mm long, flowers 3−8 in the axil of each bract, bracts early deciduous.

Found in central Natal, extending into eastern and central Transvaal, and across the southern border into Transkei and the eastern Cape Province; among rocks in thorn scrub and dry woodland. Map 81.

Vouchers: *Codd* 5939; 8596; 8597; *Strey* 4472; 6420.

The leaves are less densely pubescent and often smaller than in var. *hadiensis* and var. *tomentosus*. There is some overlapping in leaf size with *P. madagascariensis* var. *madagascariensis* (below) which, however, usually has smaller leaves, trailing stems and white flowers. The flowers of var. *woodii* are usually mauve.

19. **Plectranthus madagascariensis** *(Pers.) Benth.*, Lab. 37 (1832); in E. Mey., Comm. 230 (1837); Drège, Zwei Pfl. Doc. 153, 160 (1843); Benth. in DC., Prodr. 12: 68 (1848); Blake in Contr. Queensl. Herb. 9: 39, 110 (1971); Ross, Fl. Natal 305 (1972); Compton, Fl. Swaziland 503 (1976). Type: Mauritius or Reunion, *Commerson* (Herb. Juss., P, holo.!).

Ocimum madagascariense Pers., Syn. Pl. 2: 135 (1807). *Coleus madagascariensis* (Pers.) A. Chev. in Rev. Bot. appl. Agric. trop. 33: 338 (1953).

Perennial, often semi-succulent herb; stems procumbent, up to 1 m long (typical) or decumbent to erect, 0,3−0,45 m long (vars.), sparingly to densely and shortly tomentose, often with longer hairs and glandular hairs intermingled. *Leaves* petiolate; blade slightly succulent, drying thin to thickish in texture, ovate to subrotund, 15−30(−45) × 10−25 mm, upper surface strigose, under-surface medium to densely

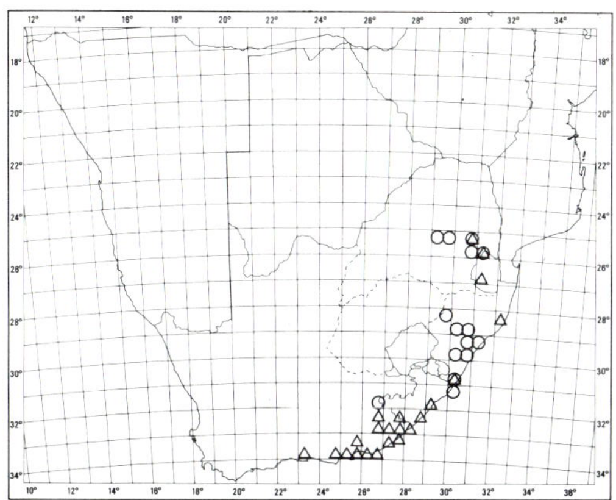

MAP 81. — ○ **Plectranthus hadiensis** var. **woodii**
△ **P. madagascariensis** var.
madagascariensis

tomentose with reddish to brown gland-dots, apex obtuse to rounded, base truncate to cuneate, margin obscurely crenate to crenate-dentate; petiole 5—35 mm long. *Inflorescence* terminal to main stem and on side branches, simple or sometimes with 1—2 pairs of branches near the base; racemes 90—250 mm long; flowers in 3—8-flowered cymes, forming 6—16-flowered verticils spaced 5—20 mm apart; bracts 3 mm long, early deciduous. *Calyx* 4—5 mm long at fruiting stage, gibbous at the base, glandular-scabrid and gland-dotted, shape as in *P. dinteri* (no. 16). *Corolla* white or mauve to purple, often reddish gland-dotted on the lips, 5—18 mm long; tube bent about the middle; lower lip boat-shaped, longer than the tube. *Stamens* free at the base, about as long as the lower corolla lip.

Found in the eastern Cape Province, Transkei, semi-coastal Natal, Swaziland and Transvaal; also in Mozambique and the Mascarenes. Grows in forest margins, dry woodland and rocky places in grassland.

Three varieties are recognized in Southern Africa and are keyed out in the key to species. In addition, a variegated form of unknown origin, with white margins to the leaves, is commonly cultivated; otherwise it has all the characteristics of var. *madagascariensis* and is not given separate taxonomic status.

(a) var. **madagascariensis.**

Codd in Bothalia 11: 403 (1975). Type: Mauritius or Reunion, *Commerson* (Herb. Juss., P, holo.!).

Ocimum tomentosum Thunb., Prodr. 2: 96 (1800); Fl. Cap. edn Schult. 448 (1823), non *Plectranthus tomentosus* Benth. Type: Cape, "Houteniquas", *Thunberg* (UPS, holo.!).

Plectranthus hirtus Benth., Lab. 38 (1832); in E. Mey., Comm. 230 (1837); Drège, Zwei Pfl. Doc. 153, 160 (1843); Benth. in DC., Prodr. 12: 68 (1848); Cooke in F.C. 5,1: 284 (1910), partly; Blake, Contr. Queensl. Herb. 9: 39 (1971); Codd in Mitt. bot. StSamml., Münch. 19: 248 (1971). Type: Cape, *Masson* (BM!).

P. mauritianus Boj., Hort. Maurit. 254 (1837). Type: *Sieber*, Fl. Maurit. exs. 152 (G!; K!; M!; P!).

Stems decumbent to procumbent, sparingly branched, rooting at the nodes, ascending at the ends producing inflorescences, fairly densely to densely tomentose; leaves drying thin to fairly thick in texture, somewhat sparsely to densely and shortly appressed tomentose; inflorescence usually simple; corolla usually white, rarely mauve or bluish, 7—18 mm long.

Found in the Cape, Transkei and Natal, in semi-coastal areas from Knysna to KwaZulu and extending into Swaziland; also in Mozambique and the Mascarenes. It grows in dry woodland and bush, among rocks or in sandy soil. Map 81.

Vouchers: *Galpin* 6466; 10829; *Pegler* 1516; *Story* 2144.

As mentioned above, there is a variegated form of unknown origin which is popular as a garden plant.

(b) var. **aliciae** *Codd* in Bothalia 11: 404 (1975). Type: Transkei, near Kentani, *Pegler* 909 (PRE, holo.!).

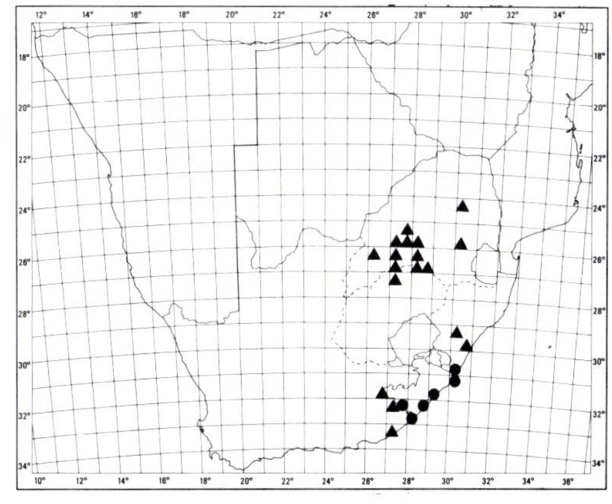

MAP 82. — ● **Plectranthus madagascariensis** var.
aliciae
▲ **P. madagascariensis** var. **ramosior**

Stems decumbent to erect, 0,2—0,4 m tall; leaves thin-textured, blade broadly ovate, 25—40 × 22—40 mm, sparingly strigose, margin shallowly crenate with 3—4 pairs of rounded teeth; inflorescence usually simple; corolla small, white to cream, 5—6 mm long.

Found in Transkei and southern Natal in semi-coastal woodland, usually in moist places. Map 82.

Vouchers: *Van Jaarsveld* 2205; 3103; 3781.

(c) var. **ramosior** *Benth.* in DC., Prodr. 12: 68 (1848); Codd in Bothalia 11: 404 (1975). Lectotype: Transvaal, "Vaal and Mooy Rivers", *Burke* (K, lecto.!; BM!).

Stems erect to decumbent, 0,2—0,35 m tall, rarely procumbent, sparingly to freely branched, fairly densely to densely tomentose; leaves medium to thick in texture, blade ovate to broadly ovate, 20—35 × 15—30 mm, rather coarsely crenate-dentate; inflorescence usually simple, several per plant; corolla mauve or bluish, rarely white, 8—12 mm long.

Concentrated in central and southern Transvaal, extending to Swaziland and the inland districts of central and northern Natal, Transkei and eastern Cape Province; usually in grass among rocks. Map 82.

Vouchers: *Codd* 8630; *Louw* 1672; *Strey* 2823; *Thode* A453.

In its typical form it is a small, erect, branched plant about 0,3 m tall with stems often arising annually from a burnt base. In the eastern Transvaal some plants may have straggly stems, which grade into *P. mutabilis* (below) while, in the eastern Cape, it is not always easy to separate it from var. *madagascariensis*. Its nearest affinity appears to be *P. hadiensis* var. *hadiensis* (no. 18a), and the two are separated mainly on size of leaf, with some specimens from Natal and the Waterberg being intermediate between the two.

20. **Plectranthus mutabilis** *Codd* in Bothalia 11: 404 (1975). Type: Transvaal, Blouberg, *Codd* 7953 (PRE, holo.!).

Perennial semi-succulent herb; stems procumbent up to 0,4 m long, sparingly branched, tomentose. *Leaves* petiolate; blade softly semi-succulent, drying fairly thin in texture, broadly ovate to subrotund, 15—50 × 15—50 mm, sparsely to densely pubescent, under-surface with yellowish gland-dots, apex obtuse to rounded, base truncate to cordate, margin deeply scalloped with few large teeth; petiole 14—30 mm long. *Inflorescence* usually simple or

with a pair of branches near the base; racemes 100—250 mm long; flowers in sessile 3—6-flowered cymes forming 6—12-flowered verticils 10—20 mm apart; bracts early deciduous. *Calyx* 4 mm long at fruiting stage, gibbous at the base. *Corolla* blue, purple-blue or lilac, 8—12 mm long; tube bent about the middle; lower lip boat-shaped, 4—5 mm long. *Stamens* free to the base, 4—6 mm long.

Found mainly on the Blouberg and Soutpansberg, extending along the eastern escarpment and inland to the Pretoria district; on rocky hillsides often in semi-shade. Map 83.

Vouchers: *Codd* 8340; 8692; *Rodin* 4011; *Strey & Schlieben* 8473.

A variable species with trailing stems, leaves with few large rounded teeth, and blue to blue-purple corolla.

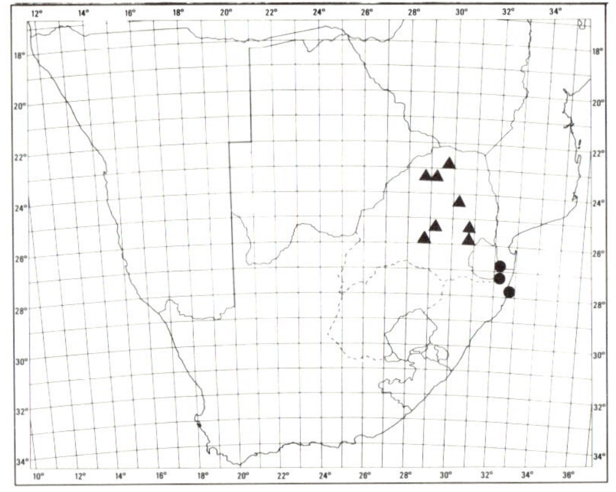

MAP 83. — ▲ **Plectranthus mutabilis**
● **P. psammophilus**

21. **Plectranthus psammophilus** *Codd* in Bothalia 11: 405 (1975). Type: Natal, Makatini Flats, *Strey* 5779 (PRE, holo.!).

Perennial semi-succulent herb; stems slender, branching, decumbent to procumbent, up to 0,5 m long, glandular-hirsute. *Leaves* petiolate; blade soft, drying fairly thin in texture, ovate-triangular, 20—40 × 18—40 mm, sparingly pubescent, under-surface with reddish brown gland-dots, apex obtuse, base truncate, margin obscurely crenate-dentate; petiole 10—20 mm long. *Inflorescence* simple or with a pair of branches near the base; racemes slender, 100−200 mm long; flowers in sessile 3−6-flowered cymes, forming 6−12-flowered verticils 1−4 mm apart; bracts often persisting to the flowering stage. *Calyx* 3 mm long at fruiting stage, gibbous at the base. *Corolla* blue-mauve, 5 mm long; tube nearly straight, expanding near the base; lower lip boat-shaped, 2,5 mm long. *Stamens* free to the base, 2,5−3 mm long.

Recorded only from northern KwaZulu; in coastal woodland on sandy flats. Map 83.

Vouchers: *Vahrmeijer & Dryfhout* 1961; *Ward* 3100.

Allied to *P. madagascariensis* var. *madagascariensis* (no. 19a) but the inflorescence tends to be denser and the flowers smaller, blue-mauve in colour, not white.

22. **Plectranthus verticillatus** *(L.f.) Druce* in Rep. botl Soc. Exch. Club Br. Isl. 1916: 640 (1917); Ross, Fl. Natal 305 (1972); Codd in Bothalia 11: 407 (1975); Compton, Fl. Swaziland 504 (1976). Type: erroneously recorded as coming from India but probably a *Thunberg* specimen from the Cape (LINN 749.4, iso.).

Ocimum verticillatum L. f., Suppl. 276 (1781), as "Ocymum"; Willd., Sp. Pl. 3: 163 (1800). *P. thunbergii* Benth., Lab. 37 (1832); in E. Mey., Comm. 229 (1837); Drège, Zwei Pfl. Doc. 125, 147 (1843); Benth. in DC., Prodr. 12: 67 (1848); Schinz in Mém. Herb. Boissier 10: 60 (1895); Cooke in F.C. 5,1: 280 (1910); Codd in Mitt. bot. StSamml., Münch. 10: 247 (1971); nom. illegit. Type: as above.

O. racemosum Thunb., Prodr. 2: 96 (1800), as "Ocymum"; Fl. Cap. edn Schult. 448 (1812). Type: Cape, "Houteniquas", *Thunberg* s.n. (UPS, holo.!; SBT!).

P. nummularius Briq. in Bull. Herb. Boissier sér. 2, 3: 1072 (1903); Cooke, l.c. 284 (1910), partly; Letty, Wild Flow. Transv. 289, t.144: 2 (1962); Compton, Check-list Fl. Swaziland 67, 158 (1966); Codd, l.c. 247 (1971); Ross, l.c. 305 (1972). Type: Natal, Camperdown, *Rehmann* 7702 (Z, holo.!).

Perennial semi-succulent herb; stems procumbent to ascending, branching, up to 1,2 m long, often rising to 0,25 m above the ground, glabrous to shortly pubescent. *Leaves* petiolate; blade softly to distinctly succulent, ovate to rotund, 16−40 × 12−40 mm, subglabrous to pubescent, undersurface with red to brownish gland-dots, apex acute to rounded, base truncate to cuneate, margin crenate-dentate to shallowly crenate with 3−6 pairs of teeth; petiole 6−30 mm long. *Inflorescence* simple or with

a pair of branches near the base, racemes 40−220 mm long (usually about 100−150 mm); flowers in sessile 1−3-flowered cymes forming 2−6-flowered verticils about 6−15 mm apart; bracts persisting beyond flowering stage. *Calyx* up to 7 mm long in fruiting stage, erect, not gibbous at the base; uppermost tooth erect, ovate, acute; 4 lower teeth spreading, linear-subulate, the lower pair the longer. *Corolla* white to pale mauve with a few mauve spots on the upper lip or freely speckled with purplish spots, 9−25 mm long; tube deflexed and expanded to a slightly saccate base, scarcely narrowing to the throat; upper lip 5−8 mm long; lower lip shallowly boat-shaped, horizontal, 5−7 mm long. *Stamens* free to the base, curved in the lower lip, 5−7 mm long. Fig. 28: 3.

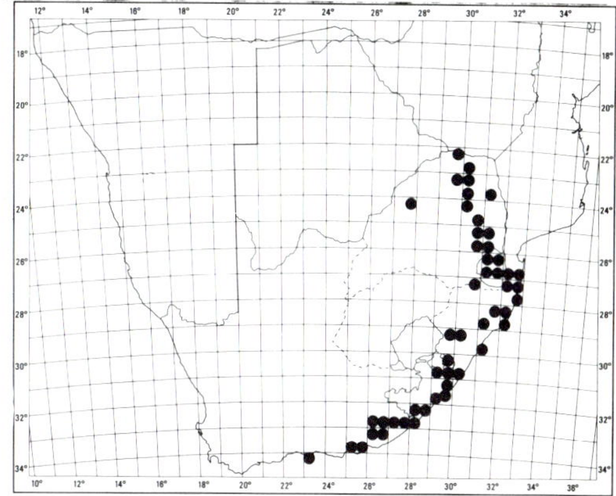

MAP 84. — **Plectranthus verticillatus**

Distributed from about Knysna through the semi-coastal parts of the eastern Cape Province, Transkei and Natal to Swaziland, eastern and northern Transvaal; also in southern Mozambique. Usually in fairly moist, stony places in forest margins, scrub forest and dry woodland. Map 84.

Vouchers: *Acocks* 9530; 10239; *Codd* ‘6100; *Flanagan* 1722; *Galpin* 10672; 10956; *Ward* 1015; 3099.

There is a good deal of variation in leaf shape and pubescence and in flower colour but it has not been possible to subdivide the material into meaningful infraspecific groups. The typical form of the eastern Cape has small ovate leaves, somewhat cuneate at the base with crenate-dentate margins, and the flowers are whitish with a few pale mauve markings on the upper lip. This grades into the Natal and Transvaal form in which the leaves are somewhat larger, glabrous or pubescent and rounded with shallowly crenate margins, and the flowers are slightly larger with freely speckled corolla. Some forms of the latter are often cultivated as ground covers or pot plants. The typification of *P. verticillatus* is discussed in Bothalia 11: 408 (1975).

In *P. verticillatus*, *P. strigosus* (no. 23) and *P. purpuratus* (no. 24) there is a red gland-dot situated between the anther cells and it is evident that these three species are closely related. In *P. verticillatus*, however, the corolla tends to be larger with the lower lip 5−7 mm long and the stamens equally long (5−7 mm), while the tube is not conspicuously narrowed near the throat as in the other two species.

23. **Plectranthus strigosus** *Benth.* in E. Mey., Comm. 229 (1837); Drège, Zwei Pfl. Doc. 153 (1843); Benth. in DC., Prodr. 12: 68 (1848); Cooke in F.C. 5,1: 280 (1910); Codd in Mitt. bot. StSamml., Münch. 10: 247 (1971); Bothalia 11: 409 (1975); Compton, Fl. Swaziland 504 (1976). Lectotype: Cape, Olifantshoek Forest (Alexandria), *Ecklon* (K, lecto.!).

P. strigosus var. *lucidus* Benth. in DC., Prodr. 12: 68 (1848); Cooke, l.c. 280 (1910). Type: Cape, Bathurst, *Burchell* 3924 (K, holo.).

P. parviflorus Gürke in Kuntze, Rev. Gen. 3,2: 261 (1898); Cooke, l.c. 281 (1910); nom. illegit. *P. kuntzeanus* Domin., Biblioth. Bot. 89: 1118 (1928). Type: Cape, East London, *Kuntze* s.n. (NY, holo.).

Perennial, semi-succulent herb; stems decumbent to ascending, up to 0,3 m long, rusty-hispid usually with multicellular purplish hairs, or greyish-strigose. *Leaves* petiolate; blade fairly thick-textured, broadly ovate to subrotund, 13−35 × 8−30 mm, strigose, under-surface often purple-tinged, with grey to rusty multicellular hairs and numerous red gland-dots, apex obtuse to rounded, base truncate to abruptly cuneate, margin obscurely crenate; petiole 5−15 mm long. *Inflorescence* usually simple; racemes 40−150 mm long; flowers in sessile 1−3-flowered cymes forming 2−6-flowered verticils 5−12 mm apart. *Calyx* up to 6 mm long in fruiting stage, sparingly strigose. *Corolla* whitish to mauve with a few darker markings on the upper lip, 6−9 mm long, tube usually narrowing distinctly between the middle and the throat; upper lip 4−5 mm long; lower lip concave, 3−4 mm long. *Stamens* free, 1,5−3 mm long. Fig. 28:1.

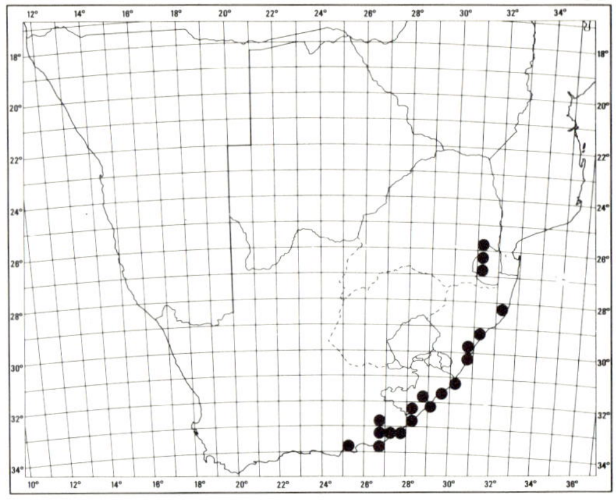

MAP 85. — **Plectranthus strigosus**

Distributed from Uitenhage through semi-coastal Transkei to the Natal border with a gap to Swaziland and the adjoining Barberton area of Transvaal; in shady rocky places and in scrub forest. Map 85.

Vouchers: *Codd* 9269; *Galpin* 278; *Pegler* 910.

In the typical form of the eastern Cape, the stems tend to be rusty-strigose but, at the other end of the distribution, in southern Natal and in Swaziland, the tomentum tends to be shorter and greyish, rather like that of *P. purpuratus* (below). The flowers of the two species are practically identical and the question arises whether subspecific status would not be more appropriate for *P. strigosus*.

24. **Plectranthus purpuratus** *Harv.*, Thes. Cap. 1: 53, t.83 (1859); Cooke in F.C. 5,1: 282 (1910); Codd in Mitt. bot. StSamml., Münch. 10: 247 (1971); Ross, Fl. Natal 305 (1972); Codd in Bothalia 11: 410 (1975). Type: ex Hort. Kew, from seed sent from Port Natal by R. Vause (K!).

Perennial succulent herb forming small mats; stems several, branched, about 0,25 m tall, densely velvety tomentulose. *Leaves* crowded, shortly petiolate; blade succulent, drying thick-textured, broadly ovate to subrotund or broadly obovate, appressed grey velvety, under-surface purple tinged, copiously red gland-dotted, apex obtuse to rounded, base obtuse to shortly cuneate, margin obscurely crenate to subentire; petiole 3−8 mm long. *Inflorescence* simple

FIG. 28. — 1, **Plectranthus strigosus,** flowering stem, × 1; 1a, mature calyx, × 4; 1b, section through corolla, × 4 (*Van Jaarsveld*, Lomati Gorge). 2, **P. purpuratus,** flowering stem, × 1; 2a, section through corolla, × 4 (*Van Jaarsveld*, Nshongweni dam). 3, **P. verticillatus,** section through corolla, × 4. 4, **P. oertendahlii,** section through corolla, × 4 (3 and 4, living plants, BRI garden).

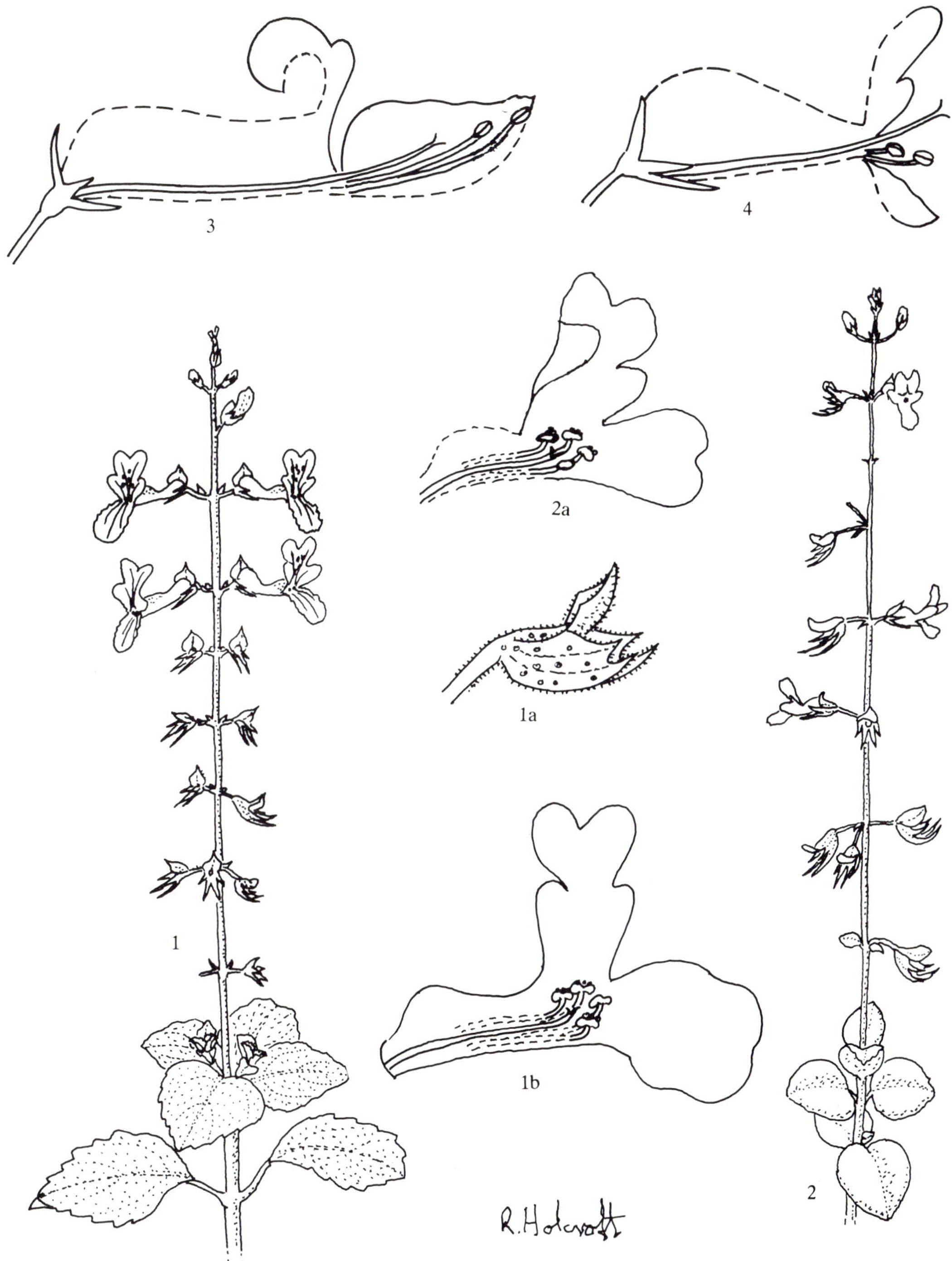

or occasionally branched near the base, racemes 40−100 mm long; flowers in sessile 1−2 (rarely 3)-flowered cymes forming 2−4-flowered verticils 4−8 mm apart. *Calyx* 5−6 mm long at fruiting stage, puberulous and freely red gland-dotted. *Corolla* white with a few blue-mauve marks, 6−9 mm long; tube narrowing about the middle; upper lip 3−4 mm long; lower lip slightly concave, 3 mm long. *Stamens* free, up to 1,5 mm long. Fig. 28:2.

Distributed from Pietermaritzburg to Pinetown; usually in moist shallow soil in rock crevices. Map 86.

Vouchers: *Eshuis* s.n.; *Killick* 504; *Strey* 5208.

See note under *P. strigosus* (above).

25. **Plectranthus oertendahlii** *Th. Fries jun.* in Acta Hort. Gothoburg. 1: 253 (1924); Codd in Bothalia 11: 411 (1975); Flower. Pl. Afr. 44: t.1729 (1977). Type: Cult. Uppsala (UPS, holo.).

Perennial semi-succulent herb, freely branched, up to 0,2 m tall; stems decumbent, rooting at the lower nodes, glandular-tomentose. *Leaves* petiolate; blade semi-succulent, broadly ovate to suborbicular, 30−40 (−45) × 25−40 (−45) mm, sparingly villous, upper surface light-veined, lower surface purple with colourless gland-dots, apex acute to obtuse, base abruptly cuneate, margin crenate-dentate, ciliate; petiole 15−40 mm long. *Inflorescence* simple or branched, racemes 70−200 mm long,

flowers in sessile, usually 3-flowered cymes, forming 6-flowered verticils 10−15 mm apart. *Calyx* up to 8 mm long in fruiting stage, glandular-hispidulous. *Corolla* whitish or suffused with pale mauve; tube 8−13 mm long, expanding and forming a saccate base 4 mm deep then narrowing gradually to about 1,75 mm at the throat; upper lip 5 mm long; lower lip concave, 4−5 mm long. *Stamens* free, 2−3 mm long. Fig. 28:4.

Recorded only from the Port Shepstone district, Natal; in wooded river valleys near the coast. Map 86.

Vouchers: *Codd* 10669; 10782; *Nicholson* 1401; *Strey* 11063.

Described originally from a cultivated plant in Sweden, said to have been introduced in the early part of this century. Some plants from the wild state may tend to have smallish leaves and the corolla tube about 8−10 mm long and may be confused with *P. verticillatus* (no. 22) but they may be distinguished by the shorter lower lip of the corolla and the very shortly exserted stamens. Also the leaves are distinctly lighter veined on the upper surface and the under-surface has colourless gland-dots, not red as in *P. verticillatus. P. oertendahlii* makes an attractive, free-flowering pot-plant.

26. **Plectranthus praetermissus** *Codd* in Flower. Pl. Afr. 45: t.1791 (1979). Type: Transkei, Port St Johns, *Stutterheim* sub PRE 57330 (PRE, holo.!).

Perennial herb, freely branched, 0,2−0,5 m tall; stems decumbent, rooting at the lower nodes, shortly pubescent. *Leaves* petiolate; blade drying thin-textured, ovate to subrotund, 40−65 × 40−60 mm, sparingly pubescent, dark green, under-surface with numerous brown to reddish gland-dots, apex obtuse to rounded, base truncate to abruptly attenuate, margin obscurely crenate-dentate; petiole 20−30 mm long. *Inflorescence* simple or sparingly branched near the base; flowers in sessile, usually 3-flowered cymes, forming 2−6-flowered verticils 10−30 mm apart. *Calyx* up to 10 mm long in fruiting stage, glandular-hispidulous. *Corolla* mauve to violet with darker blotches on the lips; tube 12−15 mm long, expanding abruptly and forming a saccate base about 4 mm deep; narrowing gradually to 1,5−2 mm at the throat; upper lip 5 mm long; lower lip concave, 4 mm long. *Stamens* free, distinctly of 2 lengths, upper pair 1,5−2 mm long, lower pair 4−5 mm long.

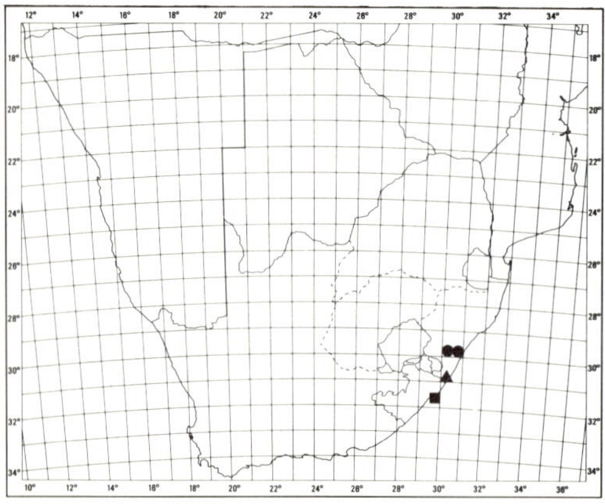

MAP 86. — ● **Plectranthus purpuratus**
▲ **P. oertendahlii**
■ **P. praetermissus**

Recorded only from Port St Johns in the Transkei; in open glades in forest. Map 86.

Voucher: *Van Jaarsveld* 3812.

27. **Plectranthus ernstii** *Codd* in Flower. Pl. Afr. 47: 1855 (1982). Type: Natal, Oribi Gorge, *Van Jaarsveld* 2196 (PRE, holo.!).

Perennial semi-succulent herb up to 0,25 m tall, branching from the base; stems thickened at the base, up to 20 mm or more in diameter, becoming brown and potato-like with age. *Leaves* petiolate; blade semi-succulent, ovate to broadly ovate, 12−30 × 10−25 mm, sparingly pubescent, under-surface often glandular-puberulous, with pale to reddish brown gland-dots, apex obtuse, base truncate, margin with few, fairly distinct teeth; petiole 6−13 mm long. *Inflorescence* simple, 30−120 mm long; flowers in sessile, 1−3-flowered cymes, forming 2−6-flowered verticils 10−20 mm apart. *Calyx* 5−6 mm long in fruiting stage, glandular-hispidulous. *Corolla* pale bluish mauve to whitish; tube 4−8 mm long, expanding abruptly and forming a saccate base 4−5 mm deep, narrowing gradually to 2 mm deep at the throat; upper lip 4−5 mm long, lower lip concave, 3−4 mm long. *Stamens* free, of two lengths, upper pair 1,5 mm long, lower pair 3 mm long.

Recorded only from Oribi Gorge, Natal, in rock crevices and steep south-facing cliffs, in humus-rich pockets of soil. Map 87.

Voucher: *Van Jaarsveld* 3876.

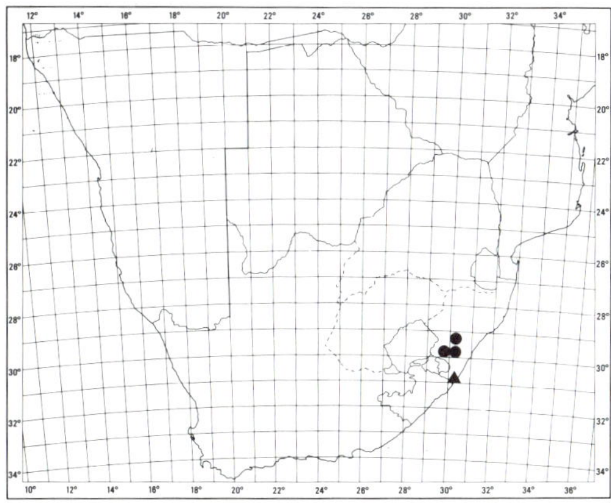

MAP 87. — ▲ **Plectranthus ernstii**
● **P. elegantulus**

The swollen, almost potato-like stem bases distinguish *P. ernstii* from all other species. The flowers are rather like those of *P. praetermissus* (above) but are smaller and pale bluish mauve in colour.

28. **Plectranthus elegantulus** *Briq.* in Bull. Herb. Boissier sér. 2,3: 1005 (1903); Cooke in F.C. 5,1: 286 (1910); Ross, Fl. Natal 305 (1972); Codd in Bothalia 11: 412 (1975). Type: Natal, Karkloof, *Rehmann* 7368 (Z, holo.!).

Perennial herb up to 0,2 m tall; stems straggling, sparingly branched, up to 0,3 m long. *Leaves* petiolate; blade thin-textured, broadly ovate, 25−40 × 20−35 mm, subglabrous to sparingly pubescent, under-surface with honey-coloured gland-dots, not suffused with purple, apex acute to obtuse, base abruptly cuneate, margin regularly crenate-dentate, finely ciliate; petiole 15−40 mm long. *Inflorescence* usually simple, rarely with a pair of branches near the base, racemes 40−120 mm long; flowers in sessile, 1−3-flowered cymes, forming 2−6-flowered verticils 10−20 mm apart. *Calyx* 7−8 mm long in fruit, glandular-puberulous with a few fringing multicellular hairs. *Corolla* whitish with a few purple spots on the lip, 7−8 mm long; tube expanding and saccate at the base, narrowing slightly towards the throat; upper lip 4−5 mm long; lower lip boat-shaped, 3 mm long. *Stamens* free, 2−2,5 mm long.

Recorded only from Natal Midlands and southern Natal; in the herb layer on forest floors. Map 87.

Vouchers: *Codd* 8582; *Marais* 827.

Related to *P. ciliatus* (below) but has smaller and less pubescent leaves, and smaller, less spotted flowers.

29. **Plectranthus ciliatus** *E. Mey. ex Benth.* in E. Mey., Comm. 227 (1837); Drège, Zwei Pfl. Doc. 150 (1843); Benth. in DC., Prodr. 12: 62 (1848); Cooke in F.C. 5,1: 275 (1910); Verdoorn in Flower. Pl. Afr. 27: t.1051 (1949); Ross, Fl. Natal 305 (1972); Codd in Bothalia 11: 414 (1975); Compton, Fl. Swaziland 502 (1976). Type: Transkei, "Omsamwubo" (Umzimvubu River), *Drège* (K, ex Herb. Benth. No. 4777, holo.!; MO!; P!; S!).

P. natalensis Gürke in Bull. Herb. Boissier 6: 552 (1898); Cooke, l.c. 283 (1910), partly, excl. *Tyson* 1793, *Wood* 558. Type: Natal, Camperdown, *Rehmann* 7701 (Z, holo.!).

Soft, branched herb up to 0,6 m tall; stems decumbent to ascending, glandular-pilose with long and short hairs having purple sap, giving a purplish colour to the stems. *Leaves* petiolate; blade thin to thickish and rugose in texture, broadly elliptic, broadly ovate or rarely subrotund, (35−) 40−80 × 30−55 mm, sparingly to freely strigose, under-surface usually suffused with purple, with honey-coloured gland-dots, apex acute to obtuse, base attenuate or abruptly cuneate, margin regularly and shallowly crenate-dentate, conspicuously ciliate; petiole 15−35 mm long. *Inflorescence* simple or with a pair of branches near the base, racemes 60−200 mm long; flowers in sessile, usually 3-flowered cymes, forming 4−6-flowered verticils 10−20 mm apart. *Calyx* 8−10 mm long in fruiting stage, hispid to glandular-puberulous, fringed with multicellular hairs. *Corolla* with whitish background freely speckled with purple, 8−14 mm long; tube expanding, saccate and slightly deflexed at the base, narrowing slightly towards the throat; upper lip 5−7 mm long; lower lip boat-shaped, 3−6 mm long, horizontal or deflexed. *Stamens* free, exceeding the lower lip.

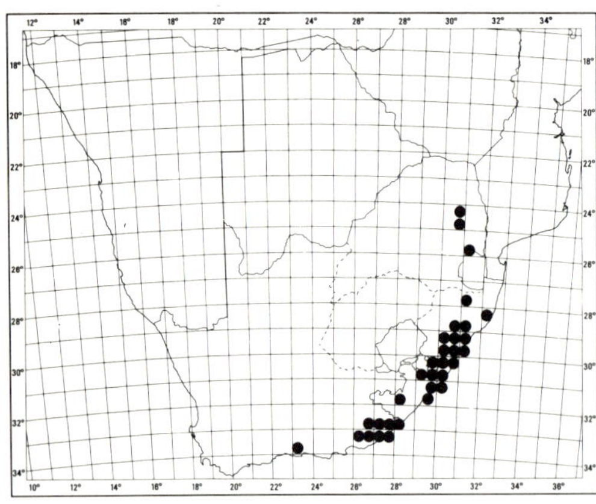

MAP 88. — **Plectranthus ciliatus**

Extending from Uniondale and Knysna in the Cape, along the semi-coastal areas of eastern Cape and Transkei to Natal, Swaziland and the mountains of eastern Transvaal; in glades in forest and in moist, shady places. Map 88.

Vouchers: *Galpin* 14743; *MacOwan* 762; *Pegler* 352; *Strey* 4938; 8085.

30. **Plectranthus fruticosus** *L'Hérit.*, Stirp. Nov. fasc. 4: 85, t.41 (March 1788); Ait., Hort. Kew. 2: 322 (1789); Willd., Sp. Pl. 3: 168 (1800); Thunb., Fl. Cap. edn Schult. 448 (1823); Benth., Lab. 32 (1832); in DC., Prodr. 12: 62 (1848); S. Moore in J. Bot., Lond. 41: 406 (1903); Cooke in F.C. 5,1: 271 (1910); Burtt in Curtis's bot. Mag. t.9616 (1940); Dyer & Bruce in Flower. Pl. Afr. 28: t.1101 (1951); Bailey, Stand. Cycl. Hort. 3: 2712 (1963); Bullock & Killick in Taxon 6: 239 (1957); Courtenay-Latimer et al., Flower. Pl. Tsitsikama t.54 (1967); Blake in Contr. Queensl. Herb. 9: 3 (1971); Codd in Mitt. bot. StSamml., Münch. 10: 247 (1971); Ross, Fl. Natal 305 (1972); Codd in Bothalia 11: 415 (1975); Compton, Fl. Swaziland 503 (1976). Type: t.41 of L'Hérit., Stirp. Nov. fasc. 4 (1788).

Germanea urticifolia Lam., Encycl. 2: 690 (April 1788); Tabl. Encycl. 3: t.514 (1819). *Plectranthus urticifolius* (Lam.) Salisb., Prodr. 88 (1796). Type: a cultivated plant as illustrated in Tabl. Encycl. 3: t.514 (1819).

P. galpinii Schltr. in J. Bot., Lond. 34: 393 (1896); Cooke, l.c. 282 (1910); Phillips in Flower. Pl. S. Afr. 8: t.294 (1928). Type: Transvaal, Barberton, Rimer's Creek, *Galpin* 939 (GRA!; NH!; PRE!).

P. arthropodus Briq. in Bull. Herb. Boissier sér. 2, 3: 1073 (1903); Cooke, l.c. 273 (1910). Type: Transvaal, Houtbosch, *Rehmann* 6151 (Z, holo.!).

P. charianthus Briq. in Bull. Herb. Boissier sér. 2,6: 824 (1906). Type: Transvaal, Houtbosch, *Rehmann* 6157 (Z, holo.!).

P. peglerae T. Cooke in Kew Bull. 1909: 378 (1909); in F.C. 5,1: 283 (1910); Bews, Plant Forms and Evol. in S. Afr. 98 (1925). Type: Transkei, Kentani, *Pegler* 377 (K, holo.!; GRA!; PRE!).

P. behrii Compton in Jl S.Afr. Bot. 11: 122 (1945); Lewis in Flower. Pl. Afr. 28: t.1109 (1951); Batten & Bokelmann, Wild Flow. E. Cape 126, t.100 (1966). Type: Transkei, Lusikisiki, *Behr* sub NBG 1252/31 (NBG, holo.!).

Soft shrub 0,6−2 m tall, freely branched; roots fibrous; branches ascending or rarely decumbent, usually purplish, sparingly pubescent to glandular-pubescent, with hairs longer and spreading at the

FIG. 29. — 1, **Plectranthus fruticosus**, inflorescence, × 1; a, leaf, × 1; b, flower, × 3; c, section through corolla, × 3; d, mature calyx, × 5 (*Van Jaarsveld* in BRI garden 26215).

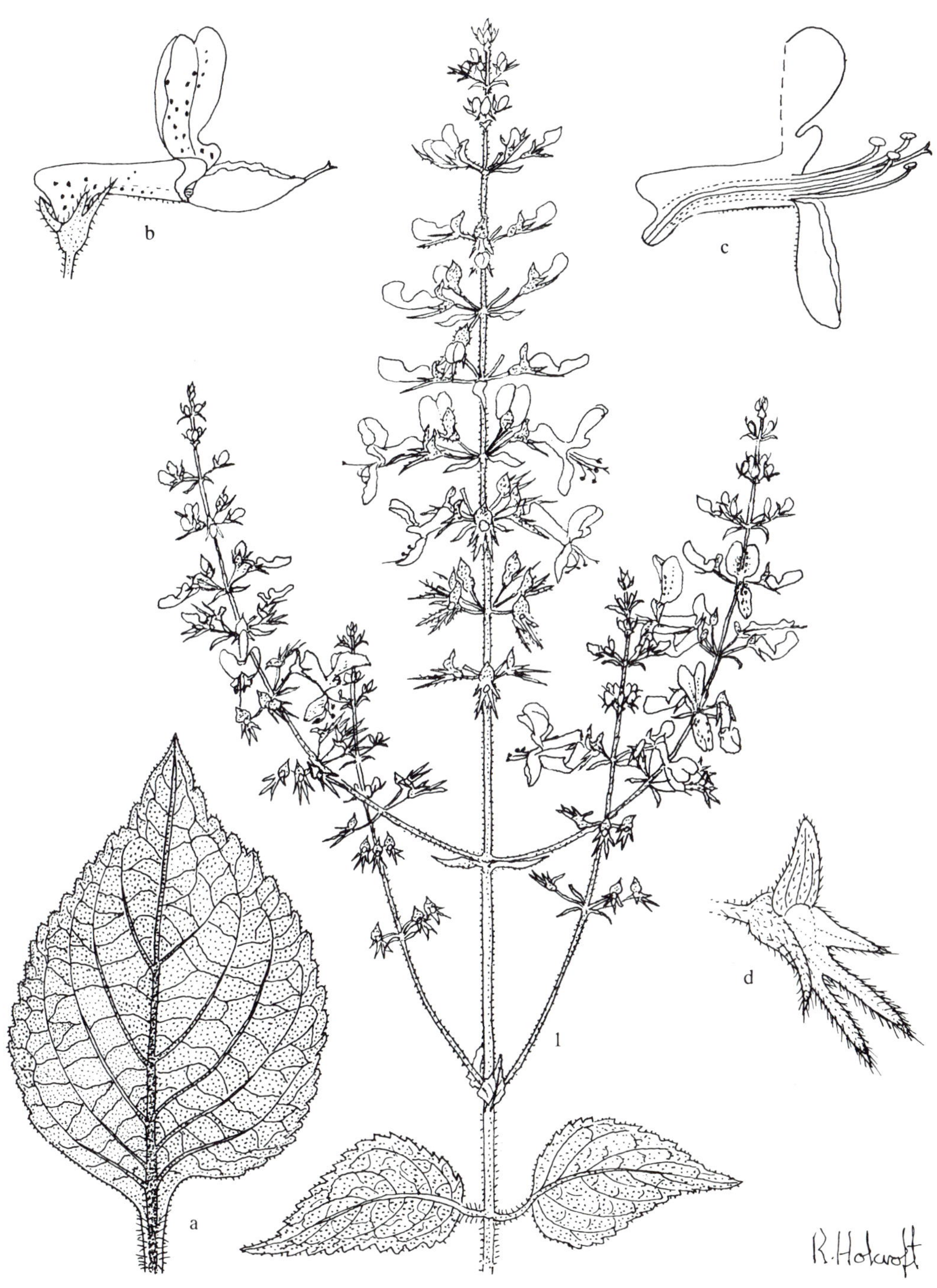

nodes. *Leaves* petiolate; blade broadly ovate to ovate-elliptic or rarely lanceolate-elliptic, 40−140 × 35−110 mm, sparingly pubescent or glandular-hispidulous, under-surface with honey-coloured gland-dots and usually suffused with purple, apex obtuse to acute, base obtuse or truncate and often abruptly attenuate, margin regularly crenate-dentate; petiole 20−50 mm long. *Inflorescence* paniculate, 80−250 mm long; flowers in sessile (1−)3-flowered cymes forming 2−6-flowered verticils 5−25 mm apart. *Calyx* 7−8 mm long in fruiting stage, glandular-hispid with scattered multicellular hairs. *Corolla* bluish mauve, rarely pink or pale blue, speckled with purple on the upper lip, 5−13 mm long; tube deflexed, saccate to distinctly spurred at the base, narrowing slightly towards the throat; upper lip 2,5−6 mm long; lower lip boat-shaped, 2−5 mm long, later deflexed. *Stamens* free, up to 8 mm long. Fig. 29.

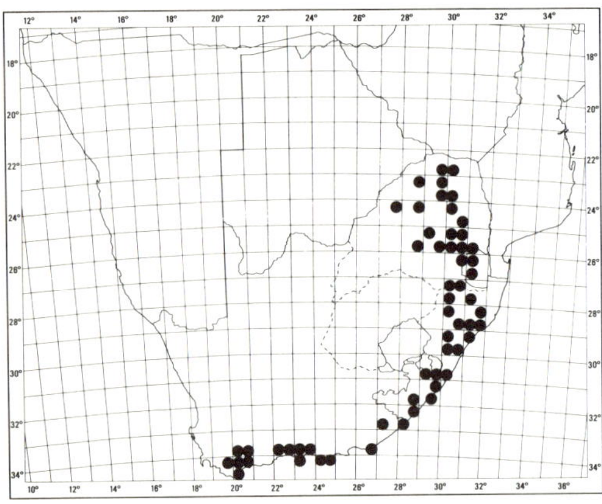

MAP 89. — **Plectranthus fruticosus**

Extending from Caledon district in south-western Cape along the semi-coastal southern and eastern Cape to the Transkei, eastern Natal, Swaziland and the mountains of eastern, central and northern Transvaal; in forest, scrub forest and shady places among rocks. Map 89.

Vouchers: *Codd* 7869; 8183; *Compton* 25785; *Dyer* 4350; *Galpin* 939; 4423; 13750; *Marloth* 2463.

One of the commonest species in Southern Africa which varies in stature and degree of pubescence according to growing conditions. See notes after *P. grallatus* (no. 32) and *P. rubropunctatus* (no. 33). Flower colour is usually pale to deep mauve with purple flecks on the upper lip. A form with pinkish flowers from the Transkei was described as *P. behrii* and is popular as a garden plant for shady places. In other respects it is not separable from *P. fruticosus*.

31. **Plectranthus oribiensis** *Codd* in Flower. Pl. Afr. 46: t.1809 (1980). Type: Natal, Oribi Gorge, *Van Jaarsveld* 2198 (PRE, holo.!).

Herb or soft shrub up to 1,5 m tall, erect, branched; roots tuberous; stems ascending, shortly and densely pubescent. *Leaves* petiolate; blade ovate-orbicular, 50−100 × 50−90 mm, densely tomentose, under-surface reticulate-veined with whitish gland-dots, apex obtuse, base cordate, margin crenate-dentate; petiole 40−70 mm long. *Inflorescence* usually with 1 or 2 pairs of branches near the base, racemes up to 200 mm long; flowers in sessile 3−5-flowered cymes forming 6−10-flowered verticils 10−25 mm apart. *Calyx* 6−7 mm long in fruiting stage, purple-tinged, glandular-hispid. *Corolla* mauve, not speckled, villous and dotted with white gland-dots on the back, 10−12 mm long; tube deflexed, expanding and spurred near the base, narrowing slightly towards the throat; upper lip 5−6 mm long; lower lip boat-shaped, 5−7 mm long, later deflexed. *Stamens* free, 2−3 mm long.

Known only from Oribi Gorge and Umtamvuna River in southern Natal; at forest margins and in wooded kloofs. Map 90.

Vouchers: *Nicholson* 1054; 1207; 1942; *Van Jaarsveld* 3875.

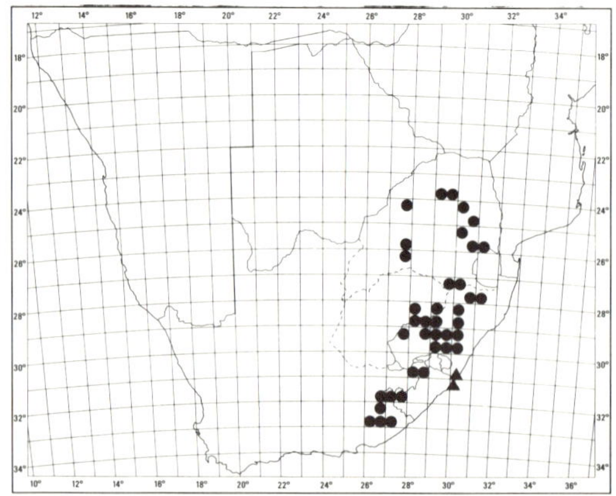

MAP 90. — ▲ **Plectranthus oribiensis**
● **P. grallatus**

The species was first collected by Mr H. B. Nicholson in 1971 and 1972. The corolla resembles that of *P. fruticosus* (above) in colour but lacks the purple spots on the upper lip. From *P. fruticosus* and *P. grallatus* (below) it differs in the villous, gland-dotted corolla and the cordate-based leaves, distinctly tomentose on the under-surface with colourless gland-dots.

32. **Plectranthus grallatus** *Briq.* in Bull. Herb. Boissier sér. 2,3: 1004 (1903); Cooke in F.C. 5,1: 287 (1910); Trauseld, Wild Flow. Drakensberg 160 (1969); Jacot Guillarmod, Fl. Lesotho 239 (1971); Ross, Fl. Natal 305 (1972); Codd in Bothalia 11: 418 (1975). Type: Transkei, Mount Frere, *Schlechter* 6415 (Z, holo.!; GRA!; PRE!).

P. transvaalensis Briq., l.c. 1005 (1903); Cooke l.c. 288 (1910); Phillips in Ann. S. Afr. Mus. 16: 241 (1917). Type: Transvaal, Houtbosch, *Rehmann* 6154 (Z, holo.!).

P. krookii Gürke ex Zahlbr. in Annln naturh. Mus. Wien 20: 48 (1905); Cooke, l.c. 274 (1910). Type: Cape, East Griqualand, *Krook* in Pl. Penther 1698 (W, holo.!; K!).

P. praetervisus Briq. in Bull. Herb. Boissier sér.2,6: 825 (1906). Type: Natal, Mt Prospect, *Rehmann* 6965 (Z, holo.!).

P. transvaalensis var. *grandifolia* T. Cooke in F.C. 5,1: 275 (1910). Type: Cape, East Griqualand, near Kokstad, *Tyson* 1793 (K, holo.!).

P. cooperi T. Cooke in Kew Bull. 1909: 377 (1909); in F.C. 5,1: 278 (1910), partly, as to *Cooper* 2982 (K!).

P. natalensis forma *glandulosa* Phillips in Ann. S. Afr. Mus. 16: 241 (1917). Syntypes: several, including Leribe, *Dieterlen* 417 (PRE!); near Witzieshoek, *Flanagan* 1927 (PRE!).

P. ciliatus and *P. fruticosus* sensu Jacot Guillarmod, Fl. Lesotho 239 (1971).

Herb 0,4—1,5 m tall with 1—3 stems arising annually from a tuberous rootstock; stems usually erect or ascending, sparingly branched, pubescent. *Leaves* petiolate; blade thin and smooth to medium-thick and somewhat rugose, broadly ovate, 50—160 × 35—140 mm, thinly pilose to fairly densely pubescent, under-surface with red to brownish gland-dots, not suffused with purple, apex acute to abruptly acuminate, base abruptly to gradually cuneate, rarely truncate, margin rather irregularly crenate-dentate, the teeth usually with small secondary teeth; petiole 20—100 mm long. *Inflorescence* paniculate, 100—260 mm long; flowers in sessile 3-flowered cymes, forming usually 6-flowered verticils 6—20 mm apart. *Calyx* 7—8 mm long in fruiting stage, glandular-hispid, usually with some fringing hairs. *Corolla* white with a flush of pink and a few spots on the upper lip, 9—13 mm long; tube slightly deflexed, expanding and saccate at the base, narrowing slightly towards the throat; upper lip 4—6 mm long; lower lip boat-shaped, 4—5 mm long. *Stamens* free, up to 7 mm long.

Found in inland areas in the eastern Cape Province and Transkei to Natal, mainly along the Drakensberg escarpment and into neighbouring parts of Lesotho and Orange Free State, extending to the higher parts of central and eastern Transvaal; in forest and scrub forest and among rocks in shady places. Map 90.

Vouchers: *Acocks* 12546; 20132; *Codd* 2419; 8513; 9546; *Galpin* 8042; 13358.

Like *P. oribiensis* (above) this species has tuberous roots. It is often confused in the herbarium with *P. fruticosus* (no. 30) which has honey-coloured gland-dots on the under-surface of the leaves, whereas *P. grallatus* has red to brownish gland-dots. There is also a difference in flower colour and, to some extent, in the toothing of the leaf margin, with *P. grallatus* having usually larger teeth with small secondary teeth often present. See also *P. rubropunctatus* (below).

There is very little overlapping in the distributions of *P. grallatus* and *P. fruticosus*. *P. grallatus* occupies the more inland and higher areas and the two meet only in the Woodbush area of north-eastern Transvaal.

33. **Plectranthus rubropunctatus** *Codd* in Bothalia 11: 420 (1975). Type: Transvaal, Nelshoogte Forestry Station, *Strey* 4081 (PRE, holo.!).

P. arthropodus sensu Compton, Fl. Swaziland 500 (1976).

Perennial herb or soft shrub; stems erect, up to 2 m tall or procumbent up to 2 m long, usually purplish, glandular-pubescent. *Leaves* petiolate; blade thin to somewhat thick-textured, broadly elliptic or broadly ovate to subrotund, 40—150 × 30—100 mm, finely to coarsely pubescent, under-surface with numerous red to brownish gland-dots, apex obtuse to rounded, base subcordate or truncate to attenuate or decurrent on the petiole, margin regularly crenate; petiole 20—80 mm long. *Inflorescence* usually paniculate, 100—250 mm long; flowers in sessile 3-flowered cymes, forming usually 6-flowered verticils 5—12 mm apart. *Calyx* 6—8 mm long in fruit, glandular-hispid with scattered long hairs. *Corolla* white, scarcely or noticeably flushed with pinkish mauve, 5,5—8 mm long; tube slightly deflexed, expanding and saccate at the base, narrowing slightly towards the throat; upper lip 2,5—4 mm long; lower lip boat-shaped 2,5—4 mm long. *Stamens* free, up to 6 mm long.

Found at relatively high altitudes of 1 200 to 2 000 m in the northern and eastern Transvaal and in Swaziland; in forest, scrub forest or in shady places among rocks or grass, extending above the forest zone. Map 91.

Vouchers: *Codd* 8143; 9478; *Compton* 26748; 29986; *Galpin* 14484; *Schlieben* 9547.

Closely related to *P. grallatus* (above) but does not have tuberous roots. The main distinction is in the leaf margin, with *P. rubropunctatus* having more shallowly crenate leaves and the teeth do not show the tendency to have smaller secondary teeth as in *P. grallatus*.

In the Mariepskop area both *P. fruticosus* (no. 30) and *P. rubropunctatus* apparently occur in exposed situations above the forest zone and herbarium specimens from this area are sometimes difficult to identify with certainty. Normally the two can be readily separated on corolla colour and on the basis of the honey-coloured gland-dots in *P. fruticosus* and the red to brownish gland-dots in *P. rubropunctatus*. In these exposed plants the leaves tend to be more rugose with denser tomentum and there appear to be intermediates between the two species.

On the Soutpansberg and Blouberg a form occurs with long trailing stems and mauve flowers which is included in *P. rubropunctatus* on the basis of its reddish gland-dots, but further investigation is necessary to determine whether it should be given separate status.

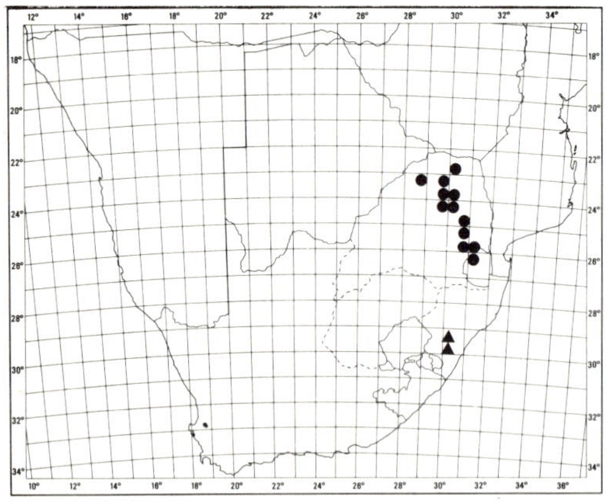

MAP 91. — ● Plectranthus rubropunctatus
▲ P. rehmannii

34. **Plectranthus rehmannii** *Gürke* in Bull, Herb. Boissier 6: 553 (1898); Cooke in F.C. 5,1: 274 (1910); Ross, Fl. Natal 305 (1972); Codd in Bothalia 11: 421 (1975).

Type: Natal, Karkloof, *Rehmann* 7359 (Z, holo.!: K!).

Erect, branched herb or subshrub 0,6—1,2 m tall; stems ascending, finely tomentose. *Leaves* petiolate; blade thickish textured, ovate to oblong-ovate, 80—140 × 50—80 mm, sparingly pubescent mainly on the nerves, under-surface with orange-brown to dark gland-dots and small subsessile glands, apex acute to acuminate, base rounded to truncate, margin finely and regularly crenate-serrate; petiole 15—60 mm long. *Inflorescence* paniculate, 250—350 mm long; flowers in few-flowered cymes consisting of a central flower and 2 short lateral branchlets in the axil of each bract, the peduncles of the lateral cymes up to 7 mm long. *Calyx* up to 9 mm long in fruiting stage, finely glandular-puberulous. *Corolla* white, about 7 mm long, covered with a tomentum of white multicellular hairs, deflexed, expanded and markedly saccate at the base, narrowing somewhat towards the throat; upper lip 2 mm long; lower lip boat-shaped, curved upwards, 4 mm long. *Stamens* free, about 2,5 mm long.

Distribution limited to the Natal Midlands where it is often locally common in forest margin scrub. Map 91.

Vouchers: *Codd* 8587; *Hilliard* 4852; *Medley Wood* 6313; 10268.

A clear-cut species with finely toothed leaf margins and small white tomentose flowers with a very short upper lip.

35. **Plectranthus swynnertonii** *S. Moore* in J. Linn. Soc., Bot. 40: 176 (1911); Codd in Bothalia 11: 422 (1975). Type: Zimbabwe, Chirinda Forest, *Swynnerton* 337 (K!).

Perennial, soft, branched herb 0,4—0,75 m tall; stems spreading-ascending, finely glandular-puberulous with a fringe of longer hairs at the nodes. *Leaves* petiolate; blade thin-textured, broadly ovate to subrotund, 50—150 × 45—140 mm, with scattered multicellular hairs, under-surface with yellowish to brownish gland-dots, often slightly sunken, apex obtuse to acute, base truncate to subcordate, margin coarsely and deeply serrate-dentate, teeth 6—16 mm long, usually with small secondary teeth; petiole 40—130 mm long. *Inflorescence* simple or with a pair of branches near the base, racemes 60—150 mm long; flowers in sessile or subsessile (1−) 3-flowered cymes, forming 2−6-flowered verticils 10−30 mm apart. *Calyx* up to 8 mm long in fruiting stage, puberulous. *Corolla* 8−10 mm long, white flushed with mauve-pink and with a few purple dots on the upper lip, fringed with white hairs; tube scarcely deflexed, expanding and saccate at the base, narrowing towards the throat; upper lip 4−6 mm long; lower lip boat-shaped, 4−5 mm long.

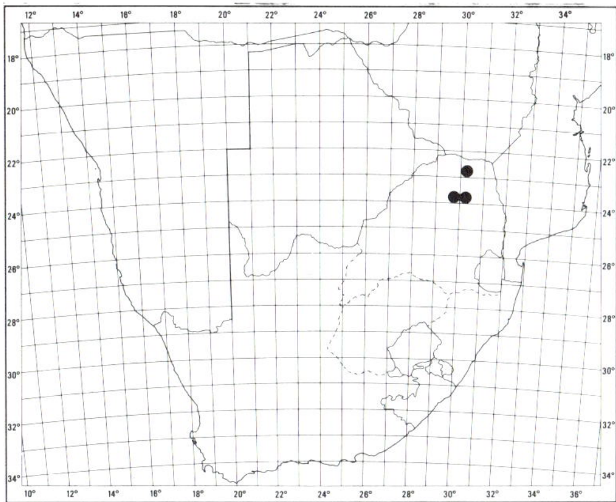

MAP 92. — **Plectranthus swynnertonii**

Found in north-eastern and northern Transvaal and eastern Zimbabwe as a ground-layer herb or soft shrub in moist humus-rich soil in mountain forests. Map 92.

Vouchers: *Codd* 8388; 9420; *Galpin* 10249; *Scheepers* 947.

Characterized by the thin, deeply toothed leaves.

36. **Plectranthus dolichopodus** *Briq.* in Bull. Herb. Boissier sér. 2,3: 1069 (1903); Cooke in F.C. 5,1: 287 (1910), partly, excluding *Flanagan* 740; Ross, Fl. Natal 305 (1972); Codd in Bothalia 11: 423 (1975). Type: Natal, Karkloof, *Rehmann* 7383 (Z, holo.!).

P. cooperi sensu Cooke in Kew Bull. 1909: 377 (1909); in F.C., 5,1: 279 (1910); partly, as to *Wood* 1843 and *Gerrard* 1673.

Erect or straggling, probably perennial herb, 0,25−1 m tall, branched; stems glandular-pilose with long multicellular hairs and gland-tipped hairs. *Leaves* petiolate; blade thin-textured, broadly ovate to

ovate-deltoid, 30−100 × 25−80 mm, sub-glabrous to thinly pubescent, under-surface with colourless gland-dots, apex acute to abruptly acute, base broadly truncate and shortly attenuate to the petiole, margin coarsely crenate-dentate; petiole 20−60 mm long. *Inflorescence* simple or paniculate, 100−200 mm long; flowers in sessile, usually 3-flowered cymes forming 2−6-flowered verticils 10−30 mm apart. *Calyx* 6 mm long in fruiting stage, sparsely glandular-puberulous. *Corolla* sky-blue to deep blue, 8−10 mm long, sparingly pubescent, tube deflexed and expanding but not markedly saccate at the base, enlarging slightly towards the throat; upper lip 1,5−2 mm long; lower lip boat-shaped, curved upwards, 4−5 mm long. *Stamens* free, 4 mm long.

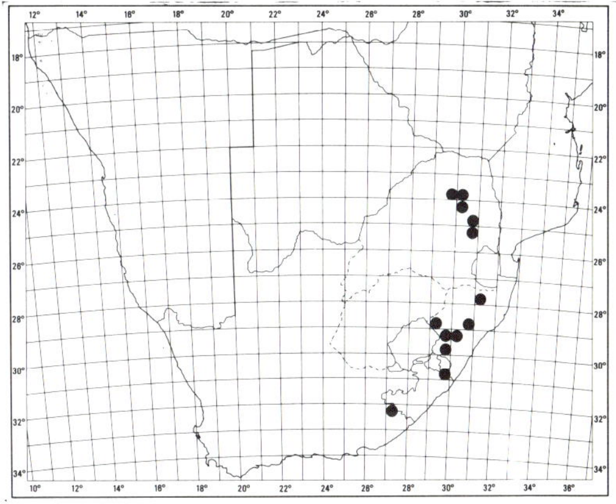

MAP 93. — **Plectranthus dolichopodus**

Known distribution somewhat disjunct, being recorded from the Stutterheim district in the Cape, the Natal Midlands and foothills of the Drakensberg, and from the Mariepskop-Woodbush area in the Transvaal; often locally common as a ground-layer herb in cool moist forests. Map 93.

Vouchers: *Codd* 7862; 8581; 9674; *Galpin* 11839; *Killick* 1676; 1965.

Characterized by the bright blue corolla which widens slightly towards the throat and has a very short upper lip.

37. **Plectranthus zuluensis** *T. Cooke* in Kew Bull. 1909: 379 (1909); in F.C. 5,1: 281 (1910); E. A. Bruce in Flower. Pl. Afr. 28: t.1110 (1951); Codd in Mitt. bot. StSamml.,

Münch. 10: 247 (1971); Ross, Fl. Natal 305 (1972); Codd in Bothalia 11: 424 (1975). Type: Natal, *Gerrard* 1675 (K, holo.!).

Erect soft shrub 1−2 m tall, freely branched; stems ascending, shortly and finely tomentose when young. *Leaves* petiolate; blade softly semi-succulent, ovate to broadly ovate, 30−70 × 25−55 mm, thinly and shortly pubescent on both surfaces, under-surface with colourless gland-dots, apex acute, base truncate to shortly cuneate, margin regularly and coarsely crenate; petiole 25−60 mm long. *Inflorescence* simple or rarely branched near the base, racemes 40−80 mm long; flowers in sessile 3-flowered cymes, forming 6-flowered verticils 5−12 mm apart. *Calyx* 7 mm long in fruiting stage, purple-tinged, glandular-pubescent. *Corolla* 10−16 mm long, pale blue-mauve to almost white with usually six rows of mauve dots on the upper lip; tube deflexed, expanding to a large saccate base and narrowing towards the throat; upper lip 5−6 mm long; lower lip concave, 5−6 mm long, soon deflexed. *Stamens* free, lower 2 fertile, 5−7 mm long, upper 2 reduced to staminodes 1−2 mm long. Fig. 30: a.

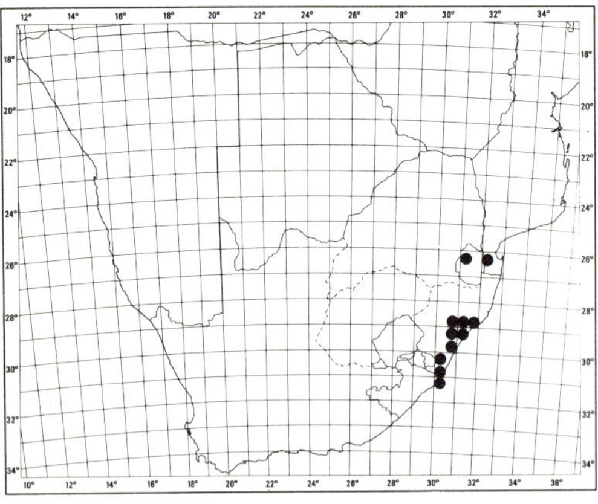

Map 94. — **Plectranthus zuluensis**

Found in semi-coastal Natal from Port Shepstone district to KwaZulu and in southern Swaziland; in forest margins, often common along streams. Map 94.

Vouchers: *Codd* 6969; 9691; *Strey* 6242.

May be recognized by the softly velvety young twigs and leaves and the medium-sized pale blue flowers which are unusual in having only 2 fertile stamens.

38. **Plectranthus saccatus** *Benth.* in E. Mey., Comm. 227 (1837). Type: Transkei, Umzimvubu River, probably near Port St Johns, *Drège* (K, ex Herb. Benth., holo.!; G!; P!; S!).

Erect to spreading soft shrub 0,5−1,2 m tall, freely branched; stems semi-succulent to somewhat woody, purple-tinged, glandular-puberulous. *Leaves* petiolate; blade herbaceous to semi-succulent, drying thin-textured, broadly ovate to ovate-deltoid, 20−70 × 15−50 mm, sub-glabrous to glandular-puberulous, under-surface with colourless gland-dots, apex acute, base truncate to obtuse or shortly cuneate, margin dentate with few, fairly large teeth; petiole 15−50 mm long. *Inflorescence* simple or occasionally branched near the base, racemes 50−120 mm long with relatively few but large flowers; flowers in sessile 1−3-flowered cymes forming 2−6-flowered verticils 10−20 mm apart. *Calyx* up to 8 mm long in fruiting stage, subglabrous to puberulous. *Corolla* mauve to pale blue or rarely white, varying in length (see vars.) from 13−30 mm; tube deflexed, enlarged and markedly saccate at the base, parallel-sided or narrowing slightly towards the throat; upper lip 10−16 mm long; lower lip boat-shaped, 5−12 mm long, horizontal or deflexed. *Stamens* free, 5−10 mm long.

Distributed from the Kentani district in the Transkei to semi-coastal Natal as far north as the Ingwavuma district; in forests or semi-shady rocky places not far from the coast.

P. saccatus may be distinguished from all other species by the large corolla, the upper lip of which is 10−16 mm long and equally broad. There is a good deal of variation in the degree of succulence of the leaves and in corolla length and colour. Two varieties are based on apparent discontinuity in the lengths of the corolla tube.

For key to varieties, see key to species.

(a) var. **saccatus.**

Codd in Bothalia 11: 427 (1975).

Plectranthus saccatus Benth. in E. Mey., Comm. 227 (1837); in DC., Prodr. 12: 62 (1848); Wood & Evans, Natal Pl. 1: t.85 (1899); Hook. f. in Curtis's bot. Mag. t. 7841 (1902); Cooke in F.C. 5,1: 273 (1910); Batten & Bokelmann, Wild Flow. E. Cape 127, t.101 (1967);

Codd in Flower. Pl. Afr. 41: t.1601 (1970); Ross, Fl. Natal 305 (1972).

The leaf blade tends to be smaller (20—50 mm long) and the corolla tube is distinctly shorter (8—16 mm) than in var. *longitubus*. Fig. 30: b.

Distribution as for the species with the exception of the Ingwavuma district. Map 95.

Vouchers: *Codd* 8574; 9296; 9351; *Galpin* 2840; 11460; *Medley Wood* 5300; 7382; 10208.

There is a good deal of variation in var. *saccatus* in degree of woodiness, in leaf texture and in flower colour. *De Winter* 8200 from Qudeni Forest has a relatively short corolla tube and the upper lip is strongly marked with purple.

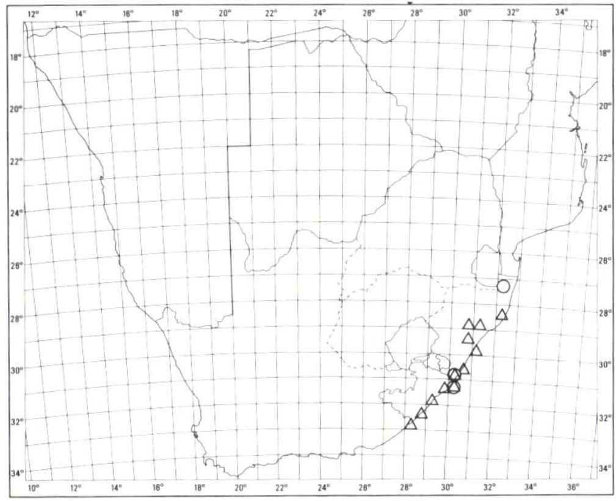

MAP 95. — △ **Plectranthus saccatus** var. **saccatus**
○ **P. saccatus** var. **longitubus**

(b) var. **longitubus** *Codd* in Bothalia 11: 428 (1975). Type: Natal, Ingwavuma District, Gwalaweni Forest, *Edwards* 2930 (PRE, holo.!).

Leaves tend to be larger, 30—70 × 25—50 mm as against 20—50 × 15—40 mm in var. *saccatus;* the corolla tube is longer (20—26 mm) but tends to be narrower (4—5 mm deep at the base as against 5—6 mm in var. *saccatus*), and the corolla lips tend to be smaller.

Recorded from the Gwalaweni Forest at the southern end of the Lebombo Range and occasionally further south.

Vouchers: *Vahrmeijer* 1913; *Vahrmeijer & Hardy* 1699.

The flower colour is normally pale blue-mauve but occasional white-flowered plants may be encountered.

39. **Plectranthus hilliardiae** *Codd* in Bothalia 11: 282 (1974); *ibid.* 11: 428 (1975). Type: Natal, near Umtamvuna River, *Hilliard & Burtt* 6767 (PRE, holo.!; NU).

Erect semi-succulent, branched perennial herb 0,3—0,4 m tall; stems ascending, shortly pilose with longer hairs at the nodes. *Leaves* petiolate; blade semi-succulent, drying membranous, broadly elliptical to obovate-elliptical, 55—90 × 40—60 mm, dark green, sparingly strigose, undersurface with colourless gland-dots, apex acute to obtuse, base cuneate, margin shallowly crenate-dentate above the middle,

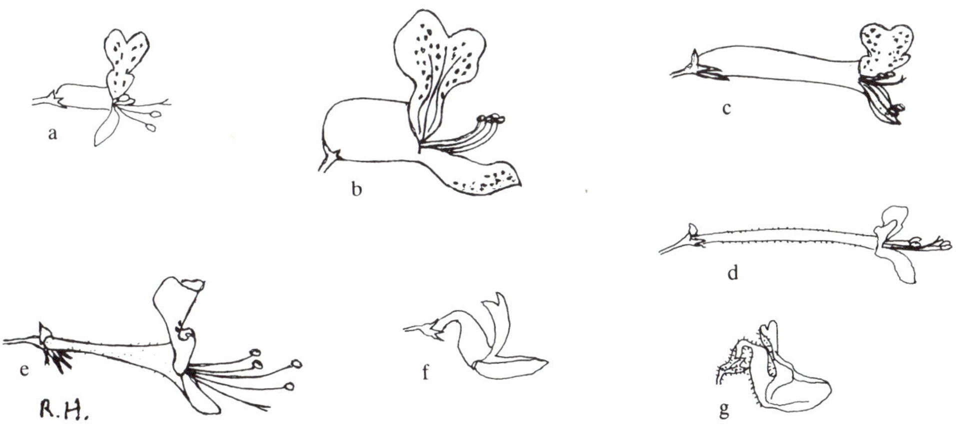

FIG. 30. — Flowers of: a, **Plectranthus zuluensis**; b, **P. saccatus** var. **saccatus**; c, **P. hilliardiae**; d, **P. ambiguus**; e, **P. ecklonii**; f, **P. petiolaris**; g, **P. laxiflorus**; all × 1.

ciliate; petiole 15—35 mm long. *Inflorescence* simple or with 1 or 2 pairs of branches near the base, racemes 80—150 mm long; flowers in sessile, usually 3-flowered cymes forming 4—6-flowered verticils 15—25 mm apart. *Calyx* 7—8 mm long in fruiting stage, glandular-hispidulous near the base. *Corolla* pale bluish with purple flecks on the lobes, 26—30 mm long; tube deflexed, expanding and saccate at the base and parallel-sided or narrowing slightly towards the throat; upper lip 5—6 mm long; lower lip concave, 4 mm long. *Stamens* free, up to 6 mm long. Fig. 30: c.

Known only from an area on each side of the Umtamvuna River in southern Natal and adjoining Transkei, not far from the coast; among rocks near and in the margins of scrub forest. Map 96.

Vouchers: *Nicholson* s.n.; *Van Jaarsveld* 3892.

The corolla is reminiscent of *P. saccatus* var. *longitubus* (above) but the leaves are larger, more fleshy with shallow toothing only in the upper half.

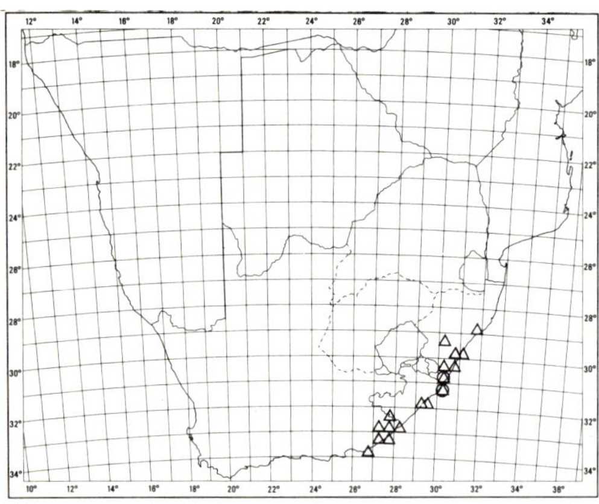

MAP 96. — ○ **Plectranthus hilliardiae**
△ **P. ambiguus**

40. **Plectranthus ambiguus** *(H. Bol.) Codd* in Bothalia 8: 159 (1964); Batten & Bokelmann, Wild Flow. E. Cape 125 (1975). Type: Cape, near Grahamstown, *MacOwan* 987 (BOL, holo.!; SAM!).

P. coloratus E. Mey. ex Benth. in E. Mey., Comm. 228 (1837), non Don (1825); Cooke in F.C. 5,1: 279 (1910), partly, excluding *Gerrard* 1671, *Wood* 3036, 3977, *Gueinzius* s.n. *P. dregei* Codd in Flower. Pl. Afr. 32: t.1244 (1957). Syntypes. Transkei, between Umgazana and Umzimvubu Rivers, *Drège a* (BM!; K!; MO!; P!; S!); Natal, between Umzimkulu and Umkomaas Rivers, *Drège b* (P!).

Orthosiphon ambiguus H. Bol. in J. Linn. Soc., Bot. 18: 394 (1881). Type: as for *P. ambiguus*.

Perennial herb or soft shrublet, branching from near the base, 0,4—1,2 m tall; stems erect or decumbent, shortly and densely to sparingly pubescent with tufts of longer hairs at the nodes. *Leaves* petiolate; blade thin to thickish and slightly rugose in texture, ovate to broadly ovate, 25—120 × 20—90 mm, subglabrous to thinly pubescent, under-surface with honey-coloured to brown gland-dots, apex obtuse to acute, base abruptly cuneate to somewhat decurrent on the petiole, margin shallowly crenate; petiole 10—70 mm long. *Inflorescence* a congested panicle, rarely simple, 40—170 mm long; flowers in sessile, (1—) 3-flowered cymes forming (2—) 6-flowered verticils 2—6 mm apart. *Calyx* up to 8 mm long in fruiting stage, glandular-hispidulous, usually suffused with purple. *Corolla* violet to purple, 23—30 mm long; tube not deflexed nor expanded at the base, straight, almost parallel-sided, increasing gradually to about 2 mm deep at the throat; upper lip 4—5 mm long; lower lip concave, 3—5 mm long. *Stamens* free, up to 6 mm long. Fig. 30: d.

Distributed from the Albany and Bathurst districts of the Cape along the semi-coastal areas of the Transkei to Ngoye Forest in Natal; in forest margins and on shady, rocky slopes. Map 96.

Vouchers: *Acocks* 13311; *Codd* 8574; 9296; 9351; *Pegler* 907.

There is considerable variation in size and texture of leaves, probably according to growing conditions. Specimens with shorter corollas have been seen which are difficult to separate with certainty from *P. ecklonii* (below), but whether this is due to hybridization between the two is not known.

41. **Plectranthus ecklonii** *Benth.* in DC., Prodr. 12: 64 (1848); Cooke in F.C. 5,1: 279 (1910); Batten & Bokelmann, Wild Flow. E. Cape 126, t.101 (1966); Ross, Fl. Natal 305 (1972); Codd in Bothalia 11: 431 (1975); in Flower. Pl. Afr. 47: t.1854 (1982). Type: Cape, Katberg, *Ecklon* s.n. (K, holo.!).

Erect soft shrub 0,7—2,5 m tall, freely branched, woody below; stems ascending, strigose, with longer hairs at the nodes. *Leaves* petiolate; blade firm-textured, often slightly rugose, ovate to oblong-elliptic,

60−170 × 40−100 mm, subglabrous to thinly pubescent, under-surface with reddish brown gland-dots, apex acute, base cuneate to rarely obtuse, margin conspicuously crenate-dentate; petiole 20−50 mm long. *Inflorescence* paniculate, 120−250 mm long; flowers in sessile, usually 3-flowered cymes forming usually 6-flowered verticils 5−15 mm apart. *Calyx* 10−11 mm long in fruiting stage, glandular-puberulous towards the base, often suffused with purple. *Corolla* pale blue or mauve to bluish purple, rarely white or pink, 16−21 mm long; tube not deflexed nor expanded at the base, straight, increasing gradually to about 3 mm deep at the throat; upper lip 5−6 mm long; lower lip concave, 4−5 mm long. *Stamens* free, up to 15 mm long. Fig. 30: e.

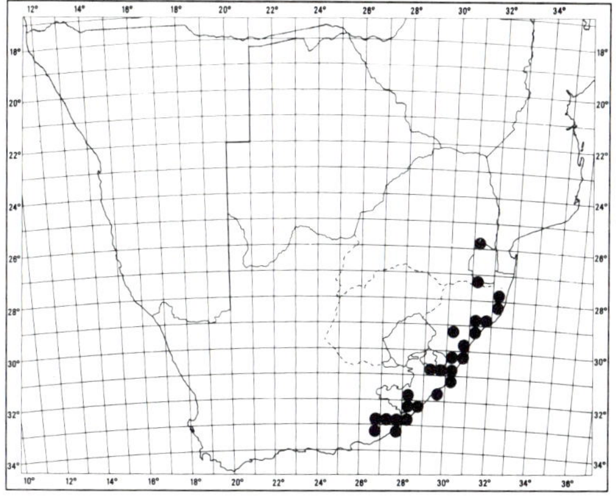

MAP 97. — **Plectranthus ecklonii**

Distributed from Somerset East and Albany districts in the Cape through coastal and midland areas of the Transkei and Natal to Barberton in Transvaal; a locally common under-storey soft shrub at forest margins or wooded stream banks. Map 97.

Vouchers: *Codd* 6973; 8578; 9246; *MacOwan* 500; *Pegler* 376; *Strey* 7350.

P. ecklonii is an attractive shrub for semi-shady places where frost is not too severe. The corolla is shorter, wider at the mouth and paler in colour than in *P. ambiguus* (above) though occasional specimens are difficult to identify with certainty. A white-flowered and a pink-flowered form are known.

42. **Plectranthus dolomiticus** *Codd* in Bothalia 15: 142 (1984). Type: Transvaal, near Penge Mine, *Van Jaarsveld* 7052 (PRE, holo.!).

Perennial semi-succulent herb up to 0,3 m tall and of equal spread; roots tuberous; stems decumbent, greyish tomentulose. *Leaves* petiolate; blade semi-succulent, broadly ovate, 20−30 × 18−30 mm, subglabrous, under-surface with colourless gland-dots, apex rounded, base truncate, margin crenate; petiole 15−30 mm long. *Inflorescence* simple or sparingly branched, 70−130 mm long; flowers in sessile, 1−3-flowered cymes forming 2−4-flowered verticils 10−25 mm apart. *Calyx* 5−6 mm long in the fruiting stage, broadly toothed, the uppermost slightly larger than the rest, minutely puberulous. *Corolla* mauve, 9−10 mm long; tube 7 mm long, somewhat sigmoid, narrow and ascending for 3 mm then decurved and expanding to about 2 mm deep at the throat; upper lip 2 mm long; lower lip concave, curved upwards, 2 mm long. *Stamens* free, 2 mm long.

Known only from the eastern Transvaal; in dry bushveld on dolomite formation, in rock crevices in full sun. Map 98.

Voucher: Only the type specimen seen.

The shape of the corolla suggests a relationship to *P. petiolaris* (below) but it differs considerably from that species in its decumbent habit, smaller leaves and smaller corolla.

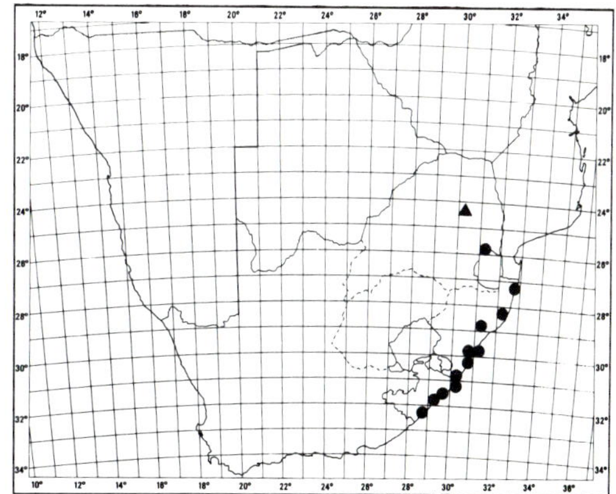

MAP 98. — ▲ **Plectranthus dolomiticus**
● **P. petiolaris**

43. **Plectranthus petiolaris** *E. Mey. ex Benth.* in E. Mey., Comm. 228 (1837); in DC., Prodr. 12: 66 (1848); Cooke in F.C. 5,1: 272 (1910); Ross, Fl. Natal 305 (1972); Codd in Bothalia 11: 431 (1975). Lectotype:

Transskei, between Umtata and Umzimvubu Rivers, *Drège* (No. 4773b in K, lecto.!; MO!; P!; S!).

P. kuntzei Gürke in Kuntze, Rev. Gen. 3,2: 260 (1898); Cooke, l.c. 277 (1910). Type: Natal, Clairmont, *Kuntze* s.n. (K!).

Perennial branched herb up to 1 m tall; stems ascending to spreading, fairly densely pubescent with tufts of longer hairs at the nodes. *Leaves* petiolate; blade thin-textured, broadly ovate-deltoid, 40—140 × 35—110 mm, thinly to fairly densely strigose, under-surface with colourless gland-dots, apex obtuse to acute, base truncate to subcordate, margin coarsely crenate-dentate, the teeth often having small secondary teeth; petiole 20—150 mm long. *Inflorescence* simple or with 1 or 2 pairs of branches near the base, racemes 100—250 mm long; flowers in sessile, 1–3-flowered cymes forming 2–6-flowered verticils 10—30 mm apart. *Calyx* up to 8 mm long in fruiting stage, glandular-puberulous. *Corolla* deep violet-purple, often with bluish lips, 12—15 mm long; tube somewhat sigmoid, not expanding at the base, narrow and ascending for 3 mm, then deflexed and expanding to about 3 mm deep at the throat; upper lip 6—8 mm long; lower lip shallowly boat-shaped, 7—9 mm long. *Stamens* free, 4—5 mm long. Fig. 30: f.

Distributed from the Port St Johns area in the Transkei, usually not far from the coast, to KwaZulu and inland to the Kaap River valley of south-eastern Transvaal. Map 98.

Vouchers: *Codd* 9295; *Medley Wood* 3390; 5754.

Can be recognized by the characteristic curved corolla tube and the large, coarsely toothed leaves.

44. Plectranthus laxiflorus *Benth.* in E. Mey., Comm. 228 (1837); Drège, Zwei Pfl. Doc. 145, 149, 157 (1843); Benth. in DC., Prodr. 12: 63 (1848); Cooke in F.C. 5,1: 276 (1910); Hulme, Wild Flow. Natal t.26, f. 2 (1954); Ross, Fl. Natal 305 (1972); Codd in Bothalia 11: 434 (1975); Compton, Fl. Swaziland 503 (1976). Lectotype: Natal, between Umzimkulu and Umkomaas Rivers, *Drège* (No. 3586 in K, lecto.!; P!; S!).

Germanea laxiflora (Benth.) Hiern, Cat. Afr. Pl. Welw. 1,4: 861 (1900).

P. hylophilus sensu Cooke, l.c. 277 (1910).

Perennial freely-branched herb or soft shrub 0,7—1,5 m tall; stems ascending or spreading, sparingly to densely glandular-pubescent. *Leaves* petiolate; blade thin-textured to somewhat rugose, broadly ovate-deltoid, 60—100 × 40—60 mm, thinly pubescent, under-surface with reddish gland-dots (sometimes not easily visible), apex acute to acuminate, base cordate, margin regularly and finely crenate-dentate; petiole 25—80 mm long. *Inflorescence* simple or laxly branched, racemes 100—300 mm long; flowers in usually 3-flowered pedunculate (rarely subsessile) cymes forming usually 6-flowered verticils. *Calyx* up to 7 mm long in fruiting stage, glandular-pubescent. *Corolla* whitish to pale mauve, 12—14 mm long, with 4—5 purple vertical lines on the upper lip; tube more or less sigmoid, not expanding at the base, narrow and ascending for 2,5 mm then deflexed and expanding to 2,5 mm deep at the throat; upper lip 6—7 mm long; lower lip boat-shaped, somewhat ascending, 5—7 mm long. *Stamens* free, about 5 mm long. Fig. 30: g.

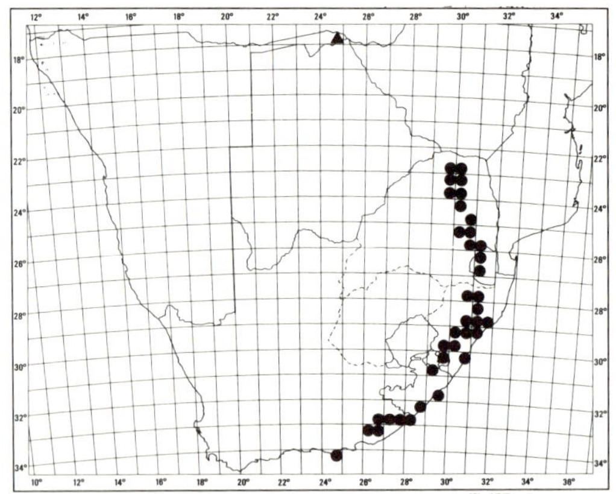

MAP 99. — ● **Plectranthus laxiflorus**
▲ **Holostylon baumii**

Distributed from Humansdorp in the Cape in semi-coastal and adjacent regions through the Transkei to Natal coast and midlands, Swaziland, eastern and northern Transvaal, extending into tropical Africa; often locally common in forest margins and on shady stream banks. Map 99.

Vouchers: *Codd* 7829; 8185; 8579; *Galpin* 10109; *Pegler* 161; *Schlechter* 4762.

The leaves have a sharp citronella-like scent unlike that of any other Southern African species.

It is probable that several tropical species names will be placed in synonymy (see Bothalia 11: 435, 1975).

7350b   **24. HOLOSTYLON**

Holostylon *Robyns & Lebrun* in Annals Soc. scient. Brux. sér. B, 49: 103 (1929); Codd in Mitt. bot. StSamml., Münch. 10: 251 (1971); R.A. Dyer, Gen. 532 (1975). Lectotype species: *H. gracilipedicellatum* Robyns & Lebrun.

Herbs or subshrubs with 1 or more erect, virgate stems arising from a perennial base. *Leaves* opposite, often quite large, margin toothed. *Inflorescence* a terminal panicle, often occupying 1/3 or more of the plant; bracts minute, each subtending a single flower. *Calyx* not or slightly gibbous at the base, subequally 5-toothed, accrescent; tube campanulate; teeth short, deltoid to deltoid-lanceolate. *Corolla* bilabiate; tube short, declinate, bent and expanding just beyond the calyx; upper lip erect, short, obscurely 4-lobed; lower lip large, boat-shaped. *Stamens* 4, declinate, attached at the mouth of the corolla tube and lying in the lower lip; filaments united at the base into a sheath open above. *Style* slightly exceeding the stamens; stigma entire. *Nutlets* subrotund, slightly compressed, triquetrous, glabrous.

Species probably 4, in tropical Africa, 1 of which extends into the Flora area. They are closely allied to *Plectranthus* (no. 23) but may be separated on a combination of characters: the minute bracts subtending solitary flowers, the subequally 5-toothed calyx, stamens united at the base, and the entire style. They come closest to *P. esculentus* N.E. Br. and, like that species, have erect, sparingly branched stems which tend to flower in winter after the leaves have dropped, but the inflorescences, except in *H. robustum* (Hiern) G. Tayl., are more diffusely branched and the flowers are usually shades of blue (rarely white).

Holostylon baumii *(Gürke) G. Tayl.* in J. Bot., Lond. 69, Suppl. 2: 161 (1931). Type: Angola, Kubango, Massaca, *Baum* 283 (K).

Plectranthus baumii Gürke in Warb., Kunene-Samb. Exped. 356 (1903).

H. gracilipedicellatum Robyns & Lebrun in Annls Soc. scient. Brux. sér. B, 49: 103 (1929). Type: Zaire, *Robyns* 2196 (BR, holo.; K).

Stems 1—several, 1,2—2,5 m tall, woody below, from a perennial rootstock, puberulous. *Leaves* sessile or subsessile; blade ovate to ovate-deltoid, 50—70 × 25—40 mm, puberulous, apex acute, base truncate, margin coarsely to shallowly serrate. *Inflorescence* a lax panicle up to 500 mm long, 150—250 mm broad, the lower panicle branches unbranched for 30—100 mm, then branching and finally flexuose for the ultimate 30—50 mm; rhachis and inflorescence branches finely puberulous; pedicels glabrous, persistent, shorter towards the ends of the panicle branches, 6—12 mm long. *Calyx* hispidulous, c. 8 mm long at maturity, with a hispidulous stipe 1,5—2 mm long and thicker than the pedicel. *Corolla* blue, 15 mm long; lower lip 10 mm long.

Recorded from the Chobe area in Botswana. Also in Angola, Zambia, Zimbabwe, Malawi and Zaire. Usually associated with *Brachystegia* woodland on sandy flats and streambanks. Map 99.

Voucher: *Robertson & Elffers* 60.

The species is characterized by the stipitate calyx and the unusual branching of the inflorescence, in which the ends of the branches are flexuose. The ultimate branchlets of the inflorescence are puberulous whereas the persistent pedicels are glabrous, from which it is possible to see how the flexuose branchlets are derived from a basically cymose structure.

Fig. 31. — 1, **Rabdosiella calycina,** inflorescence, × 1; a, habit, much reduced; b, flower, × 3; c, section through corolla, × 3; d, mature calyx, × 3; e, nutlet, × 9 (living plant, BRI garden, from Machadodorp).

7350c   **25. RABDOSIELLA**

Rabdosiella *Codd* in Bothalia 15: 9 (1984). Type species: *R. calycina* (Benth.) Codd.

Plectranthus sect. *Pyramidium* Benth., Lab. 44 (1832); in DC., Prodr. 12: 61 (1848).

Rabdosia sensu Codd in Bothalia 11: 117 (1973); ibid. 11: 436 (1975); sensu R.A. Dyer, Gen. 532 (1975).

Allied to *Plectranthus* L'Hérit., but the stems tend to be more woody, the leaves are often ternate and the bracts are leaf-like, becoming progressively smaller towards the apex of the inflorescence. Also allied to *Isodon* (Schrad. ex Benth.) Spach, a mainly Asiatic genus with 1 species widespread in tropical Africa, but the inflorescence of *Rabdosiella* is a dense terminal panicle, the mature calyx is erect, tubular and distinctly 10-nerved, and the corolla is saccate at the base.

A genus of 2 species, 1 of which occurs in Southern Africa and the other in India.

Rabdosiella calycina *(Benth.) Codd* in Bothalia 15: 10 (1984). Type: Transkei, between St Johns and Umsikaba Rivers, *Drège* 3584 (K, lecto.! = *Drège* b in G!; MO!; P!; S!).

Plectranthus calycinus Benth. in E. Mey., Comm. 230 (1837); Drège, Zwei Pfl. Doc. 148, 152 (1843); Benth. in DC., Prodr. 12: 61 (1848); Cooke in F.C. 5,1: 270 (1910); Trauseld, Wild Flow. Drakensberg 160 (1969); Codd in Mitt. bot. StSamml., Münch. 10: 250 (1971). *Rabdosia calycina* (Benth.) Codd in Bothalia 11: 117 (1973); ibid. 11: 436 (1975).

P. pyramidatus Gürke in Bull. Herb. Boissier 6: 522 (1898). Type: Transvaal, Houtbosch, *Rehmann* 6179 (Z, holo.!).

P. pachystachyus Briq. in Bull. Herb. Boissier sér. 2,3: 1003 (1903). *P. calycinus* var. *pachystachyus* (Briq.) T. Cooke in F.C. 5,1: 271 (1910). Type: Natal, Umkomaas, *Medley Wood* 4621 (K!).

A soft shrub branched above, or stems 1—several arising annually from a woody rootstock, 0,6—1,5 m tall, ribbed towards the base, glandular-puberulous to densely tomentose. *Leaves* subsessile or shortly petiolate; blade ovate-lanceolate to broadly ovate, 40—100 × 20—45 mm, subcoriaceous, upper surface subglabrous to strigose, under-surface reticulately veined, subglabrous to densely tomentose and freely dotted with orange gland-dots, apex acute to obtuse, base cuneate to rounded, margin crenate-dentate. *Inflorescence* a terminal panicle 100—300 mm long; flowers in dense, somewhat scorpioid cymes. *Calyx* 7—9 mm long and erect at fruiting stage; teeth subequal, narrowly deltoid. *Corolla* white to cream, flushed with mauve on the lips, 8—11 mm long; tube 4 mm deep at the base, narrowing gradually to 3 mm at the throat; upper lip erect, 2 mm long; lower lip spreading, shallowly boat-shaped, 4—5 mm long. *Stamens* declinate, attached at the mouth of the corolla tube, 2,5—4,5 mm long, enclosed in the lower corolla lip; filaments free. *Stigma* minutely bifid. Fig. 31.

Distributed from the Blouberg and Soutpansberg in the northern Transvaal along the eastern escarpment to Swaziland, eastern Orange Free State, Natal, Transkei and into the eastern Cape Province to around King William's Town. Map 100.

Vouchers: *Codd* 9266; 9433; *Galpin* 8161; 10113; 12019; *Tyson* 2749.

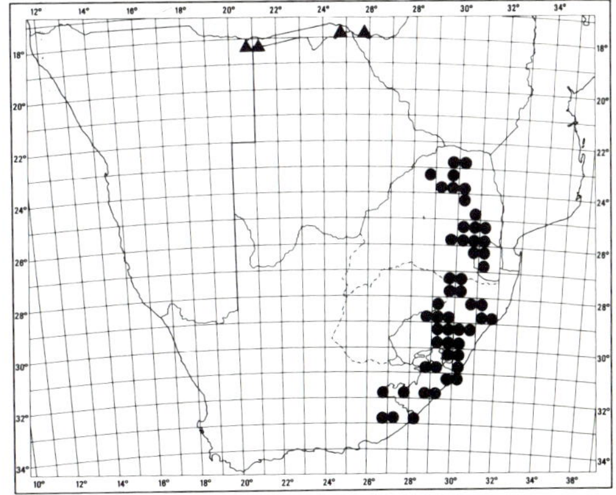

MAP 100. — ● **Rabdosiella calycina**
▲ **Englerastrum schweinfurthii**

26. ENGLERASTRUM

Englerastrum *Briq.* in Bot. Jb. 19: 178, t.3, fig. A (1894); in Natürl. PflFam. 4,3a: 358 (1897); Bak. in F.T.A. 5: 445 (1900); T.C.E. Fries in Notizbl. bot. Gart. Mus. Berl. 9: 61 (1924), partly; Alston in Kew Bull. 1926: 295 (1926), partly; Hutch. & Dandy in Kew Bull. 1926: 479 (1926); Morton in J. Linn. Soc., Bot. 58: 257 (1962): in F.W.T.A. edn 2,2: 465 (1963); Launert & Schreiber in F.S.W.A. 123: 11 (1969); Codd in Mitt. bot. StSamml., Münch. 10: 250 (1971); R.A. Dyer, Gen. 532 (1975). Type species: *E. schweinfurthii* Briq.

Soft herbs, annual or perennial, decumbent or erect. *Leaves* membranous, subsessile or petiolate. *Inflorescences* slender, racemose or paniculate, borne singly or in pairs in the axils of the leaves for almost the entire length of the stem as well as terminally; flowers small, solitary (or rarely 2) in the axils of minute bracteoles. *Calyx* campanulate, equally 5-toothed. *Corolla* small, bilabiate, usually blue; upper lip erect, short, subequally 4-lobed, lower lip slightly longer, patent, keel-shaped. *Stamens* 4, declinate, attached at the mouth of the corolla tube, included in the lower corolla lip; filaments shortly united at the base. *Style* slightly exceeding the stamens, shortly forked at apex.

Probably 5 species, 1 from Ceylon, the remainder in tropical Africa, with 1 extending into the Flora region. They are delicate herbs with flowers reduced to 1 or 2 per floral bract, arranged in short slender racemes or panicles, which appear to have evolved by reduction from *Isodon* (Schrad. ex Benth.) Spach, a mainly Asiatic genus with one species in Africa, in which the flowers are arranged in axillary, well-branched dichasia. A similar evolutionary trend toward solitary flowers may be noted in *Plectranthus*, where the calyx is normally more or less bilabiate, with the uppermost tooth larger than the rest. T.C.E. Fries included such species in *Englerastrum* but the resemblance is essentially a superficial one and Hutchinson & Dandy, l.c., restricted the genus to those species with equally 5-toothed calyces, as in *Isodon*.

Englerastrum schweinfurthii *Briq.* in Bot. Jb. 19: 178, t.3, fig. A (1894); Bak. in F.T.A. 5: 445 (1900); Morton in F.W.T.A. edn 2,2: 465 (1963); Launert & Schreiber in F.S.W.A. 123: 11 (1969). Type: Sudan, *Schweinfurth* 2532 (PRE, iso.!).

Annual or perennial herb, stems erect or straggling, sparingly branched, up to 0,6 m long, sparingly pilose. *Leaves* subsessile or petiolate; blade ovate to ovate-oblong, 30—50 × 15—25 mm, subglabrous, apex acute to obtuse, base obtuse, shortly decurrent on the petiole, margin shallowly crenate; petiole up to 20 mm long. *Inflorescences* slender, terminal or axillary along almost the entire length of stem, racemose or sparingly branched, 25—70 mm long, on long peduncles; pedicels 1—2 mm long. *Calyx* pilose, 3 mm long at fruiting stage; teeth deltoid, 1 mm long. *Corolla* mottled blue and white, 5 mm long.

Recorded from the Caprivi area in S.W.A./Namibia, along the Okavango and Zambesi Rivers, on river banks and islands, usually in muddy places. Widespread in tropical Africa. Map 100.

Vouchers: *Hardy* 5618; *Müller & Giess* 541.

7354   **27. SOLENOSTEMON**

Solenostemon *Thonn.* in Schumach., Beskr. Guin. Pl. 271 (1827), emend. J. K. Morton in J. Linn. Soc., Bot. 58: 251 (1962); in F.W.T.A. edn 2,2: 462 (1963); Codd in Mitt. bot. StSamml., Münch. 10: 249 (1971); Blake in Contr. Queensl. Herb. 9: 6 (1971); Codd in Bothalia 11: 437 (1975); R. A. Dyer, Gen. 533 (1975). Type species: *S. ocymoides* Schumach.

Coleus sect. *Solenostemon* (Thonn.) Benth., Lab. 52 (1832); in DC., Prodr. 12: 72 (1848).

— sect. *Solenostemoides* Briq. in Natürl. PflFam. 4,3a: 359 (1897); *Solenostemon* sect. *Coleoidea* J. K. Morton, l.c., descr. angl. *Solenostemon* subgen. *Solenostemoides* (Briq.) Codd in Bothalia 11: 437 (1975). Lectotype species: *S. latifolius* (Hochst. ex Benth.) J. K. Morton.

Perennial, erect or spreading herbs or subshrubs. *Leaves* often blotched on the upper surface or variegated (cultivars). *Inflorescence* terminal, racemose or paniculate, lax or dense; flowers in pedunculate or sessile, often somewhat glomerate, dichasia; bracts differentiated from the leaves, early deciduous. *Calyx* bilabiate, 5-toothed; uppermost tooth the largest, forming an ovate erect lobe; two lowermost teeth fused, forming a strap-shaped emarginate lip; two lateral teeth short and truncate to rounded. *Corolla* bilabiate; tube more or less sigmoid, narrow and ascending at the base then deflexed about the middle and expanding to the throat; upper lip obscurely 4-lobed; lower lip larger than the upper, boat-shaped. *Stamens* 4, declinate, attached at the mouth of the corolla tube and lying in the lower lip; filaments usually shortly united at the base, occasionally free. *Style* lying with the stamens in the lower lip of the corolla. *Nutlets* ovoid-triquetrous, glabrous.

In the typical species (the West African *S. ocymoides* Schumach.), which does not occur in Southern Africa, the lower lip of the calyx is entire and eventually bends upwards closing the mouth of the calyx tube. The concept of the genus was enlarged by Morton, l.c. to include his Section *Coleoidea* in which the lower lip of the calyx is emarginate or bifurcate and does not bend upwards in the mature stage. This group, which is considered to be worthy of subgeneric rank, includes 60 or more species described from Africa, Asia and Malesia, of which 2 species are recognized in Southern Africa.

One of the best known members of the group is the commonly cultivated "Coleus" with variegated and often incised leaves. Taking a broad view of species limits, the correct name for it would be *S. scutellarioides* (L.) Codd (Bothalia 11: 439, 1975).

1 Roots tuberous, potato-like; corolla 5—7 mm long, lower lip pale bluish mauve.................. 1. *S. rotundifolius*

1 Roots fibrous; corolla 8—15 mm long, lower lip violet.. 2. *S. latifolius*

1. **Solenostemon rotundifolius** *(Poir.) J. K. Morton* in J. Linn. Soc., Bot. 58: 272 (1962); in F.W.T.A. edn 2,2: 463 (1963); Ross, Fl. Natal 305 (1972); Codd in Bothalia 11: 438 (1975). Type: Mauritius, *Commerson* (P, holo.).

Germanea rotundifolia Poir. in Lam., Encycl. 2: 763 (1812). *Plectranthus rotundifolius* (Poir.) Spreng., Syst. 2: 690 (1825); Benth., Lab. 34 (1832); in DC., Prodr. 12: 65 (1848). *Coleus rotundifolius* (Poir.) A. Chev. & E. Perrot., Veg. Util. Trop. Franc. 1: 101, 119 (1905).

Coleus dysentericus Bak. in Kew Bull. 1894: 10 (1894); in F.T.A. 5: 437 (1900). Type: Niger region, *Barter* 846 (K, holo.).

Perennial herb, branching at the base, producing ovoid to roundish potato-like tubers; stems ascending to decumbent, 0,3—0,6 m long, semi-succulent, puberulous to shortly pubescent. *Leaves* petiolate; blade fairly thick-textured, ovate, 25—50 × 20—30 mm, upper surface subglabrous, lower surface puberulous, sparingly red gland-dotted, apex acute, base truncate to abruptly attenuate, margin crenate-dentate; petiole 20—30 mm long. *Inflorescence* 60—100 mm long, simple or with a pair of branches at the base; verticils many-flowered, dense. *Calyx* 3 mm long in fruit, glandular-hispid. *Corolla* 5—7 mm long, gland-dotted; upper lip whitish, lower lip pale bluish mauve. *Stamens* shortly united at the base.

Cultivated for the potato-like tubers in eastern Transvaal and KwaZulu; probably of tropical African origin. Map 101.

Vouchers: *Clarke* 65; *Scheepers* 931.

There are probably further synonyms among the species listed by Chevalier & Perrottet, l.c., but no attempt has been made to sort these out. Three varieties are also maintained by these authors.

Considered to be a delicacy and known as Matabala (Sepedi) amaTabhane, or amaData (Zulu). The tubers are prepared by boiling them in water and squeezing the edible inside part out of the skin.

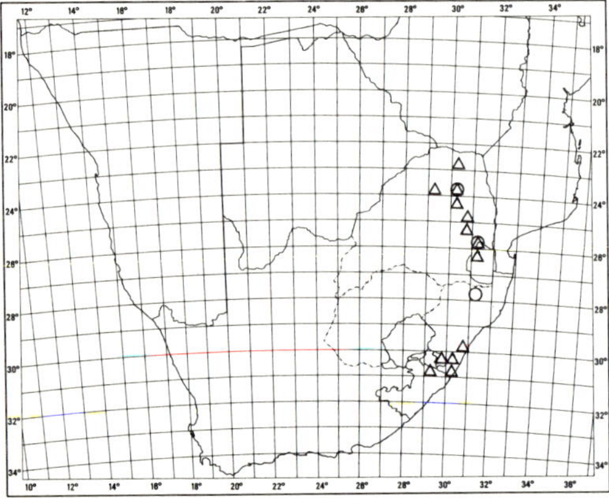

MAP 101. — ○ **Solenostemon rotundifolius**
△ **S. latifolius**

2. **Solenostemon latifolius** *(Hochst. ex Benth.) J. K. Morton* in J. Linn. Soc., Bot. 58: 271 (1962); in F.T.A. edn 2,2: 463 (1963); Ross, Fl. Natal 305 (1972); Codd in Bothalia 11: 439 (1975). Syntypes: Ethiopia, *Schimper* 825; 1828 (K!).

Coleus latifolius Hochst. ex Benth. in DC., Prodr. 12: 74 (1848); A. Rich., Tent. Fl. Abyss. 2: 184 (1851); Bak. in F.T.A. 5: 437 (1900).

Plectranthus tysonii Gürke in Bot. Jb. 24: 77 (1898); Cooke in F.C. 4,1: 276 (1910); Compton, Fl. Swaziland 506 (1976). Type: Cape, Griqualand East, Clydesdale, *Tyson* 2769 (G!; K!; PRE!).

Coleus rehmannii Briq. in Bull. Herb. Boissier sér. 2,3: 1075 (1903); Cooke, l.c. 289 (1910). Type: Transvaal, Houtbosch, *Rehmann* 6156 (Z, holo.!).

Perennial herb; stems semi-erect to procumbent, 0,2−1,5 m long, pubescent, usually with longish multicellular hairs. *Leaves* petiolate; blade thin to medium-thick in texture, broadly ovate-deltoid, 25−80 × 20−65 mm, sparingly to densely pubescent, sometimes with a dark V-shaped blotch on the upper surface, lower surface freely dotted with reddish gland-dots, apex acute, base truncate, margin crenate; petiole 15−60 (−100) mm long. *Inflorescence* usually simple, 100−350 mm long; flowers in opposite, sessile or pedunculate, usually compact dichasia or, in depauperate specimens, reduced to few-flowered sessile cymes. *Calyx* up to 7 mm long in fruit, glandular-hispidulous. *Corolla* 8−15 mm long, gland-dotted, lower lip violet to purple, upper lip paler. *Stamens* usually united at the base, occasionally free. Fig. 32.

Distributed from East Griqualand, through Natal and Swaziland to the mountains of eastern and northern Transvaal, in forest margins and open woodland among rocks; widespread in tropical Africa. Map 101.

Vouchers: *Codd* 7915; 8408; *Schlieben* 9520; *Strey* 10649.

A variable species with stems semi-erect to trailing and leaves varying in texture according to growing conditions.

FIG. 32. — 1, **Solenostemon latifolius**, flowering stem, × 1; a, leaf, × 1; b, flower, × 3; c, section through corolla, × 3; d, flowering calyx, from front, × 8; e, flowering calyx, × 8; f, mature calyx, × 3 (*Codd* 7820).

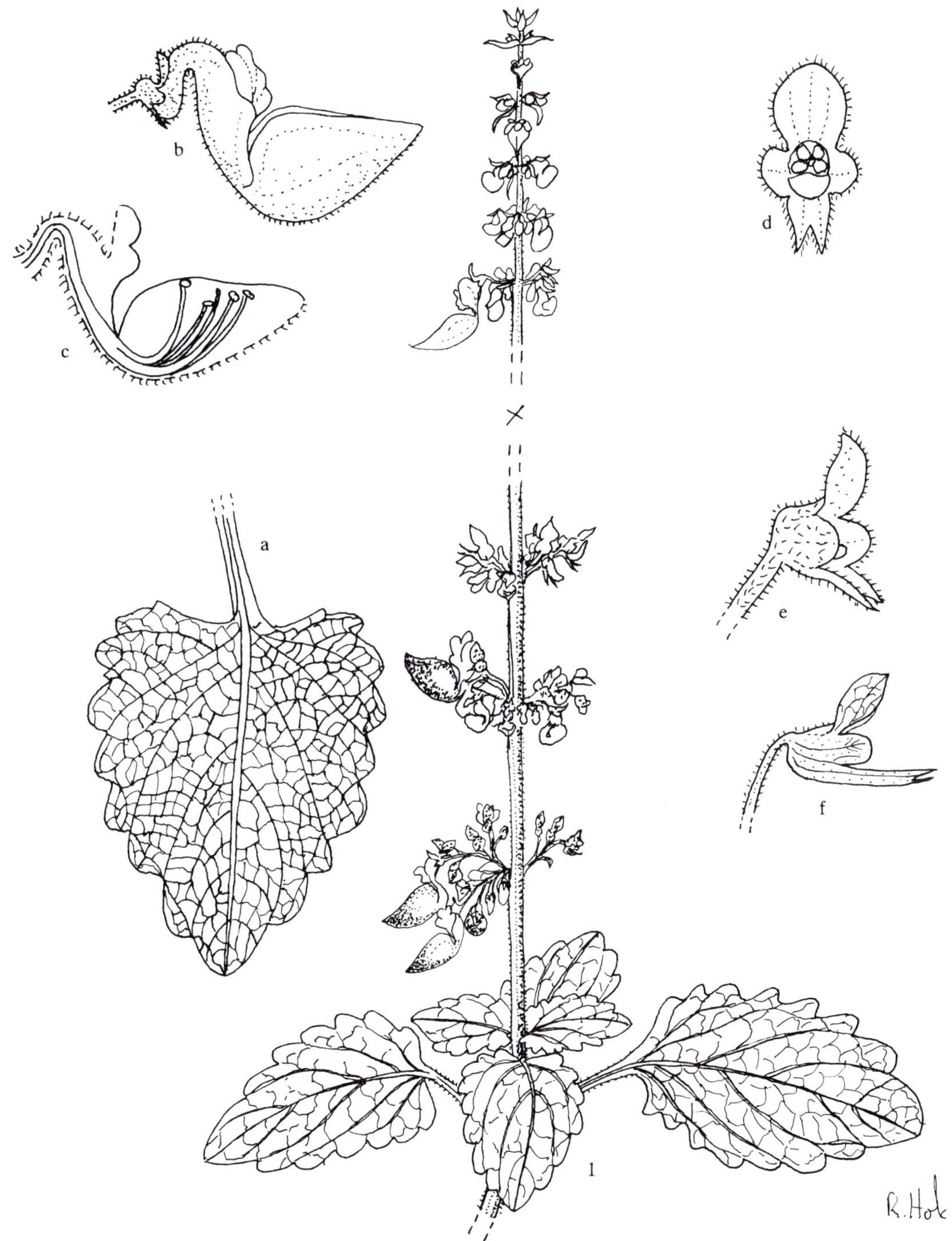

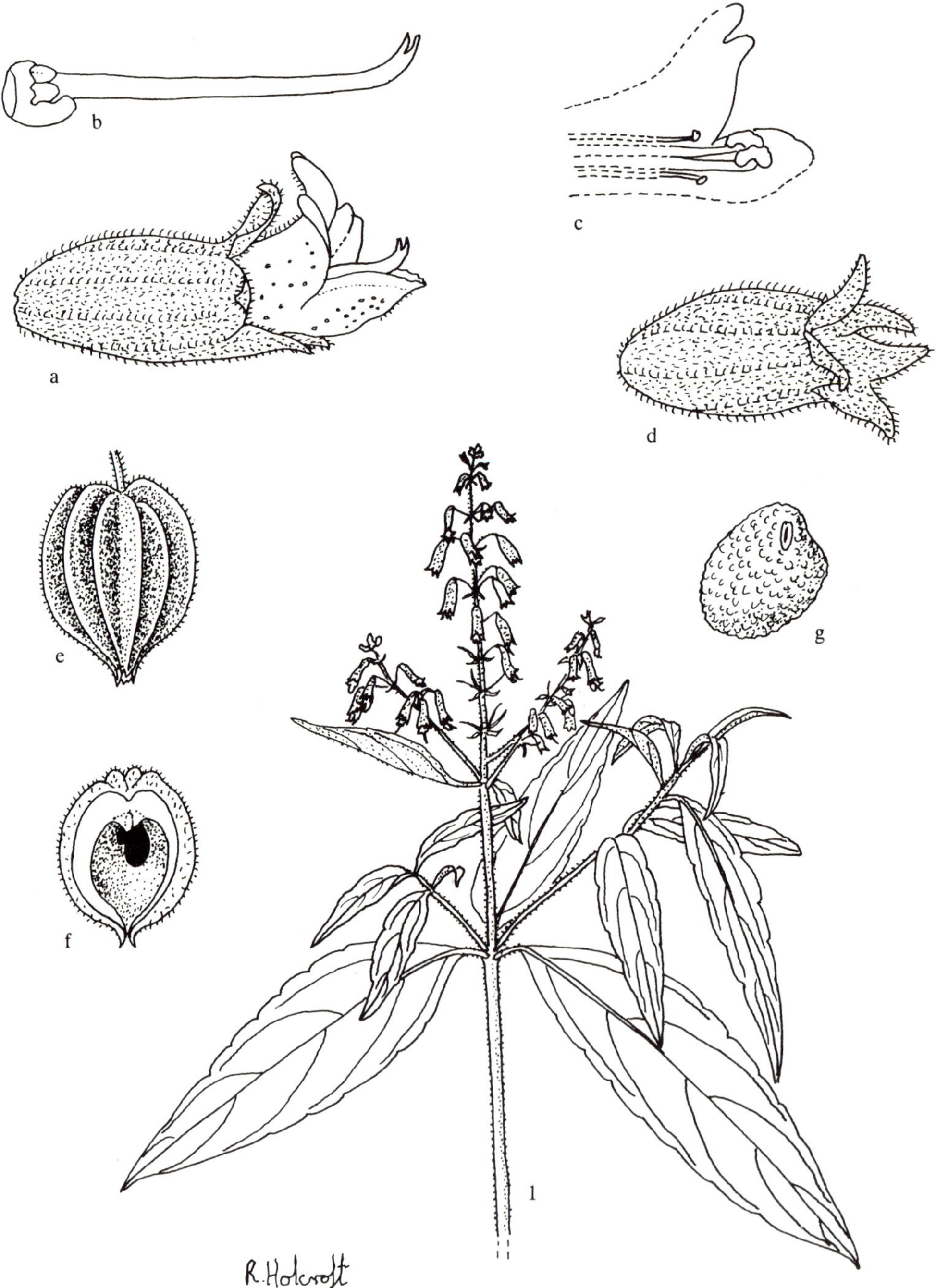

Fig. 33. — 1, **Hoslundia opposita,** flowering branch, × 1; a, flower, × 9; b, gynoecium, × 9; c, section through apex of corolla, × 9; d, flowering calyx, × 9; e, mature calyx, × 2; f, section through mature calyx, × 2; g, nutlet, × 9 (*Mrs Jenkins* s.n., living plant).

7357 **28. HOSLUNDIA**

Hoslundia *Vahl,* Enum. Pl. 1: 212 (1804); Benth., Lab. 706 (1835); in DC., Prodr. 12: 54 (1848); Benth. & Hook. f., Gen. Pl. 2,2: 1174 (1876); Briq. in Natürl. PflFam. 4,3a: 363 (1897); Bak. in F.T.A. 5: 377 (1900); Cooke in F.C. 5,1: 297 (1910); Morton in F.W.T.A. edn. 2,2: 456 (1963); Launert & Schreiber in F.S.W.A. 123: 13 (1969); R. A. Dyer, Gen. 533 (1975). Lectotype species: *H. opposita* Vahl.

A monotypic African genus characterised by having only 2 fertile stamens and a fleshy, berry-like fruit.

Hoslundia opposita *Vahl,* Enum. Pl. 1: 212 (1804); Benth., Lab. 706 (1835); in DC., Prodr. 12: 54 (1848); Bak. in F.T.A. 5: 377 (1900); Morton in J. Linn. Soc., Bot. 58: 241 (1962); in F.W.T.A. edn 2,2: 456 (1963); Launert & Schreiber in F.S.W.A. 123: 14 (1969); Compton, Fl. Swaziland 506 (1976). Type: Guinea, *Thonning* s.n.

H. verticillata Vahl, Enum. Pl. 1: 213 (1804); Benth., Lab. 706 (1835); in DC. Prodr. 12: 54 (1848); Hiern, Cat. Afr. Pl. Welw. 1,4: 860 (1900). *H. opposita* var. *verticillata* (Vahl) Bak. in F.T.A. 5: 377 (1900). Type: Senegal, *Dupuis* s.n.

H. decumbens Benth. in DC., Prodr. 12: 54 (1848); Briq. in Bull. Herb. Boissier sér. 2,3: 661 (1903); Cooke in F.C. 5,1: 298 (1910). *H. opposita* var. *decumbens* (Benth.) Bak. in F.T.A. 5: 377 (1900); Ross, Fl. Natal 305 (1972). Type: Delagoa Bay, *Forbes* s.n.

Orthosiphon physocalycinus A. Rich., Tent. Fl. Abyss. 2: 180 (1851). Type: Ethiopia, *Petit* s.n.

Spreading, erect or subscandent herb or soft shrub, 0,6−1,2 m tall. *Leaves* petiolate, opposite or sometimes ternate; blade grey-green, ovate-lanceolate to ovate-elliptic, 35−65 (−75) × 18−25 mm, sparingly pubescent to fairly densely appressed tomentose, apex acute, base cuneate, margin crenate-dentate; petiole 3−8 mm long. *Inflorescence* consisting of rather lax panicles or racemes; verticils 2−4-flowered, 3−5 mm apart; bracts minute, linear-lanceolate, 1,5−2 mm long; pedicels slender, 2,5−4 mm long, articulate at the top. *Calyx* subequally 5-toothed, 4−4,5 mm long at flowering; teeth narrow, 1−1,5 mm long; tube cylindric, becoming globose and fleshy in fruit. *Corolla* bilabiate, white or cream, 6−7 mm long; tube straight, subcylindric; upper lip short, erect, 1 mm long, 3-lobed; lower lip patent, 1,5 mm long. *Stamens* didynamous, only the lower 2 fertile, declinate, attached near the throat, exserted by 3−5 mm; the upper pair minute, attached in the throat, included. *Disc* produced on one or two sides, one lobe often exceeding the ovary. *Style* exserted, shortly 2-lobed. *Fruit* subglobose, berry-like, 4−5 mm in diam., orange-coloured, usually enclosing 2 or 3 nutlets; nutlets ellipsoid-orbicular, compressed, 2 × 1,5 mm. Fig. 33.

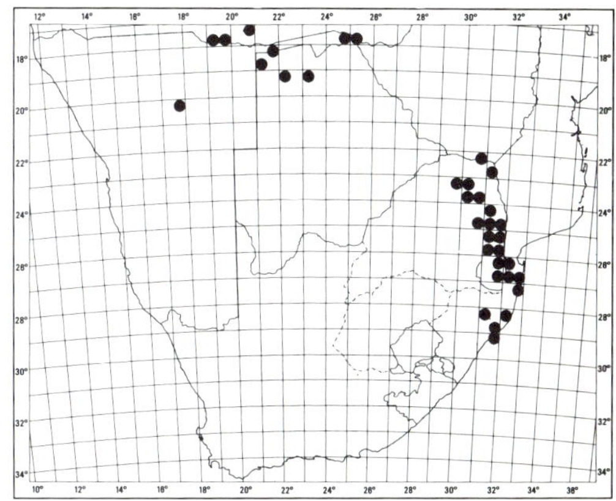

MAP 102. — **Hoslundia opposita**

Found in northern S.W.A./Namibia and Botswana, northern and eastern Transvaal, Swaziland and coastal Natal as far south as Port Shepstone, in tropical and subtropical open woodland. Widespread throughout tropical Africa to Senegal, Sudan and Ethiopia. Map 102.

Vouchers: *Codd* 4741; 5138; *De Winter* 4050; *Galpin* 1246; *Medley Wood* 10204.

The orange-coloured fleshy fruits are edible and are relished by birds. The leaves have a strong and rather unpleasant smell, said to repel bees, and have been recorded as being used in the collection of honey.

7359

29. SYNCOLOSTEMON

Syncolostemon *E. Mey. ex Benth.* in E. Mey., Comm. 230 (1837); in DC., Prodr. 12: 53 (1848); Benth. & Hook. f., Gen. Pl. 2,2: 1174 (1876); Briq. in Natürl. PflFam. 4,3a: 364 (1897); N.E. Br. in F.C. 5,1: 261 (1910); R. A. Dyer, Gen. 534 (1975); Codd in Bothalia 12: 21 (1976). Lectotype: *S. parviflorus* E. Mey. ex Benth. (Codd, l.c.).

Perennial herbs or soft shrubs. *Leaves* opposite, entire or toothed. *Inflorescence* paniculate or racemose, crowded or lax; flowers in 2—6-flowered verticils; bracts small, caducous. *Calyx* subequally 5-toothed or the uppermost tooth somewhat larger than the remaining 4; tube campanulate or cylindrical, scarcely enlarging but becoming suborbicular in some species at fruiting stage, glabrous or pubescent in the throat. *Corolla* bilabiate; tube cylindrical to cylindric-campanulate, widening slightly to the truncate mouth; upper lip short, erect, obscurely 3- or 4-lobed; lower lip spreading to deflexed, concave, longer than the upper. *Stamens* 4, didynamous, exserted; upper pair attached near or below the middle of the corolla tube with filaments free, glabrous or pubescent near the base; lower pair attached at the corolla mouth, filaments united for almost their entire length; anthers 1-thecous. *Disc* lobed, produced in front. *Style* exserted, minutely bilobed. *Nutlets* oblong, sometimes slightly frilled at the base.

A Southern African genus of 9 species, closely related to *Hemizygia* but the uppermost tooth of the calyx is not broadly ovate.

1 Corolla tube 6—10 mm long:

  2 Pubescence on leaves of stellate hairs .... 1. *S. concinnus*

  2 Pubescence on leaves dense or sparse but not stellate:

    3 Leaves greenish, sparsely to densely pubescent but not sericeous:

      4 Inflorescence lax, verticils up to 20 mm apart; bracts ovate-lanceolate, not chartaceous:

        5 Leaves obovate or elliptic to lanceolate, up to 3 times as long as broad .... 2(a). *S. parviflorus* var. *parviflorus*

        5 Leaves narrowly lanceolate to linear-lanceolate, 4—8 times as long as broad .... 2(b). *S. parviflorus* var. *lanceolatus*

      4 Inflorescence dense, verticils 2—3 mm apart; bracts broadly ovate, chartaceous............. 4. *S. comptonii*

    3 Leaves grey, densely sericeous:

      6 Leaves 12—25 × 2—8 mm, flat; inflorescence terminal, lax .... 3. *S. argenteus*

      6 Leaves 4—10 × 1,5—3 mm, margin revolute; inflorescence terminal or on short lateral shoots, dense, densely villous .... 5. *S. eriocephalus*

1 Corolla tube 14—30 mm long:

  7 Calyx teeth subequal:

    8 Verticils usually 4—6-flowered; calyx teeth linear-subulate, 3—5 mm long .... 6. *S. densiflorus*

    8 Verticils 2-flowered; calyx teeth narrowly deltoid, 1,5—2 mm long .... 7. *S. rotundifolius*

  7 Calyx with the uppermost tooth elliptic to obovate, larger than the lower 4:

    9 Corolla tube 20—30 mm long; rhachis glandular-puberulous .... 8. *S. macranthus*

    9 Corolla tube 18—20 mm long; rhachis hispid... 9. *S. latidens*

1. **Syncolostemon concinnus** *N.E. Br.* in F.C. 5,1: 264 (1910); Codd in Bothalia 12: 22 (1976); Compton, Fl. Swaziland 508 (1976). Type: Transvaal, Steenkampsberg, *Schlechter* 3891 (K, holo.; PRE!).

Herb about 0,6 m tall; stems several, softly woody, arising annually from a woody rootstock, slender, sparingly branched, hispidulous, with tufts of leaves along the stem. *Leaves* subsessile; blade obovate-oblong to narrowly elliptic, 10—18 × 2—8 mm, stellately pubescent on both surfaces, freely gland-dotted, apex rounded, base acute, margin often with a few teeth near the apex. *Inflorescence* a lax panicle 120—200 mm long; verticils 2-flowered, up to 20 mm apart; bracts 2—3 mm long. *Calyx*

5—6 mm long, glandular-hispid, becoming subrotund; tube setose in the throat. *Corolla* white, 12—13 mm long; tube 8—9 mm long; lower lip 3—4 mm long. *Stamens* well exserted; 2 upper filaments attached near base of tube, glabrous. *Style* well exserted.

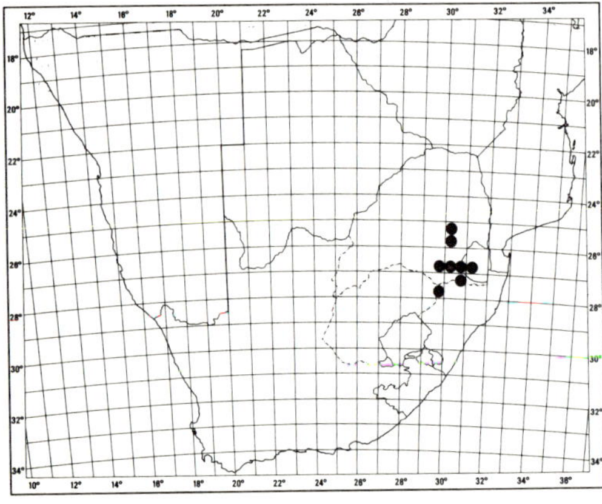

Map 103. — **Syncolostemon concinnus**

Found in eastern and south-eastern Transvaal, Swaziland and north-eastern Orange Free State, in mountain grassland, usually among rocks. Map 103.

Vouchers: *Acocks* 23803; *Codd* 9847; 10207.

This is one of the most northerly members of the genus, extending to the Lydenburg district in Transvaal. The stellate pubescence distinguishes it from the other related species.

2. Syncolostemon parviflorus *E. Mey. ex Benth.* in E. Mey., Comm. 231 (1837). Lectotype: Transkei, near Umsikaba River, *Drège* (K, lecto.!).

Herb or soft shrublet 0,4—1 m tall; stems few to several arising annually from a woody rootstock, slender, sparingly branched, with tufts of leaves along the stem. *Leaves* subsessile; blade greenish or drying blackish, elliptic-obovate to lanceolate or linear-lanceolate, 12—32 × 2—12 mm, hispidulous to fairly densely appressed pubescent, apex rounded to acute, base cuneate, margin with occasional teeth near the apex of larger leaves. *Inflorescence* a lax panicle 120—250 mm long; verticils 2-flowered, up to 20 mm apart; bracts 3—4 mm long. *Calyx* 5—6 mm long, becoming

subrotund; tube setose in the throat. *Corolla* white or flushed with pink, rarely reddish pink, 10—12 mm long; tube 7—9 mm long; lower lip 3 mm long. *Stamens* well exserted; 2 upper filaments attached near the base of the corolla tube, with a few hairs near the base. *Style* well exserted.

Found from Transkei through Natal and Swaziland to the Barberton district in Transvaal; in dense grassland, often among rocks.

Two varieties are recognized; for key to varieties see key to species.

(a) var. **parviflorus.**
Codd in Bothalia 12: 23 (1976).

S. parviflorus E. Mey. ex Benth. in E. Mey., Comm. 231 (1837); in DC., Prodr. 12: 54 (1848); N.E. Br. in F.C. 5,1: 263 (1910); Ross, Fl. Natal 306 (1972), as "parvifolius"; Compton, Fl. Swaziland 508 (1976).

— var. β Benth. in E. Mey., Comm. 231 (1837); *S. dissitiflorus* Benth. in DC., Prodr. 12: 54 (1848). *S. parviflorus* var. *dissitiflorus* (Benth.) N.E. Br., l.c. 264 (1910). Type: Natal, Port Natal, *Drège* (K, holo.!).

Leaves obovate to oblanceolate, elliptic or lanceolate, 12—26 × 4—12 mm.

Distribution and ecology as for the species. Map 104.

Vouchers: *Codd* 9510; 9530; 9664; *Compton* 27412; *Galpin* 10961; *Schlieben* 9573.

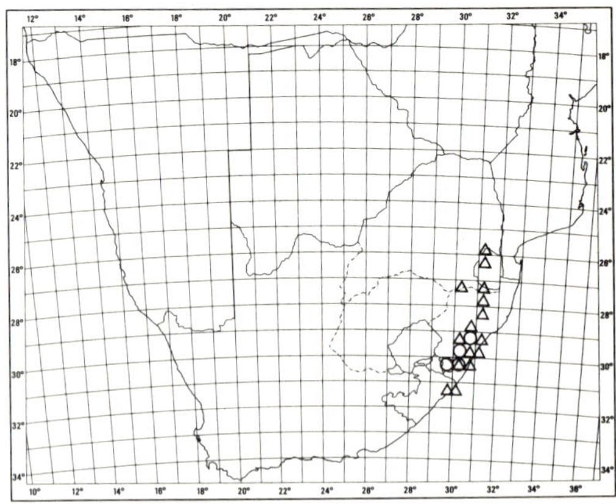

Map 104. — △ **Syncolostemon parviflorus** var. **parviflorus**
○ **S. parviflorus** var. **lanceolatus**

(b) var. **lanceolatus** (*Gürke*) *Codd* in Bothalia 12: 23 (1976). Lectotype: East Griqualand, Mt Malowe, *Tyson* in Herb. Norm. 1294 (K, lecto.!; PRE!).

S. lanceolatus Gürke in Bot. Jb. 26: 77 (1898); N.E. Br. in Fl. Cap. 5,1: 262 (1910); Ross, Fl. Natal 306 (1972).

S. cooperi Briq. in Bull. Herb. Boissier sér. 2, 3: 979 (1903). *S. lanceolatus* var. *cooperi* (Briq.) N.E. Br., l.c. 262 (1910). Syntypes: Natal, *Cooper* 1151; 2895.

S. lanceolatus var. *grandiflorus* N.E. Br., l.c. 262 (1910). Type: Natal, near Enon, *Wood* 1882 (K, holo.).

Leaves narrowly lanceolate to linear-lanceolate, 12—32 × 2—4 mm, often appressed-pubescent on both sides.

Recorded from East Griqualand and the central Natal Midlands. Although N.E. Brown records *Cooper* 2895 as coming from the Orange Free State, this is unlikely in view of its known distribution. Map 104.

Vouchers: *Galpin* 12030; 14691; *Schlechter* 6616.

Grades into var. *parviflorus* and so varietal status appears appropriate. Occasional specimens may be difficult to separate from *S. argenteus* (below) where the two meet in the Pietermaritzburg district, but *S. argenteus* is a more robust species with a mainly semi-coastal distribution and with markedly sericeous leaves.

3. Syncolostemon argenteus *N.E. Br.*

in F.C. 5,1: 263 (1910); Ross, Fl. Natal 306 (1972); Codd in Bothalia 12: 23 (1976). Type: Natal, near Inyezaan, *Medley Wood* 3875 (K, holo.!; NH!).

Herb or soft shrublet 0,6—1,3 m tall; stems solitary or few from a woody rootstock, slender, sparingly branched, sericeous, with tufts of leaves along the stems. *Leaves* subsessile; blade linear-lanceolate to elliptic-obovate, 12—25 × 2—8 mm, densely sericeous, apex obtuse, base

cuneate, margin entire. *Inflorescence* a fairly lax panicle 90—250 mm long; verticils 2-flowered, up to 15 mm apart; bracts 2,5—3 mm long. *Calyx* 5—6 mm long, becoming subrotund; tube villous in the throat. *Corolla* white to pinkish, 10—12 mm long; tube 8 mm long; lower lip 3 mm long. *Stamens* well exserted; 2 upper filaments attached near the middle of the tube, sparsely hairy. *Style* well exserted.

Found in Natal midlands and semi-coastal areas at altitudes of 300 to 1 000 m, in dense grassland, often adjoining forest. Map 105.

Vouchers: *Medley Wood* 9361; 10359; *Strey* 4592; 6475.

4. Syncolostemon comptonii *Codd* in

Bothalia 12: 23 (1976); Compton, Fl. Swaziland 507 (1976). Type: Swaziland, near Komati Bridge, *Compton* 28839 (PRE, holo.!).

Soft shrub up to 1,6 m tall; stems few or solitary from the base, sparingly branched, shortly pilose. *Leaves* subsessile; blade oblanceolate to narrowly elliptic, 20—35 × 3—6 mm, sparingly hispid, gland-dotted, apex acute, base narrowly cuneate, margin usually with a few teeth at the apex. *Inflorescence* a compact panicle 50—80 mm long; verticils 2-flowered, 2—3 mm apart; bracts broadly ovate, 2,5 mm long. *Calyx* 6 mm long, glandular-hispid; throat setose. *Corolla* white, 9—10 mm long; tube 6—7 mm long; lower lip about 3 mm long. *Stamens* well exserted; 2 upper filaments attached near the middle of the corolla tube, sparingly pubescent near the base. *Style* well exserted.

Known only from the type gathering, in tall grass in open woodland, at Komati Bridge, Swaziland. Map 105.

Related to *S. parviflorus* (no. 2) but the plants are more robust, the inflorescence more compact and the bracts are broadly ovate with a fringe of woolly hairs.

5. Syncolostemon eriocephalus *Verdoorn* in Kew Bull. 1937: 447 (1937); Codd

in Bothalia 12: 25 (1976). Type: Transvaal, Pilgrims Rest, *Morisse* 51 (PRE, holo.!).

Shrub 0,6—2 m tall, much branched; stems with tufts of leaves along the stem, densely villous. *Leaves* sessile; blade linear to narrowly oblong, 4—10 × 1,5—3 mm, densely silvery sericeous on both surfaces, apex rounded, base shortly cuneate, margin

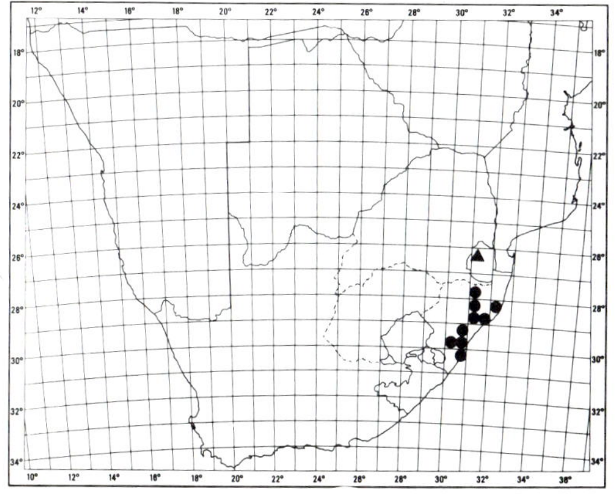

MAP 105. — ● **Syncolostemon argenteus**
▲ **S. comptonii**

entire. *Inflorescence* borne terminally and on short lateral shoots, racemose or occasionally branched, dense, 20—50 mm long, densely villous; verticils 2-flowered, 1—2 mm apart; bracts 1,5—3 mm long. *Calyx* 4 mm long, thickly covered with white to pale yellowish woolly hairs; tube villous in the throat. *Corolla* cream, yellow or brownish yellow, 7—9 mm long; tube 5—7 mm long; lower lip 2 mm long. *Stamens* exserted; 2 upper filaments attached about the middle of the corolla tube, pubescent near the base. *Style* well exserted.

Found along the escarpment in the eastern Transvaal from Pilgrims Rest to near The Downs, at altitudes of 1 400—2 000 m, in shallow sandy soil among quartzite rocks. Map 106.

Vouchers: *Codd & De Winter* 3349; *Galpin* 14601; *Rauh & Schlieben* 9661.

A very distinctive species because of its small grey leaves, dense small inflorescences partly obscured by woolly hairs, and small cream to yellowish brown flowers.

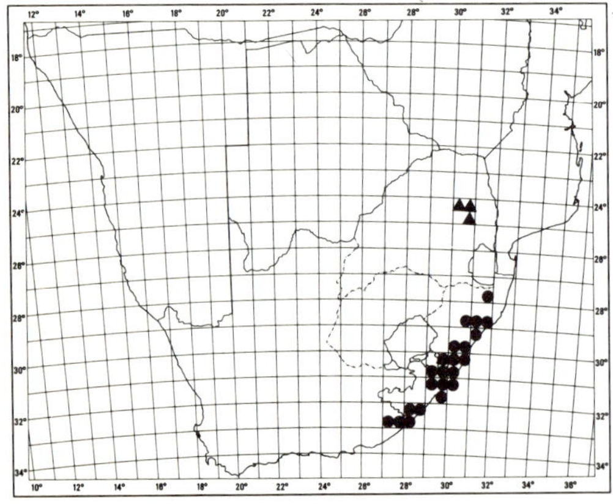

MAP 106. — ▲ Syncolostemon eriocephalus
● S. densiflorus

6. Syncolostemon densiflorus *Benth.* in E. Mey., Comm. 230 (1837); Hochst. in Flora 28: 67 (1845); Benth. in DC., Prodr. 12: 54 (1848); N.E. Br. in F.C. 5,1: 265 (1910); Codd in Flower. Pl. Afr. 32: t.1252 (1957); Ross, Fl. Natal 306 (1972); Codd in Bothalia 12: 25 (1976). Lectotype: Transkei, between

Umzimkulu and Umsikaba Rivers, *Drège* 4744c (K, lecto.!).

S. ramulosus E. Mey. ex Benth. in E. Mey., Comm. 231 (1837); Hochst. in Flora 28: 68 (1845); Benth. in DC., Prodr. 12: 54 (1848); N.E. Br., l.c. 264 (1910). Syntypes: near Morley, *Drège* 4744b (K!); Umsikaba River, *Drège*.

Shrub 1—2,2 m tall, sparingly branched; stems white tomentose. *Leaves* petiolate; blade ovate or broadly elliptic to rotund, 5—15 × 4—10 mm, scabrid to subglabrous, gland-dotted, apex obtuse to rounded, base cuneate, margin subentire or toothed above the middle; petiole 1—6 mm long. *Inflorescence* a dense, often compact, terminal panicle 50—160 mm long, 40—65 mm in diameter; verticils 4—6-flowered, 2—3 mm apart; bracts 3—5 × 3—4 mm. *Calyx* cylindrical, 10 mm long; teeth deltoid-subulate, subequal or with the uppermost shorter than the rest; tube glabrous in the throat. *Corolla* crimson, pink or rarely whitish, 18—23 mm long; tube 15—20 mm long, gradually widening to a truncate mouth 5—6 mm wide; lower lip 3—4 mm long, deflexed at maturity. *Stamens* well exserted, often coiled; 2 upper filaments attached below the middle of the corolla tube, puberulous near the base. *Style* well exserted. Fig. 34.

Distributed from Keiskammahoek through the Transkei to about Nongoma in Natal, in semi-coastal grassland and forest margins at altitudes up to about 1 000 m. Map 106.

Vouchers: *Acocks* 12250; 13778; *Codd* 9241; *Strey* 4161; 6407.

Although superficially similar to *S. rotundifolius* (below), *S. densiflorus* may be distinguished by the longer and more subulate calyx teeth, with the uppermost one usually smaller than the rest, and the 4—6-flowered verticils; in *S. rotundifolius* the verticils are 2-flowered and the calyx teeth are more deltoid.

S. densiflorus, *S. rotundifolius* and *S. macranthus* (no. 8) are attractive when in flower but have proved difficult to maintain in cultivation. They are worth persisting with as garden plants until the problems of cultivation have been overcome.

7. Syncolostemon rotundifolius *E. Mey. ex Benth.* in E. Mey., Comm. 231 (1837); in DC., Prodr. 12: 53 (1848); N.E. Br. in F.C. 5,1: 265 (1910); Ross, Fl. Natal 306 (1972); Codd in Bothalia 12: 25 (1976).

FIG. 34. — 1, **Syncolostemon densiflorus,** flowering stem, × 1; a, mature calyx, × 3; b, section through corolla, × 2; c, ovary and disc, × 5 (after Flower. Pl. Afr. 32: t.1252, 1957).

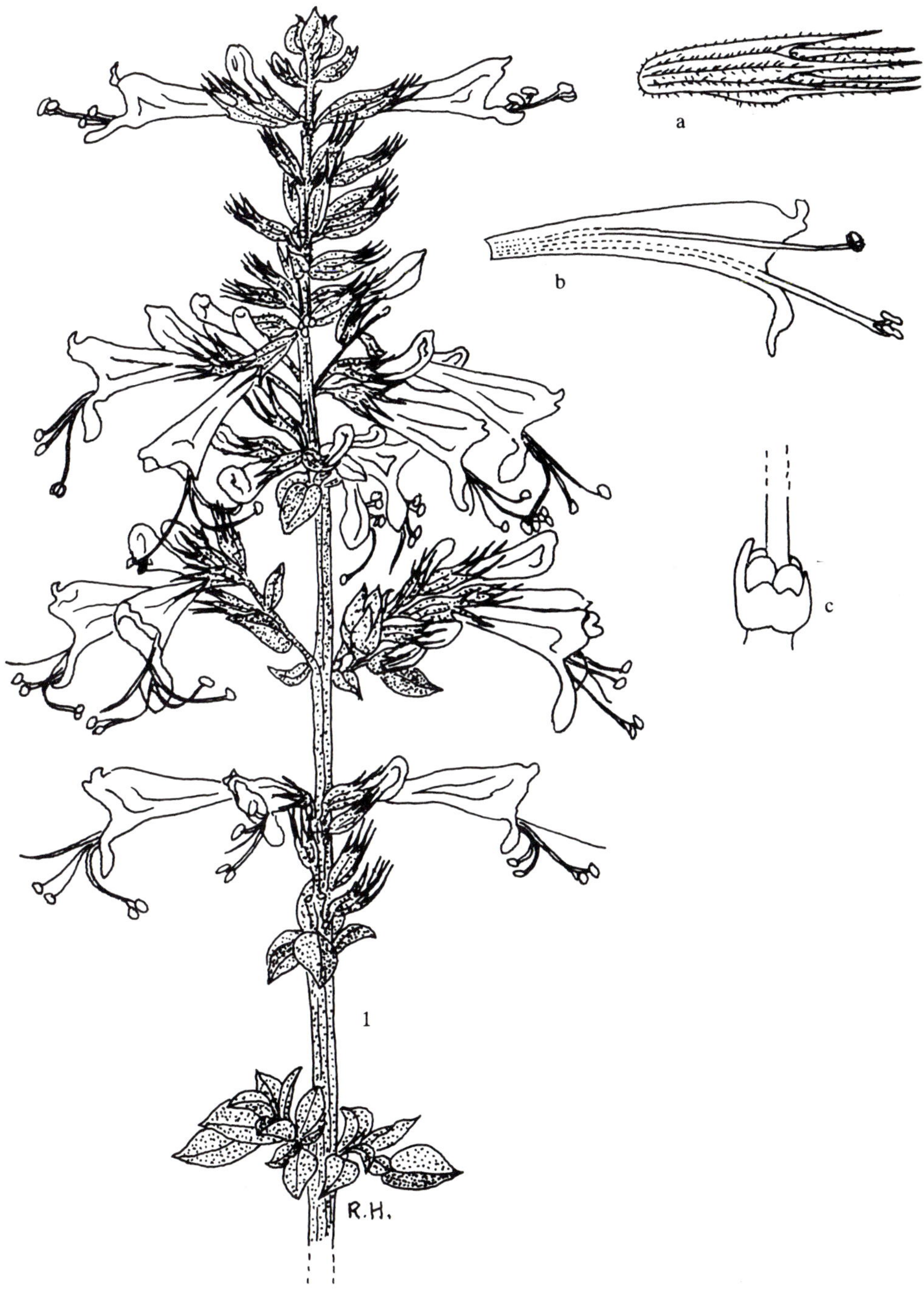

Lectotype: Transkei, between Umzimvubu and Umsikaba Rivers, *Drège* 4743a (K, lecto.!; PRE!).

Soft shrub 0,6−2 m tall, sparingly branched; stems white tomentulose. *Leaves* petiolate; blade broadly elliptic or broadly obovate to subrotund, 10−25 × 6−18 mm, tomentulose and gland-dotted, apex rounded, base cuneate to obtuse, margin entire or faintly crenate-dentate above the middle; petiole 2−5 mm long. *Inflorescence* a fairly dense panicle, rarely simple, 50−80 mm long and up to 60 mm in diameter; verticils 2-flowered; bracts 3−5 × 3−4 mm. *Calyx* cylindrical, 9−10 mm long; teeth subequal, deltoid, the uppermost often slightly shorter than the rest; tube glabrous in the throat. *Corolla* mauve, pink or magenta-pink, 23−27 mm long; tube 20−23 mm long, gradually widening to 5−6 mm at the mouth; lower lip 3−5 mm long, deflexed at maturity. *Stamens* well exserted; 2 upper filaments attached below the middle of the corolla tube, pubescent near the base. *Style* well exserted.

Found from about Port St Johns in the Transkei to Port Shepstone in Natal, in grassland and scrub on rocky slopes, usually not far from the sea at altitudes of up to 400 m. Map 107.

Vouchers: *Acocks* 10918; 13331; *Codd* 9321; *Strey* 5798; 8438; 8769.

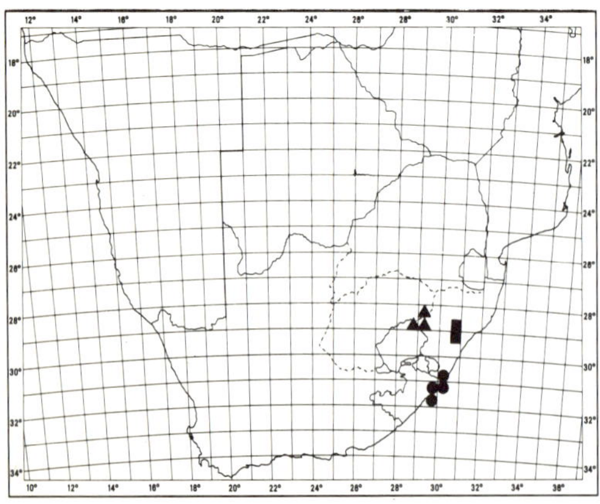

MAP 107. — ● **Syncolostemon rotundifolius**
    ▲ **S. macranthus**
    ■ **S. latidens**

8. **Syncolostemon macranthus** *(Gürke) Ashby* in J. Bot., Lond. 73: 357 (1935); Ross, Fl. Natal 306 (1972); Codd in Bothalia 12: 26 (1976). Lectotype: Natal, Van Reenens Pass, *Medley Wood* 3573, in NH 949 (K, lecto.!).

Orthosiphon macranthus Gürke in Bot. Jb. 26: 84 (1898); N.E. Br. in F.C. 5,1: 242 (1910).

Hemizygia cooperi Briq. in Bull. Herb. Bossier sér. 2,3: 992 (1903). Type: "Orange Free State", *Cooper* 1015 (K, lecto.!).

Shrub 1−2,5 m tall, much branched; stems hispidulous. *Leaves* petiolate; blade ovate to ovate-lanceolate, 20−45 × 12−20 mm, scabrid, gland-dotted, apex acute to obtuse, base obtuse to cuneate, margin obscurely crenate-dentate; petiole 2−8 mm long. *Inflorescence* usually paniculate, 80−180 mm long, fairly dense to lax; verticils 4−6 (rarely 2)-flowered, 4−18 mm apart; bracts 3−5 × 3−4 mm. *Calyx* cylindrical, 9−10 mm long, densely glandular-puberulous, with the uppermost tooth obovate-elliptic, 2−2,5 mm long, distinctly larger than the lower 4 narrowly deltoid teeth. *Corolla* pink to pale mauve or purple, 25−30 mm long; tube 20−25 mm long, gradually widening to 5−6 mm at the mouth; lower lip 3−5 mm long, usually deflexed at maturity. *Stamens* well exserted, often coiled; 2 upper filaments inserted about the middle of the corolla tube, glabrous. *Style* well exserted.

Recorded from a restricted area of the Natal Drakensberg between Cathedral Peak and Van Reenens Pass and just extending into the eastern Orange Free State. Locally frequent along streams and at forest margins at altitudes of 1 600−2 200 m. Map 107.

Vouchers: *Acocks* 11207; *Codd* 8516; *Killick* 1074; *Schlechter* 6912.

Differs from *S. rotundifolius* (above) in the larger, scabrid and more acute leaves and the upper calyx tooth being larger than the lower four. See also note after *S. latidens* (below).

9. **Syncolostemon latidens** *(N.E. Br.) Codd* in Bothalia 12: 26 (1976). Type: Natal, Umvoti District, *Gerrard* 1233 (K, holo.!; PRE, photo.!).

Orthosiphon latidens N.E. Br. in F.C. 5,1: 242 (1910). *Hemizygia latidens* (N.E. Br) Ashby in J. Bot., Lond. 73: 348 (1935); Ross, Fl. Natal 306 (1972).

Soft shrub 1−1,5 m tall, branching; stems hispidulous. *Leaves* petiolate; blade ovate to broadly ovate, 30−50 × 20−35 mm, drying dark brown, tomentulose, apex obtuse to acute, base obtuse to truncate, margin crenate-dentate; petiole 3−10 mm long. *Inflorescence* a fairly dense terminal panicle 100−200 mm long; verticils usually 6-flowered, 4−8 mm apart; bracts 5 × 4 mm. *Calyx* cylindrical, 10−11 mm long, glandular-hispid, with the uppermost tooth broadly obovate, distinctly larger than the lower 4 lanceolate-deltoid teeth. *Corolla* mauve-pink to deep pink, 22−25 mm long; tube 20−22 mm long, gradually widening to 5−6 mm at the mouth; lower lip 3−5 mm long, often deflexed at maturity. *Stamens* well exserted; 2 upper filaments attached below the middle of the corolla tube, puberulous near the base. *Style* well exserted.

Known from only a restricted area near Kranskop in central Natal, growing in and near the forest margin. Map 107.

Vouchers: *Dyer* 4353; *Strey* 4248.

On the basis of calyx shape, *S. macranthus* (no. 8) and *S. latidens* are transitional between *Syncolostemon* and *Hemizygia*, in which the uppermost calyx tooth is broadly ovate and usually decurrent on the tube. However, both these species are so obviously allied to *S. densiflorus* (no. 6) and *S. rotundifolius* (no. 7) that it is considered best to retain them in *Syncolostemon*. In *S. latidens* the uppermost calyx tooth is larger than in *S. macranthus*, but not decurrent on the tube, and the pubescence on the leaves is short and soft, not as scabrid as in *S. macranthus*.

30. HEMIZYGIA

Hemizygia *(Benth.) Briq.* in Natürl. PflFam. 4, 3a: 368 (1897); Annu. Conserv. Jard. bot. Genève 2: 247 (1898); Ashby in J. Bot., Lond. 73: 312, 343 (1935); Launert & Schreiber in F.S.W.A. 123: 11 (1969); R.A. Dyer, Gen. 535 (1975); Codd in Bothalia 12: 1 (1976). Type species: *H. teucriifolia* (Hochst.) Briq.

Ocimum sect. *Hemizygia* Benth. in DC., Prodr. 12: 41 (1848).

Orthosiphon sensu Bak. in F.T.A. 5: 365 (1900), partly; sensu N.E. Br. in F.C. 5,1: 237 (1910), partly.

Bouetia A. Chev. in Mém. Soc. bot. Fr. 8: 200 (1912). Type species: *B. ocimoides* A. Chev.

Perennial soft shrubs or annual herbs, or stems arising annually from a perennial woody rootstock. *Leaves* opposite or rarely ternate, sessile or petiolate, usually toothed. *Inflorescence* paniculate or racemose, crowded or lax; flowers in 2−6-flowered verticils; bracts small and caducous or persistent, or the terminal few pairs large and persistent as a colourful coma. *Calyx* bilabiate, 5-toothed, the uppermost tooth broadly ovate to subrotund, decurrent on the tube; 2 lower teeth subulate to spinescent, longer than the 2 lateral, deltoid-lanceolate teeth. *Corolla* bilabiate; tube subcylindrical or widening slightly to the truncate mouth; upper lip short, erect, obscurely 3- or 4-lobed; lower lip longer than the upper, concave, horizontal to deflexed. *Stamens* 4, didynamous, exserted (upper pair included in *H. pretoriae);* upper pair attached near or below the middle of the corolla tube, filaments free, usually pubescent near the base and sometimes higher as well; lower pair attached at the corolla mouth, filaments connate for all or part of their length (occasionally almost free), glabrous. *Disc* usually crenate, produced in front. *Style* exserted, minutely bifid or occasionally clavate. *Nutlets* ovoid.

Species about 35, mostly African; 28 species in Southern Africa. Closely related to *Syncolostemon* (no. 29), in which the calyx is subequally 5-toothed or the uppermost tooth, if larger than the rest, is not broadly ovate.

1 Stellate or dendroid (branched) hairs present on leaves and other parts, often intermingled with simple hairs: (second half of couplet on p. 4: 194)

  2 Verticils 3−6-flowered (2-flowered verticils may occasionally also be present):

    3 Leaf blade 30−90 × 18−30 mm; inflorescence usually paniculate:

      4 Leaf blade 50−90 × 25−30 mm, concolorous, upper surface densely pubescent, petiole 2−4 mm long; inflorescence laxly branched, up to 600 mm long, bracts greenish; calyx setose in the throat .... 1. *H. macrophylla*

      4 Leaf blade (25−)30−60× 18−30 mm, discolorous, upper surface subglabrous, petiole 6−12 mm long; inflorescence lax to dense, up to 250 mm long, bracts mauve-purple; calyx not setose in the throat .... 2. *H. obermeyerae*

    3 Leaf blade 5−30 × 2−15 mm; inflorescence usually simple or with a pair of branches near the base:

      5 Leaf margin flat, not revolute:

        6 Terminal bracts small, inconspicuous, 4−7 mm long:

          7 Calyx 8−9 mm long; corolla 12−15 (tube 10−12) mm long; leaves 15−35 × 6−12 mm, upper surface coarsely velvety .... 5. *H. incana*

          7 Calyx 5−7 mm long; corolla 8−11 (tube 6−9) mm long; leaves 7−20 × 2−7 mm, upper surface finely velvety, often darker than the lower .... 6. *H. cinerea*

        6 Terminal bracts 7−11 mm long, persistent, colourful:

          8 Leaf blade grey velvety on both surfaces; upper pair of stamens exserted from the corolla tube; stigma capitate .... 7. *H. elliottii*

          8 Leaf blade subglabrous to villous; upper pair of stamens included in the corolla tube; stigma shortly bifid .... 15(b). *H. pretoriae* subsp. *heterotricha*

      5 Leaf margin revolute:

        9 Corolla tube widening towards the mouth; stamens exserted well beyond the lower lip of the corolla:

10 Apical bracts 10−15 mm long, colourful .... 10. *H. stenophylla*

10 Apical bracts 4−7 mm long, inconspicuous:

  11 Leaves finely grey velvety on both surfacs; stem finely grey tomentose .... 6. *H. cinerea*

  11 Leaves with under-surface coarsely pubescent, upper surface darker, finely pubescent; stem villous .... 11. *H. rehmannii*

9 Corolla tube cylindrical, often slightly narrowed at the mouth; stamens exserted scarcely beyond the lower lip of the corolla:

  12 Lower internodes of main stems less than 20 mm long; leaves usually not more than 4 mm broad .... 12. *H. subvelutina*

  12 Lower internodes of main stems usually more than 20 mm long; leaves 3−6 mm or more broad, especially the lower .... 13. *H. teucriifolia*

2 Verticils all 2-flowered:

  13 Leaf blade with upper surface rugose, subglabrous to hispidulous, under-surface grey tomentose:

    14 Leaf blade lanceolate-elliptic, 15−25 mm long; apical bracts conspicuous, up to 15 mm long .... 3. *H. rugosifolia*

    14 Leaf blade ovate, 6−11 mm long; apical bracts inconspicuous, up to 3,5 mm long .... 4. *H. parvifolia*

  13 Leaf blade densely grey tomentose to grey floccose on both surfaces:

    15 Leaf blade densely grey tomentose, 12−25 × 4−12 mm; apical bracts mauve-purple, 7−11 mm long:

      16 Calyx 5 mm long; corolla 12−14 mm long; stigma capitate .... 7. *H. elliottii*

      16 Calyx 8−10 mm long; corolla 17−20 mm long; stigma shortly bifid .... 8. *H. gerrardii*

    15 Leaf blade densely grey floccose, 28−45 × 15−22 mm; apical bracts inconspicuous, 4−5 mm long .... 9. *H. floccosa*

1 (from p. 4: 193) Stellate or dendroid (branched) hairs absent:

  17 Leaves narrow, leathery, margin revolute, under-surface thickly tomentose with long white hairs, upper surface somewhat varnished .... 14. *H. albiflora*

  17 Leaves broad or narrow, margin not revolute, under-surface glabrous to tomentose but not as above:

    18 Apical bracts of the inflorescence like the lower ones, small and inconspicuous, usually caducous:

      19 Upper pair of stamens included in the corolla tube .... 15(a). *H. pretoriae* subsp. *pretoriae*

      19 Upper pair of stamens exserted from the corolla tube:

        20 Verticils 2-flowered; leaves 6−15 × 3−7 mm:

          21 Stems 0,12−0,25 m tall, usually sparingly branched, arising annually from a woody rootstock .... 16. *H. modesta*

          21 Stems 0,6−1,2 m tall, shrubby, much branched .... 17. *H. punctata*

        20 Verticils 3−6-flowered; leaves usually longer than above:

          22 Leaves elliptic-ovate to broadly ovate, obtuse to rounded at the apex, obtuse or broadly cuneate at the base; stems 0,25−0,4 m arising annually from a woody rootstock ..... 18. *H. bolusii*

          22 Leaves linear to ovate, apex acute, base cuneate; annual or perennial herbs not arising annually from a perennial woody rootstock:

            23 Stem and leaves with pubescence of short or fairly short, dense and often crisped hairs:

              24 Leaves ovate-lanceolate to ovate; petiole 6−14 mm long .... 24. *H. petiolata*

              24 Leaves linear to lanceolate or, rarely, ovate-lanceolate; petiole usually less than 5 mm long .... 25. *H. canescens*

            23 Stem villous to subglabrous, not as above; leaves subglabrous or sparingly pubescent to canescent or villous, often with long and short hairs intermingled:

              25 Leaves linear or with some leaves on a plant up to 5 mm broad, subglabrous; stems subglabrous with few long hairs, often somewhat varnished .... 26. *H. linearis*

              25 Leaves linear-lanceolate to ovate-lanceolate, usually more than 5 mm broad; stems and leaves sparingly to densely villous .... 27. *H. petrensis*

18 Apical bracts of inflorescence distinct from the lower ones, membranous, forming a persistent colourful coma (often small but coloured in *H. petiolata* and *H. petrensis*):

  26 Stamens not exserted beyond the lower lip of the corolla; filaments of upper pair of stamens pubescent from the base to near the apex .... 23. *H. persimilis*

  26 Stamens exserted beyond the lower lip of the corolla; filaments of upper pair pubescent only near the base:

    27 Verticils 2-flowered:

      28 Stems shrubby, up to 1 m tall, much branched; leaves obovate to oblanceolate, 15−25 × 6−11 mm; corolla 25−28 mm long .... 19. *H. ramosa*

      28 Stems up to 0,3 m long arising annually from a woody rootstock; leaves ovate, usually exceeding 25 mm long and 11 mm wide; corolla 12−15 mm long .... 21. *H. foliosa*

    27 Verticils 3−6-flowered:

      29 Terminal bracts ovate to linear-lanceolate, cuneate at the base, pairs of bracts often spaced 10−20 mm apart, more than twice as long as broad (sometimes less in *H. transvaalensis* but then corolla tube more than 12 mm long):

        30 Corolla tube more than 12 mm long; terminal bracts ovate to lanceolate, rarely linear-lanceolate .... 20. *H. transvaalensis*

        30 Corolla tube less than 12 mm long; terminal bracts lanceolate to linear-lanceolate .... 22. *H. thorncroftii*

      29 Terminal bracts broadly ovate, not cuneate at the base, densely crowded, usually less than twice as long as broad:

        31 Petiole of mature leaves more than 5 mm long; under-surface of leaves covered with a fine greyish white pubescence .... 24. *H. petiolata*

        31 Petiole of mature leaves less than 5 mm long or leaves sessile; under-surface of leaves glabrous to variously pubescent:

          32 Terminal bracts conspicuous, violet or white, up to 14 × 9 mm; corolla white .... 28. *H. bracteosa*

          32 Terminal bracts small, often purplish, about 5 × 3 mm; corolla mauve .... 27. *H. petrensis*

---

1. **Hemizygia macrophylla** *(Gürke)* Codd in Bothalia 12: 3 (1976). Type: Natal, Drakensberg, *Rehmann* 7016 (Z, holo.!).

Syncolostemon macrophyllus Gürke in Bull. Herb. Boissier 6: 555 (1898); Ross, Fl. Natal 306 (1972); *Orthosiphon macrophyllus* (Gürke) N.E. Br. in F.C. 5,1: 241 (1910).

Soft shrub 1−1,5 m tall, highly aromatic; stems several from a perennial woody rootstock, sparingly branched, grey pubescent. *Leaves* shortly petiolate; blade ovate-lanceolate to lanceolate, 60−90 × 25−30 mm, both surfaces densely and coarsely stellate velvety, tending to fold along the midrib and then somewhat falcate, apex acute, base cuneate, margin regularly serrate in the upper two-thirds; petiole up to 4 mm long. *Inflorescence* a large lax panicle up to 600 × 250 mm; verticils 3−6-flowered, 20 mm or more apart; bracts caducous, broadly ovate, acute, 6−8 mm long with a white margin and patches of white tomentum. *Calyx* up to 7 mm long, glandular-hispid, setose in the throat becoming swollen and narrow at the mouth when in the fruiting stage. *Corolla* purple, 17−18 mm long; tube 11−12 mm long, widening to 5−6 mm at the mouth; upper lip a small appendage, 1 mm long; lower lip 5−6 mm long, horizontal. *Stamens* exserted well beyond the lower lip, curved upwards; upper pair attached below the middle of the corolla tube, puberulous near the base; lower pair united to near the apex. *Stigma* bifid.

Recorded from the foothills of the Drakensberg in northern Natal and southern Transvaal, in dense grassland subjected to periodic burning, on slopes usually among dolerite rocks at altitudes of 1 500−1 800 m. Map 108.

Vouchers: *Codd* 9979; *Devenish* 444; 1590.

Distinguished from all other species by the large grey leaves, coarsely stellate (dendroid) pubescence on both surfaces and the large lax inflorescence with purple flowers. The calyx, which is setose in the throat and becomes swollen and narrowed at the mouth at maturity, makes this species somewhat intermediate between *Syncolostemon* (no. 29) and *Hemizygia*.

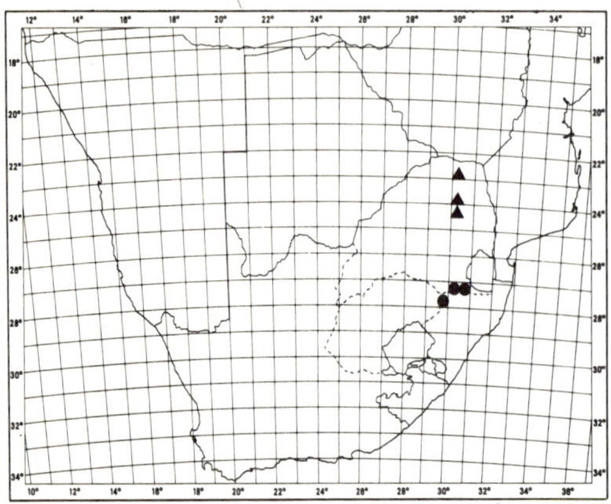

MAP 108. — ● **Hemizygia macrophylla**
       ▲ **H. obermeyerae**

2. Hemizygia obermeyerae *Ashby* in J. Bot., Lond. 73: 343 (1935); Codd in Bothalia 12: 4 (1976). Type: Transvaal, Soutpansberg, *Obermeyer* sub TRV 31556 (PRE, holo.!).

Soft shrub 1–1,5 m tall, freely branched; stems softly stellate-tomentose. *Leaves* petiolate; blade broadly ovate to ovate-lanceolate, 30–60 × 18–30 mm, upper surface subglabrous, under-surface densely grey stellate-pubescent, apex obtuse to rounded, base truncate to obtuse, margin finely crenate-dentate; petiole 6–12 mm long. *Inflorescence* usually branched, fairly dense, 80–180 mm long; verticils 4–6-flowered, 10–20 mm apart; bracts persisting at the apex, mauve-purple, ovate, acute, 10–15 × 5–10 mm. *Calyx* about 8 mm long, glandular-setulose. *Corolla* mauve-pink, 18–22 mm long; tube 15–17 mm long, widening to 6–8 mm at the mouth; upper lip 1 mm long; lower lip 4–6 mm long, horizontal to slightly deflexed. *Stamens* exserted well beyond the lower lip, curled upwards; upper pair attached below the middle of the tube, puberulous at the base; lower pair united to the apex. *Stigma* bifid.

Found in the north-eastern Transvaal on the Soutpansberg and at The Downs at altitudes of 1 400–1 800 m, with bracken and shrub on stony hillsides and forest margins. Map 108.

Vouchers: *Codd* 4188; 8593; *Hutchinson* 2238; *Schlieben & Strey* 8353.

Easily separated from other species with stellate pubescence by the relatively large petiolate leaves. It is an attractive shrub which grows under humid conditions but has not succeeded in cultivation in the drier and colder parts of the Transvaal.

3. Hemizygia rugosifolia *Ashby* in J. Bot., Lond. 73: 344 (1935); Codd in Bothalia 12: 4 (1976). Type: Transvaal, The Downs, *Junod* 4342 (PRE, holo.!).

Erect soft shrub, branched, probably about 1 m tall; stems shortly stellate-tomentose. *Leaves* shortly petiolate; blade ovate-lanceolate to elliptic, 15–25 × 6–10 mm, somewhat coriaceous, upper surface rugose, puberulous, with nerves immersed, under-surface densely greyish stellate-tomentose, apex obtuse, base cuneate, margin finely and regularly crenate-dentate; petiole 2–4 mm long. *Inflorescence* branched or simple, medium lax, 80–130 mm long; verticils 2-flowered, 10–15 mm apart; bracts persisting at the apex, purplish, ovate, acute to acuminate, about 10 × 5 mm. *Calyx* about 10 mm long at maturity. *Corolla* about 22 mm long; tube about 18 mm long, widening to 5–6 mm at the mouth; upper lip 1 mm long; lower lip 4 mm long, usually deflexed. *Stamens* well exserted beyond the lower lip; upper pair attached about the middle of the corolla tube, filaments glabrous; lower pair united to the apex. *Stigma* minutely bifid.

Known from only three gatherings near the Downs and Blyde River escarpment in north-eastern Transvaal, where it apparently grows among quartzite rocks. Map 109.

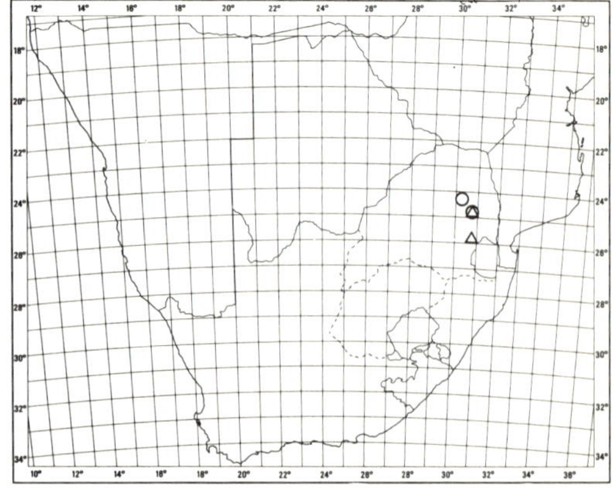

MAP 109. — ○ **Hemizygia rugosifolia**
       △ **H. parvifolia**

Vouchers: *Rogers* 20188; *Van Jaarsveld* 6038.

See the next species, *H. parvifolia*, for differences between the two.

4. Hemizygia parvifolia *Codd* in Bothalia 12: 4 (1976). Type: Transvaal, farm Belvedere, overlooking Blyde River Gorge, *Codd* 10321 (PRE, holo.!).

Twiggy shrub 0,5—1 m tall; stems stellate-floccose. *Leaves* shortly petiolate; blade ovate to broadly ovate, 6—11 × 4—9 mm, discolorous, upper surface rugose, brown, subglabrous, under-surface densely grey tomentose with dendroid hairs, apex obtuse to rounded, base obtuse to truncate, margin minutely crenate-dentate; petiole 1—2,5 mm long. *Inflorescence* simple or with a pair of branches near the base, 50—80 mm long; verticils 2-flowered, 8—15 mm apart; bracts caducous, 2,5—5 mm long, sometimes tinged with purple. *Calyx* 9—10 mm long at maturity. *Corolla* white, 15—17 mm long; tube 11—14 mm long, widening to 4 mm at the mouth; lower lip 4 mm long. *Stamens* well exserted beyond the lower lip; upper pair attached about the middle of the corolla tube, filaments pubescent near the base; lower pair united to the apex. *Stigma* shortly bifid.

Found on the eastern Transvaal escarpment from Kaapsche Hoop to Blyde River, at altitudes of 1 300 to 1 500 m, among quartzite rocks. Map 109.

Vouchers: *Codd* 9555; 10321; *Davidson* 2663.

Allied to *H. rugosifolia* (above) but has smaller, more ovate leaves, smaller, less conspicuous apical bracts and whiter, more dendroid pubescence. In *H. rugosifolia* the pubescence consists mostly of short, simple hairs with a few stellate hairs intermingled.

5. Hemizygia incana *Codd* in Bothalia 12: 5 (1976). Type: Transvaal, Kaapsche Hoop, *Codd* 5758 (PRE, holo.!).

Shrub, sparingly branched, about 0,6 m tall; stems grey tomentose with dendroid hairs and long simple hairs. *Leaves* sessile to subsessile; blade ovate or lanceolate to elliptic-lanceolate, 15—35 × 6—12 mm, densely grey stellate-velvety on both surfaces, upper surface darker grey than lower, apex obtuse to rounded, base obtuse, margin minutely crenate-dentate above the middle. *Inflorescence* simple or with a pair of branches near the base, 80—200 mm long; verticils 4—6-flowered; bracts ovate, caducous, up to 5 mm long. *Calyx* 8—9 mm long

at maturity, glandular-villous. *Corolla* mauve, 12—15 mm long; tube 10—12 mm long, widening to 3—4 mm at the mouth; lower lip 3 mm long. *Stamens* well exserted beyond the lower lip; upper pair attached near the base of the corolla tube, filaments pubescent near the base; lower pair united almost to the apex. *Stigma* shortly bifid.

Found in the neighbourhood of Kaapsche Hoop at an altitude of about 1 800 m, in sandy soil among quartzite rocks. Map 110.

Vouchers: *De Winter* 5083; *Kluge* 2663.

See note after the next species, *H. cinerea*.

6. Hemizygia cinerea *Codd* in Bothalia 12: 6 (1976). Type: Natal, Cathedral Peak Forest Research Station, *Killick* 1644 (PRE, holo.!).

Branched shrub 0,4—1,5 m tall; stems grey tomentose. *Leaves* shortly petiolate; blade lanceolate-elliptic to oblanceolate-elliptic or linear-elliptic, 7—20 × 2—7 mm, densely stellate-tomentose on both surfaces, upper surface darker grey than the lower, apex obtuse to rounded, base cuneate, margin entire or minutely crenate-dentate above the middle; petiole 1—2 mm long. *Inflorescence* simple or sparingly branched near the base, 70—150 mm long; verticils 4—6-flowered; bracts broadly ovate, acute, 4—7 mm long, caducous. *Calyx* 5—7 mm long at maturity, villous and freely gland-dotted. *Corolla* pinkish to mauve, 8—11 mm long; tube 6—9 mm long, widening to 3 mm

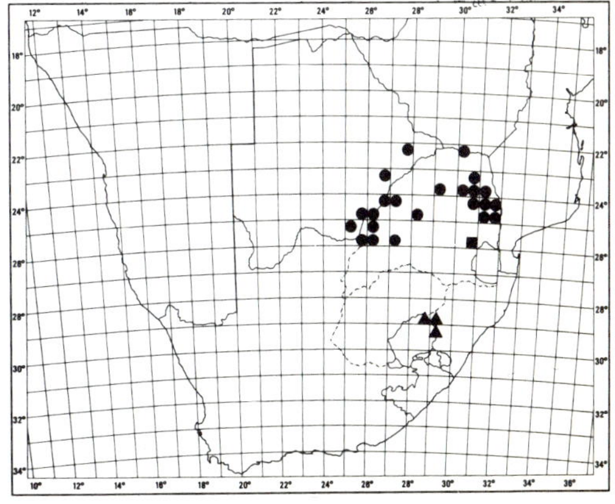

MAP 110. — ■ **Hemizygia incana**
▲ **H. cinerea**
● **H. elliottii**

at the mouth; lower lip 2,5−3 mm long. *Stamens* well exserted beyond the lower lip; upper pair attached near the base of the corolla tube, filaments minutely pubescent near the base; lower pair united almost to the apex. *Stigma* shortly bifid.

Known only from the Natal Drakensberg between Mont-aux-Sources and Cathkin Peak at altitudes of 1 700−2 300 m; often locally common along stream banks, at the foot of cliffs and on mountain sides. Map 110.

Vouchers: *Edwards* 459; 2300; *Galpin* 10168; 11846; *Sidey* 1655.

Allied to *H. incana* (above) but the leaves tend to be smaller with darker upper surfaces and the calyx and corolla are smaller. The next species, *H. elliottii,* differs from *H. cinerea* in having a distinct coma of mauve-purple bracts at the apex of the inflorescence, the calyx is stellate tomentose, not villous, and the stigma is capitate, not shortly bifid; it is also a species of the hot, dry savanna areas rather than of high altitudes. Also related to *H. stenophylla* (no. 10), which occurs further south and at lower altitudes, and which has somewhat longer leaves with revolute margins, colourful bracts at the apex of the inflorescence and a glandular-hispid calyx.

7. **Hemizygia elliottii** *(Bak.) Ashby* in J. Bot., Lond. 73: 345 (1935), partly, excluding Natal specimens; Codd in Bothalia 12: 7 (1976). Type: Matabeleland, *Elliott* s.n. (K, holo.).

Orthosiphon elliottii Bak. in F.T.A. 5: 376 (1900).

O. messinensis Good in J. Bot., Lond. 63: 173 (1925). Type: Transvaal, Messina, *Moss & Rogers* 153 (BM, holo.; PRE!).

Soft branched shrub 0,35−0,6 m tall, woody at the base; stems stellate-tomentose. *Leaves* subsessile to shortly petiolate; blade lanceolate to ovate, 15−25 × 4−12 mm, densely stellate grey velvety on both surfaces, apex acute, base obtuse, margin entire. *Inflorescence* simple or occasionally with a pair of branches at the base; verticils 2−6-flowered; bracts broadly ovate to subrotund, 7−11 × 5−8 mm, persisting as a dense, mauve-purple coma. *Calyx* 5 mm long at maturity, sparingly stellate tomentose. *Corolla* white to pale mauve, 13 mm long; tube 9 mm long, widening to 3 mm at the mouth; lower lip 4 mm long, often deflexed. *Stamens* shortly exserted, not or only slightly exceeding the lower lip; upper pair attached about the middle of the tube, filaments pubescent for about two-thirds their length; lower pair

loosely joined for about half their length. *Style* capitate.

Found in Botswana and in western, northern and eastern Transvaal at altitudes of 300 to 1 300 m, in dry subtropical savanna; also in Zimbabwe. Map 110.

Vouchers: *Codd* 5036; 8658; 8857; *Leistner* 3184; *Schlechter* 4676.

See note after *H. cinerea* (above).

An interesting variation is found in the number of flowers per verticil. In all specimens from Zimbabwe, Botswana, and western and northern Transvaal, the verticils are 2-flowered, whereas in those from the eastern Transvaal lowveld the verticils are 4−6-flowered.

8. **Hemizygia gerrardii** *(N.E. Br.) Ashby* in J. Bot., Lond. 73: 345 (1935); Ross, Fl. Natal 306 (1972); Codd in Bothalia 12: 8 (1976). Type: Natal, "near Ingoma," *Gerrard* 1239 (K, holo.; PRE, fragment!).

Orthosiphon gerrardii N.E. Br. in F.C. 5,1: 249 (1910).

Soft, branched shrub c. 1 m tall; stems stellate-pubescent, glabrescent, bark flaking in strips. *Leaves* shortly petiolate; blade ovate to broadly elliptic, c. 15 × 10 mm, thickish, densely and somewhat coarsely grey stellate-velvety on both surfaces, apex obtuse, base obtuse to truncate, margin entire; petiole 1−3 mm long. *Inflorescence* usually simple, 40−50 mm long; verticils 2-flowered; bracts broadly elliptic, c. 8 × 5 mm, persisting as a mauve-purple coma. *Calyx* 8−10 mm long at maturity, stellate-tomentose. *Corolla* mauve-pink, 20−25 mm long; tube 17−20 mm long, 3 mm wide at the mouth; lower lip 4−6 mm long. *Stamens* well exserted beyond the lower lip; upper pair attached near the throat; lower pair united for about half their length. *Stigma* entire or minutely bifid.

Known from only 2 gatherings, one from northern Natal and the other from southern Transvaal, in grass among rocks. Map 111.

Voucher: *Dyer & Verdoorn* 5829.

Resembles *H. elliottii* (above) in the entire, grey tomentose leaves and the entire (or almost entire) stigma, but the calyx and corolla are considerably larger and the stamens well exserted.

9. **Hemizygia floccosa** *Launert* in Mitt. bot. StSamml., Münch. 7: 302 (1968); Launert & Schreiber in F.S.W.A. 123: 13 (1969); Codd in Bothalia 12: 8 (1976). Type: S.W.A./Namibia, Outjo, *De Winter & Hardy* 8139 (PRE, holo.!; M).

Soft shrublet 0,4—0,8 m tall, sparingly branched; stems loosely dendroid-floccose, glabrescent and pale reddish brown with age. *Leaves* petiolate; blade ovate, 28—45 × 15—22 mm, loosely to densely floccose on both surfaces, apex subacute, base obtuse, margin obscurely and somewhat distantly crenate-dentate; petiole 5—10 mm long. *Inflorescence* simple or with a pair of branches near the base; verticils 2-flowered; bracts 5 × 2,5 mm, deciduous. *Calyx* 11 mm long at maturity, glandular-strigose. *Corolla* pale mauve, c. 20 mm long; tube c. 15 mm long, widening to 5 mm at the mouth; lower lip 5 mm long. *Stamens* shortly exserted, not exceeding the lower lip; upper pair attached near the middle of the corolla tube, filaments pubescent near the base; lower pair united for about half their length. *Stigma* capitate.

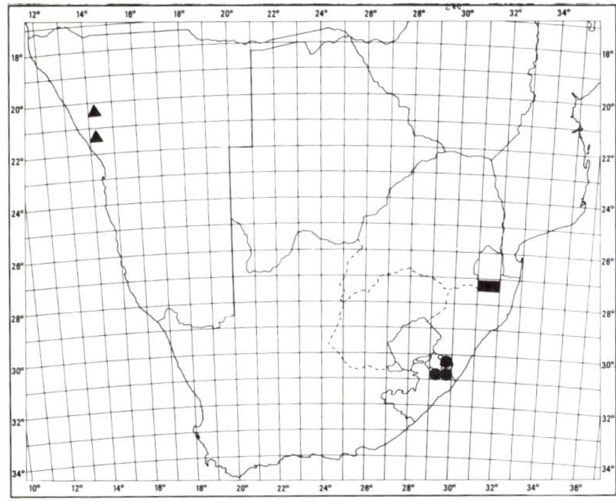

MAP 111. — ■ **Hemizygia gerrardii**
    ▲ **H. floccosa**
    ● **H. stenophylla**

A rare plant found in the central, semi-coastal part of S.W.A./Namibia, in dry watercourses. Map 111.

Vouchers: *Giess* 3929; 5003; 7900.

Related to *H. elliottii* (no. 7) but readily distinguished by the larger leaves which are petiolate, distantly toothed and floccose.

10. **Hemizygia stenophylla** *(Gürke)* Ashby in J. Bot., Lond. 73: 347 (1935); Ross, Fl. Natal 306 (1972); Codd in Bothalia 12: 8 (1976). Type: Natal, East Griqualand, near Enyembe, *Tyson* in Herb. Austr. Afr. 1293 (K, lecto.!; PRE!).

Orthosiphon stenophyllus Gürke in Bot. Jb. 26: 84 (1898); N.E. Br. in F.C. 5,1: 250 (1910).

Soft shrub 0,3—0,9 m tall, usually with several stems arising from a perennial rootstock; stems shortly stellate-tomentose. *Leaves* subsessile; blade linear-lanceolate or elliptic-lanceolate to lanceolate, 12—30 × 3—5 mm, upper surface grey to blackish and finely pubescent, under-surface densely grey stellate-velvety, apex acute, base obtuse, margin revolute, entire. *Inflorescence* simple or with 1 or 2 pairs of branches near the base, 80—180 mm long; verticils 4—6-flowered; bracts ovate-lanceolate, 10—15 mm long, persisting as a mauve-purple coma. *Calyx* 7—8 mm long at maturity, glandular-hispidulous. *Corolla* pale mauve to rosy mauve, c. 13 mm long; tube c. 10 mm long, widening to 3 mm at the mouth; lower lip 3 mm long, deflexed. *Stamens* well exserted beyond the lower lip; upper pair attached about the middle of the corolla tube, pubescent in the lower part; lower pair united to near the apex. *Stigma* minutely bifid.

Found in southern Natal, East Griqualand and the adjoining Transkei, in dense grassland often near forest and among rocks. Map 111.

Vouchers: *Codd* 8568; *Hilliard & Burtt* 6748; *Strey* 6300; 6334.

Resembles *H. rehmannii* (below) but has slightly narrower, more lanceolate leaves and a tuft of conspicuous mauve-purple bracts at the apex of the inflorescence. *H. cinerea* (no. 6), which occurs in the Natal Drakensberg at higher altitudes and also lacks the conspicuous coma of bracts, has more elliptical leaves and the calyx is distinctly villous.

11. **Hemizygia rehmannii** *(Gürke)* Ashby in J. Bot., Lond. 73: 347 (1935); Codd in Bothalia 12: 9 (1976). Type: Transvaal, Houtboschberg, *Rehmann* 6172 (Z, holo.; BM; photo of BM specimen in PRE!).

Orthosiphon rehmannii Gürke in Bull. Herb. Boissier 6: 557 (1898); N.E. Br. in F.C. 5,1: 251 (1910).

Soft shrub branching from a perennial woody rootstock, forming a round bush 0,3—0,8 m tall; stems villous with short stellate hairs intermingled. *Leaves* sessile; blade narrowly elliptic to oblong-elliptic, 10—22 × 3—8 mm, upper surface dark grey to brownish and finely pubescent, under-surface densely grey to yellowish grey stellate-velvety, apex acute to obtuse, base somewhat cuneate, margin revolute, entire or finely toothed above the middle. *Inflorescence* simple or branched, 60—220 mm long; verticils 4—6-flowered; bracts ovate, acute, 5—6 mm long, caducous. *Calyx* 9—10 mm long at maturity, glandular-hispid. *Corolla* pale mauve, c. 17 mm long; tube c. 14 mm long, widening to 4 mm at the mouth; lower lip 3 mm long, eventually deflexed. *Stamens* well exserted beyond the lower lip; upper pair attached below the middle of the tube, filaments glabrous; lower pair united to near the apex. *Stigma* minutely bifid.

Found on the Drakensberg escarpment of northeastern Transvaal from Woodbush to The Downs at altitudes of 1 500 to 2 000 m; in shallow soil among rocks in grassland, often near forest margins. Map 112.

Vouchers: *Codd* 9426; *Scheepers* 909; *Schlechter* 4442.

See note after *S. stenophylla* (above).

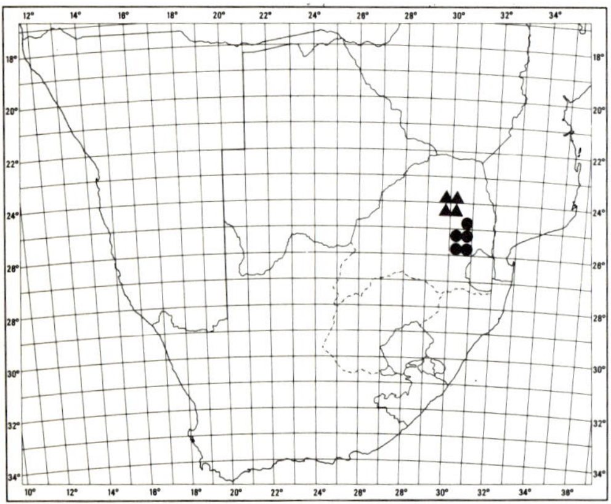

MAP 112.— ▲ **Hemizygia rehmannii**
● **H. subvelutina**

12. **Hemizygia subvelutina** *(Gürke) Ashby* in J. Bot., Lond. 73: 346 (1935); Codd in Bothalia 12: 9 (1976). Type: Transvaal, Lydenburg, near Paarde Plaats, *Wilms* 1152 (BM; K; photo of BM specimen in PRE).

Orthosiphon subvelutinus Gürke in Bot. Jb. 26: 80 (1898); N.E. Br. in F.C. 5,1: 253 (1910).

O. heterophyllus Gürke, l.c. 82 (1898). Syntypes: Transvaal, near Spitzkop, *Wilms* 1148; 1155 (BM; K).

Bushy herb or soft shrublet 0,2—0,5 (—0,8) m tall with few to many erect or ascending stems arising annually from a perennial woody rootstock; stems slender, sparingly branched, densely beset with leaves and short leafy shoots, densely stellate-pubescent, often with a yellowish tinge. *Leaves* sessile; blade somewhat ericoid, linear to linear-lanceolate (occasionally ovate near base of stem), 5—10 (—15) × 1—2 (—5) mm, coriaceous, upper surface stellate-scabrid, under-surface usually yellowish stellate-tomentose, margin revolute, entire. *Inflorescence* simple, 50—110 mm long; verticils 4—6-flowered, in the axils of persistent somewhat leaf-like bracts, 4—7 × 2—3 mm. *Calyx* 5—6 mm long at maturity, stellate-hispid. *Corolla* white, often tinged with mauve, 12—16 mm long; tube 10—12 mm long, tubular, 2,5 mm wide, often slightly constricted at the throat; lower lip 2—4 mm long. *Stamens* shortly exserted, not or scarcely exceeding the lower lip; upper pair attached below the middle of the tube, filaments puberulous near the base; lower pair united only near the base or to about half their length. *Stigma* shortly bifid.

Localized on the eastern Transvaal mountains from Lydenburg and Pilgrims Rest to Kaapsche Hoop at altitudes of 1 400 to 2 200 m; in dense grass among quartzite rocks and in rock crevices. Map 112.

Vouchers: *Codd* 5751; 8306; 9480; *Galpin* 14447.

See note after the next species, *H. teucriifolia*, to which *H. subvelutina* is closely related and of which it could be regarded as a subspecies. *H. subvelutina* tends to have narrower, more ericoid leaves, shorter internodes and a yellowish tomentum, but there appear to be some intermediates in the Kaapsche Hoop area. *H. teucriifolia* is mainly a Natal species which extends to the Barberton mountains in the Transvaal.

13. **Hemizygia teucriifolia** *(Hochst.) Briq.* in Natürl. PflFam. 4,3a: 369 (1897); in Annu. Conserv. Jard. bot. Genève 2: 247 (1898); Ashby in J. Bot., Lond. 73: 346 (1935); Ross, Fl. Natal 306 (1972); Codd in Bothalia 12: 9 (1976). Type: Natal, Table Mtn, *Krauss* 448 (BM; K; photo of BM specimen in PRE!).

Ocimum teucriifolium Hochst. in Flora 28: 66 (1845); Benth. in DC., Prodr. 12: 41 (1848). *Orthosiphon teucriifolius* (Hochst). N.E. Br. in F.C. 5,1: 254 (1910).

Orthosiphon woodii Gürke in Bot. Jb. 26: 83 (1898). Type: Natal, Entumeni, *Medley Wood* sub NH 783 (= *Medley Wood* 3964 in K; NH!).

H. galpiniana Briq. in Bull. Herb. Boissier sér. 2,3: 993 (1903). *Orthosiphon teucriifolius* var. *galpinianus* (Briq.) N.E. Br. in F.C. 5,1: 254 (1910). Type:

Transvaal, Barberton, Saddleback, *Galpin* 1217 (K; NH!; PRE!).

Bushy herb 0,15—0,3 m tall, with few to many erect or ascending stems arising annually from a perennial woody rootstock; stems slender, usually simple, greyish stellate-pubescent. *Leaves* sessile; blade subcoriaceous, linear to lanceolate or elliptic, 8—18 × 3—6 mm, upper surface blackish, stellate-scabrid; under-surface greyish stellate-tomentose, apex acute, base obtuse, margin revolute, entire. *Inflorescence* simple, 40—80 mm long; verticils 4—6-flowered, in the axils of persistent somewhat leaf-like bracts, 4—6 × 2—3 mm. *Calyx* 5—6 mm long at maturity, stellate-pubescent. *Corolla* mauve, 10—12 mm long; tube 9—10 mm long, tubular, 2,5 mm wide, slightly constricted at the throat, glabrous; lower lip 2 mm long. *Stamens* shortly exserted, not or scarcely exceeding the lower lip; upper pair attached below the middle of the tube, filaments puberulous in the lower half; lower pair united for half or more of their length. *Stigma* minutely bifid.

Found at scattered localities from Stutterheim in the eastern Cape Province to Natal and into the Barberton district of the Transvaal at altitudes of 600 to 1 500 m, not as yet recorded from Swaziland; locally frequent in dense grassland, tending to spread where overgrazing has occurred. Map 113.

Vouchers: *Acocks* 11781; *Hilliard & Burtt* 3193; 6980; 8369; *Strey* 9217; 9338.

See also the closely related *H. subvelutina* (above) for differences between the two. These two species have certain characteristics which set them apart from most other species, for example: the simple, not branched inflorescence; the leaf-like bracts which persist for the entire length of the inflorescence; the tubular corolla tube; and the shortly exserted stamens. Somewhat similar characteristics are shown by *H. pretoriae* (no. 15) in which the two upper stamens are not exserted, and *H. persimilis* (no. 23), in which the bracts, though persistent, are more colourful.

14. **Hemizygia albiflora** *(N.E. Br.) Ashby* in J. Bot., Lond. 73: 348 (1935); Codd in Bothalia 12: 10 (1976); Compton, Fl. Swaziland 509 (1976). Type: Transvaal, Pilgrims Rest district, Mac Mac, *Mudd* s.n. (K, holo.).

Orthosiphon albiflorus N.E. Br. in F.C. 5,1: 251 (1910).

O. decipiens N.E. Br., l.c. 252 (1910). Type: Transvaal, Mac Mac, *Mudd* s.n. (K, holo.).

Woody shrublet, somewhat gnarled and branching, decumbent to ascending, 0,3—1,5 m tall; stems villous. *Leaves* sessile; blade ericoid, coriaceous, linear to linear-elliptic, 10—30 × 1,5—5 mm, upper surface dark green to blackish and subglabrous to appressed villous, under-surface densely appressed villous with long white matted hairs; apex and base tapering, margin strongly revolute, entire. *Inflorescence* simple or with a pair of branches near the base, 50—100 mm long; verticils 4—6-flowered; bracts ovate, acute, 5—8 × 3—5 mm, caducous. *Calyx* 6—8 mm long at maturity, glandular-hispid. *Corolla* white, 12—15 mm long; tube 10—15 mm long, more or less tubular, 3 mm wide at the mouth; lower lip 2—3 mm long. *Stamens* exserted beyond the lower lip; upper pair attached below the middle of the tube, filaments finely puberulous in the lower half; lower pair united for almost their entire length. *Stigma* minutely bifid.

Found on the eastern Transvaal mountains from Mariepskop to Barberton and extending into northern Swaziland at altitudes of 1 800 to 2 400 m; usually in crevices in quartzite rocks. Map 113.

Vouchers: *Codd* 7853; 8207; 8270; *Galpin* 13052; 13068; 14352.

15. **Hemizygia pretoriae** *(Gürke) Ashby* in J. Bot., Lond. 73: 356 (1935). Type: Transvaal, Pretoria, *Wilms* 1151 (BM).

Bushy herb 0,1—0,3 m tall with few to many erect or ascending stems arising annually from a perennial woody rootstock;

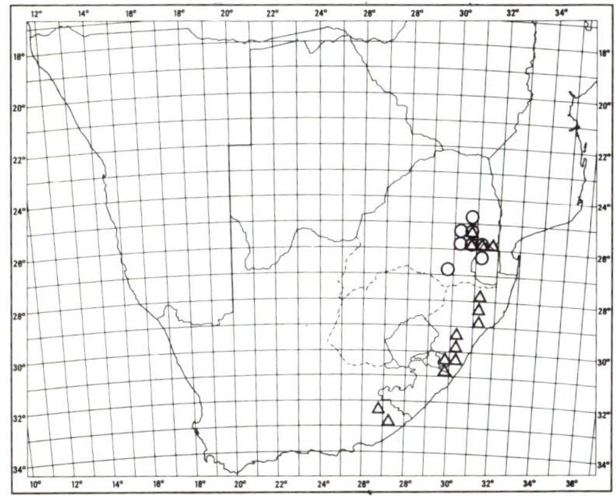

MAP 113. — △ **Hemizygia teucriifolia**
○ **H. albiflora**

stems simple, slender, hispid to villous, sometimes with branched hairs intermingled (subsp. *heterotricha*). *Leaves* subsessile to shortly petiolate; blade narrowly elliptic or oblanceolate to ovate, obovate or subrotund, 8−24 × 2−15 mm, subglabrous to villous or tomentose, sometimes with stellate or branched hairs, conspicuously glanddotted, often folded along the midrib, apex acute to obtuse, base cuneate, margin entire or rarely with a few small teeth in the upper third. *Inflorescence* simple, 40−80 mm long; verticils (2−) 3−6-flowered, borne in the axils of persistent, leaf-like bracts 6−10 × 3−4 mm. *Calyx* 9−11 mm long in fruit, glandular-hispid. *Corolla* whitish to pale mauve, 14−16 mm long; tube 10−12 mm long, narrowly tubular, widening slightly to 2 mm at the throat; upper lip narrow, 3 mm long; lower lip 4 mm long. *Stamens:* upper pair included, attached near the middle of the tube with glabrous filaments; lower pair united for more than half their length, exserted by 2−3 mm. *Stigma* minutely bifid.

Distributed from central to eastern Transvaal, Swaziland and northern Natal, in dense grassland subject to periodic burning, often among rocks, at altitudes of 1 000 to 1 800 m.

An anomalous species in which the bracts subtending the verticils are persistent and leaf-like, the corolla tube and lips are relatively long and narrow resembling *Orthosiphon* (no. 36), and the upper two stamens are included in the corolla tube. However, the united lower stamens, which are shortly exserted, and the large upper tooth of the calyx, indicate that it belongs in *Hemizygia*.

Two subspecies are recognized and are separated on the presence or absence of stellate or dendroid hairs (see key to species).

(a) subsp. **pretoriae**.

Codd in Bothalia 12: 11 (1976).

Orthosiphon pretoriae Gürke in Bot. Jb. 26: 81 (1898); N.E. Br. in F.C. 5,1: 254(1910). *Hemizygia pretoriae* (Gürke) Ashby in J. Bot., Lond. 73: 356 (1935); Ross, Fl. Natal 306 (1972); Compton, Fl. Swaziland 510 (1976), partly. Type: Transvaal, Pretoria, *Wilms* 1151 (BM).

O. natalensis Gürke, l.c. 82 (1898). Syntypes: Natal, Glencoe, *Medley Wood* 4756 (K, NH!); *Kuntze* s.n.; Coldstream, *Rehmann* 6918 (K).

No stellate or branched hairs present; leaves narrowly elliptic to oblanceolate,

obovate or ovate, subglabrous to densely pubescent. Fig. 35.

Distribution and ecology more or less as for the species but absent from the areas where subsp. *heterotricha* occurs. Map 114.

Vouchers: *Acocks* 11256; 20880; *Galpin* 9645; 12442; *C.A. Smith* 1062.

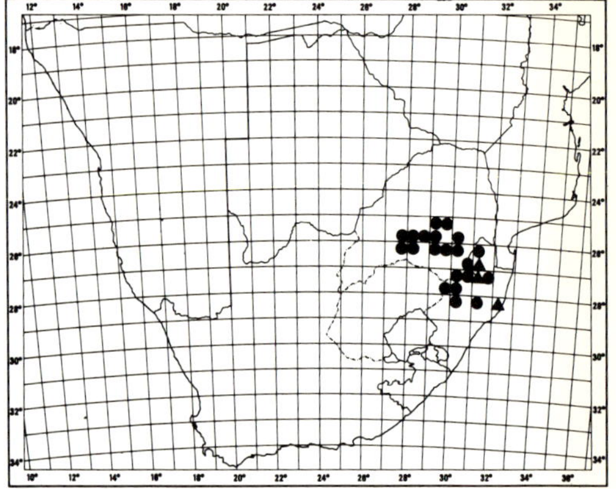

MAP 114. — ● **Hemizygia pretoriae** subsp. **pretoriae**
▲ **H. pretoriae** subsp. **heterotricha**

(b) subsp. **heterotricha** *Codd* in Bothalia 12: 11 (1976). Type: Swaziland, near Hlatikulu, *Compton* 26320 (PRE, holo.!).

Stellate or branched hairs present on stems, leaves and bracts often mixed with long simple hairs; leaves ovate to ovate-rotund, usually densely pubescent.

Found in south-western Swaziland, the Piet Retief area of Transvaal and the Hluhluwe area of Natal. Map 114.

Vouchers: *Acocks* 13154; *Compton* 28323; 30458.

16. **Hemizygia modesta** *Codd* in Bothalia 12: 12 (1976); Compton, Fl. Swaziland 510 (1976). Type: Swaziland, Mbabane, Bomvu Ridge, *Compton* 28368 (PRE, holo.!).

Herb 0,12−0,25 m tall with a few to several stems arising annually from a perennial woody rootstock; stems slender, sparingly branched, softly woody below,

FIG. 35. — 1, **Hemizygia pretoriae** subsp. **pretoriae**, portion of plant, × 1; a, flower, × 2; b, section through corolla, × 2; c, mature calyx, × 2 (*Mrs B. Clarke* s.n., Pretoria District).

hispid to villous. *Leaves* sessile or subsessile; blade lanceolate-elliptic or elliptic to broadly ovate, 6−12 × 4−6 mm, sparingly to densely hispid, apex acute to obtuse, base obtuse, margin entire. *Inflorescence* simple, 50−100 mm long; verticils 2-flowered; bracts ovate, 4−5 mm long, caducous. *Calyx* 7−8 mm long at maturity, hispid, freely gland-dotted. *Corolla* white to pale mauve, 12−15 mm long; tube 8−12 mm long, widening to 4 mm at the mouth; lower lip 3−5 mm long. *Stamens* well exserted beyond the lower lip; upper pair attached about the middle of the tube, filaments pubescent near the base; lower pair united to near the apex. *Stigma* shortly bifid.

Found in the Piet Retief and Barberton districts of Transvaal and the adjoining parts of Swaziland; in mountain grassland subjected to periodic burning. Map 115.

Vouchers: *Acocks* 12867; *Compton* 29123; 30013; *Leipoldt* s.n.

It is sometimes confused with another dwarf grassland species, *H. thorncroftii* (no. 22), but the latter usually has narrower leaves, 4−6-flowered verticils, and a persistent coma of colourful bracts; see also note after the following species.

17. **Hemizygia punctata** *Codd* in Bothalia 12: 13 (1976). Type: Transvaal, 18 km S.W. of Lydenburg, *Codd* 8038 (PRE, holo.!).

Soft shrub 0,6−1,2 m tall, branching above; stems slender, hispidulous. *Leaves* shortly petiolate; blade elliptic or elliptic-oblanceolate to obovate, 10−15 × 3−10 mm, subglabrous to hispidulous, gland-dotted, apex acute to obtuse, base cuneate, margin entire or with a few teeth in the upper half. *Inflorescence* simple or with a pair of branches near the base, 80−150 mm long; verticils 2-flowered; bracts broadly ovate, 4−6 mm long, caducous. *Calyx* 7−9 mm long at maturity, glandular-hispid. *Corolla* pale mauve, 9−12 mm long; tube 7−10 mm long, widening to 4 mm at the mouth; lower lip 2−3 mm long. *Stamens* well exserted beyond the lower lip; upper pair attached near the base of the tube, filaments pubescent near the base; lower pair united to near the apex. *Style* shortly bifid.

Recorded from the Lydenburg, Nelspruit and Barberton districts, on stony slopes in grassland. Map 115.

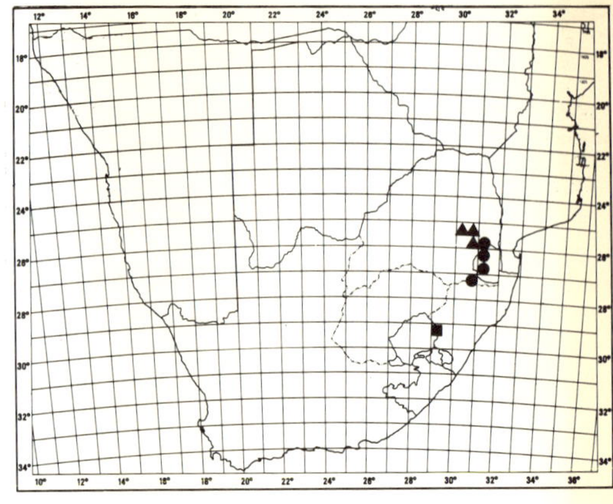

MAP 115. — ● **Hemizygia modesta**
▲ **H. punctata**
■ **H. bolusii**

Vouchers: *Edwards* 4113; *Liebenberg* 3323; *Van Jaarsveld* 1014.

Closely related to the previous species, *H. modesta*, but grows at lower altitudes forming a taller soft shrub up to 1,2 m tall, and tends to have smaller flowers. It superficially resembles the small-leaved form of *H. transvaalensis* (no. 20), but the latter has 4−6-flowered verticils and a persistent coma of colourful bracts.

18. **Hemizygia bolusii** *(N.E. Br.) Codd* in Bothalia 8: 159 (1964); Ross, Fl. Natal 306 (1972); Codd in Bothalia 12: 14 (1976). Type: Natal, Giants Castle, *A. Bolus* in Herb. Guthrie 4894 (BOL, holo.!).

Orthosiphon bolusii N.E. Br. in F.C. 5,1: 258 (1910).

Stems several, erect, 0,25−0,3 m tall arising annually from a perennial woody rootstock; stems sparingly branched, villous. *Leaves* petiolate; blade ovate, 20−25 × 14−18 mm, upper surface brownish and appressed hispid, under-surface paler, hispid to villous, apex and base obtuse to rounded, margin with a few minute teeth above the middle; petiole 2−4 mm long. *Inflorescence* simple, 100−140 mm long; verticils 4−6-flowered; bracts ovate, 4−5 × 2−2,5 mm. *Calyx* 11−12 mm long at maturity, glandular-villous. *Corolla* 14 mm long; tube 10 mm long, widening to 4−5 mm; lower lip 4 mm long, eventually deflexed. *Stamens* well exserted beyond the lower lip; upper pair attached below the middle of the tube, filaments puberulous near the base; lower pair united to near the apex.

Known from only one gathering near Giants Castle in the Natal Drakensberg at about 3 000 m; in mountain grassland. Map 115.

Voucher: only the type seen.

19. Hemizygia ramosa *Codd* in Bothalia 12: 14 (1976). Type: Natal, near Mkuze, *Moll* 3158 (PRE, holo.!).

Shrub 1−2 m tall, much branched; stems shortly tomentose. *Leaves* shortly petiolate; blade obovate to oblanceolate, 15−25 × 6−11 mm, upper surface sparingly hispid, under-surface hispid and gland-dotted, apex rounded, base obtuse to cuneate, margin obscurely crenate-dentate mainly above the middle; petiole 1−3 mm long. *Inflorescence* usually sparingly branched near the base, lax, 10−150 mm long; verticils 2-flowered; bracts ovate to broadly elliptical, 14−16 × 6−8 mm, mauve pink, persisting as a distinct coma. *Calyx* 10 mm long at maturity, sparingly glandular-hispidulous. *Corolla* mauve, 25−28 mm long; tube 20−22 mm long, widening to 4−5 mm at the mouth; lower lip 5 mm long. *Stamens* well exserted beyond the lower lip; upper pair attached about the middle of the tube, filaments pubescent near the base; lower pair united to the apex. *Stigma* shortly bifid.

Recorded from the southern end of the Lebombo Range, near Mkuze in Natal; in shallow soil among rocks in open woodland. Map 116.

Voucher: *Ward* 4074.

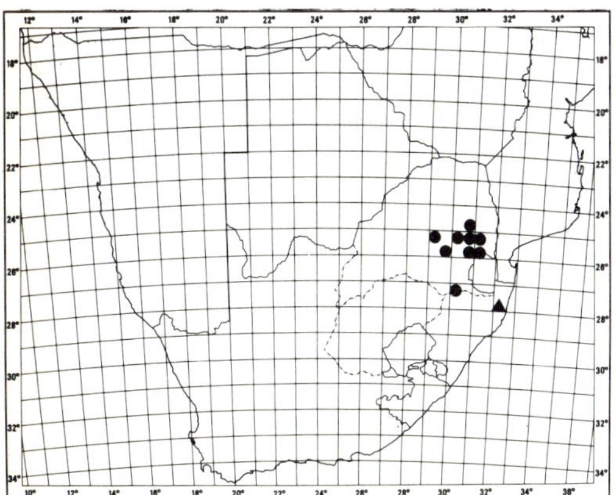

MAP 116. — ▲ **Hemizygia ramosa**
      ● **H. transvaalensis**

20. Hemizygia transvaalensis *(Schltr.) Ashby* in J. Bot., Lond. 73: 349 (1935); Letty, Wild Flow. Transv. 285, t.141, 4 (1962); Codd in Bothalia 12: 15 (1976). Type: Transvaal, Barberton, *Galpin* 468 (PRE!; SAM!).

Orthosiphon transvaalensis Schltr. in J. Bot., Lond. 35: 281 (1897); N.E. Br. in F.C. 5,1: 244 (1910).

Ocimum wilmsii Gürke in Bot. Jb. 26: 79 (1898). Syntypes: Transvaal, Lydenburg, *Wilms* 1107 (BM; K; PRE!); 1108.

Orthosiphon muddii N.E. Br., l.c. 245 (1910). Syntypes: Transvaal, Drakensberg, *Mudd* s.n. (K; PRE, fragment!); Spitzkop, *Burtt Davy* 1570 (K).

Soft shrublet 0,3−1 m tall; stems arising annually from a perennial woody rootstock, sparingly to freely branched and sometimes broom-like (in the latter case with many small leaves), sparingly to densely hispid. *Leaves* sessile or shortly petiolate; blade, in typical form, ovate to broadly ovate, 15−40 × 8−22 mm, in broom-like form ovate-elliptic to ovate, 12−20 × 4−8 mm, concolorous, sparingly to densely pubescent on both surfaces, apex acute to obtuse, base obtuse to rounded, margin serrate-dentate chiefly in the upper two-thirds, rarely with teeth obscure. *Inflorescence* paniculate, lax, 70−200 mm long; verticils (2−) 3−6-flowered; bracts ovate to lanceolate, the terminal ones pinkish purple, 12−24 × 4−10 mm, often forming a lax coma. *Calyx* 12−14 mm long at maturity, densely glandular-hispidulous. *Corolla* whitish to mauve or lilac-pink, 18−22 mm long; tube 14−17 mm long, widening to 5 mm at the mouth; lower lip 4−6 mm long, often deflexed. *Stamens* well exserted beyond the lower-lip; upper pair attached about the middle of the tube, filaments pubescent in the lower half; lower pair united to the apex or nearly so. *Stigma* bifid.

Found in the eastern Transvaal from Mariepskop and Lydenburg to Barberton, at medium altitudes of 1 000 to 1 700 m; often locally common on grassy slopes and flats, often among rocks. Map 116.

Vouchers: *Galpin* 14313; 14553; *Rogers* 23232; *Schlechter* 3916.

Some specimens branch freely and produce numerous small leaves giving them a broom-like appearance. The type of *Ocimum wilmsii* is such a specimen. However, there appear to be intermediates between this form and the typical specimens and so separate status for the small-leaved form is not considered justified. There are no floral differences to support a formal subdivision of the species.

H. transvaalensis is related to the next two species *H. foliosa* and *H thorncroftii* but can usually be distinguished by its more robust stature and the longer corolla (18−22 mm). Depauperate specimens may flower when only 0,2 m tall and these may be confused with *H. thorncroftii* which, however, usually has narrowly elliptic leaves and the corolla is 14−16 mm long. *H. foliosa* tends to have decumbent stems with larger, elliptical leaves, 2-flowered verticils, and the corolla is 12−14 mm long.

21. **Hemizygia foliosa** S. Moore in J. Bot., Lond. 43: 172 (1905); Ashby in J. Bot., Lond. 73: 348 (1935); Codd in Bothalia 12: 15 (1976); Compton, Fl. Swaziland 510 (1976). Type: Swaziland, Mbabane, *Burtt Davy* 2833 (BM, holo.; K; PRE!).

Orthosiphon foliosus (S. Moore) N.E. Br. in F.C. 5,1: 243 (1910).

O. humilis N.E. Br., l.c. 259 (1910). *Hemizygia humilis* (N.E. Br.) Ashby, l.c. 348 (1935). Type: Transvaal, Waterval Onder, *Rogers* 4375 (K, holo.; PRE!).

Perennial herb with 1−several stems from a woody rootstock; stems decumbent to ascending 0,2−0,35 m long, thinly to densely villous. *Leaves* sessile or shortly petiolate; blade ovate to ovate-elliptic or elliptic, variable in size but usually large when mature, 25−70 × 15−35 mm, concolorous, subglabrous to pilose and gland-dotted on both surfaces, apex obtuse to rounded, base obtuse to truncate, margin entire to somewhat distantly dentate. *Inflorescence* paniculate, lax, 100−200 mm long; verticils 2-flowered; bracts ovate-lanceolate, 8−18 × 3−8 mm, mauve-purple, persisting as an apical coma. *Calyx* 9−10 mm long at maturity, glandular-hispid. *Corolla* whitish to mauve, 12−14 mm long; tube 9−10 mm long, widening to 4 mm at the mouth; lower lip 3−4 mm long, often deflexed. *Stamens* well exserted beyond the lower lip; upper pair attached about the middle of the tube, filaments pubescent near the base; lower pair united to the apex. *Stigma* bifid.

Found in the south-eastern Transvaal and western Swaziland at altitudes of 1 300 to 1 700 m; in dense mountain grassland, often among rocks. Map 117.

Vouchers: *Bolus* 12250; *Codd* 2101; 4726; 9507; *Galpin* 10207.

See note after *H. transvaalensis* (above). The type of *H. humilis* has densely pubescent and smaller leaves and the bracts are smaller (about 10 mm long), but there are intermediates linking it with the typical form.

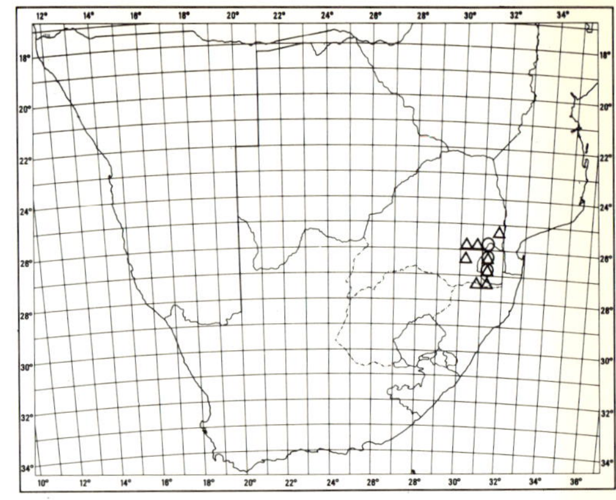

Map 117. — △ **Hemizygia foliosa**
○ **H. thorncroftii**

22. **Hemizygia thorncroftii** *(N.E. Br.) Ashby* in J. Bot., Lond. 73: 349 (1935); Codd in Bothalia 12: 16 (1976); Compton, Fl. Swaziland 510 (1976). Type: Transvaal, Barberton, *Thorncroft* sub TRV 3123 (K, lecto.; PRE, fragment!; = *Thorncroft* sub TRV 3125 in PRE!). See note below on the confusion of numbers.

Orthosiphon thorncroftii N.E. Br. in F.C. 5,1: 246 (1910).

Perennial herb 0,15−0,30 m tall with few to several erect stems arising annually from a woody rootstock; stems slender, subglabrous to glandular-hispid. *Leaves* subsessile; blade elliptic to linear-elliptic, 15−40 × 4−10 mm, concolorous, sparingly pubescent on both surfaces, apex obtuse to acute, base cuneate to attenuate, margin with a few small teeth towards the apex. *Inflorescence* simple or occasionally with a pair of branches near the base, 70−100 mm long, lax; verticils 3−6-flowered; bracts lanceolate to linear-lanceolate, 15−30 × 2−5 mm, mauve-purple, persisting as an apical coma. *Calyx* 10−11 mm long at maturity, glandular-hispid. *Corolla* mauve, 14−16 mm long; tube 10−12 mm long, widening to 4 mm at the mouth; lower lip 4 mm long, often deflexed. *Stamens* well exserted beyond the lower lip; upper pair attached about the middle of the tube, filaments pubescent in the lower part; lower pair united for their entire length. *Stigma* minutely bifid.

Found in the Barberton area of the Transvaal and in western Swaziland at altitudes of 1 000 to 1 800 m; in mountain grassland. Map 117.

Vouchers: *Codd* 9791; *Compton* 29076; 29165; *Galpin* 465.

See note after *H. transvaalensis* (no. 20). In the Transvaal Museum Herbarium register (now in PRE), two *Thorncroft* specimens were entered on the same day; no. 3123 is *H. transvaalensis* and no. 3125 is *H. thorncroftii*. When duplicates were sent to Kew the numbers appear to have become interchanged so that, on the Kew specimens, no. 3123 is *H. thorncroftii* and no. 3125 is *H. transvaalensis*.

23. **Hemizygia persimilis** *(N.E. Br.) Ashby* in J. Bot., Lond. 73: 349 (1935); Codd in Bothalia 12: 16 (1976). Lectotype: Transvaal, Barberton, *Thorncroft* sub TRV 3132 (K, lecto.; PRE!; SAM!).

Orthosiphon persimilis N.E. Br. in F.C. 5,1: 246 (1910).

O. rogersii N.E. Br., l.c. 247 (1910). Syntypes: Transvaal, Nelspruit, *Rogers* sub TRV 4740 (K; PRE!; SAM!); Devil's Kantoor, Kaapsche Hoop, *Bolus* 9742.

Bushy herb 0,15—0,3 m tall with several erect stems arising from a perennial woody rootstock; stems simple or branched, glandular-hispid. *Leaves* subsessile; blade lanceolate-elliptic, 15—20 × 7—9 mm, more or less concolorous, sparingly pubescent, the surface somewhat wrinkled and gland-dotted, apex obtuse, base cuneate, margin entire. *Inflorescence* simple, 30—120 mm long, fairly dense; verticils 2—6-flowered; apical bracts occupying the upper third of the raceme, ovate to broadly ovate, 12—15 × 7—10 mm, whitish to rose-purple. *Calyx* 7—9 mm long at maturity, glandular-villous. *Corolla* white, drying yellow-brown, 11—12 mm long; tube c. 8 mm long, not expanding towards the throat; lower lip 3—4 mm long. *Stamens* shortly exserted, not exceeding the lower lip; upper pair attached about 1 mm from the base of the tube, pubescent for their entire length, scarcely exserted; lower pair united only at the base for c. 0,5 mm, filaments sparingly pubescent. *Stigma* clavate.

Known from only the Nelspruit-Barberton-Kaapsche Hoop area at altitudes of about 1 000 m; in grassy places among rocks and in open woodland. Map 118.

Vouchers: *De Souza* 423; *Mauve* 4942.

An anomalous species with narrow corolla tube and persistent bracts similar to those of *H. pretoriae* (no. 15) but the upper bracts are colourful; the upper stamens are attached near the base of the corolla tube and the filaments are pubescent for almost their whole length; and the filaments of the lower pair of stamens are united for only about 0,5 mm at the base.

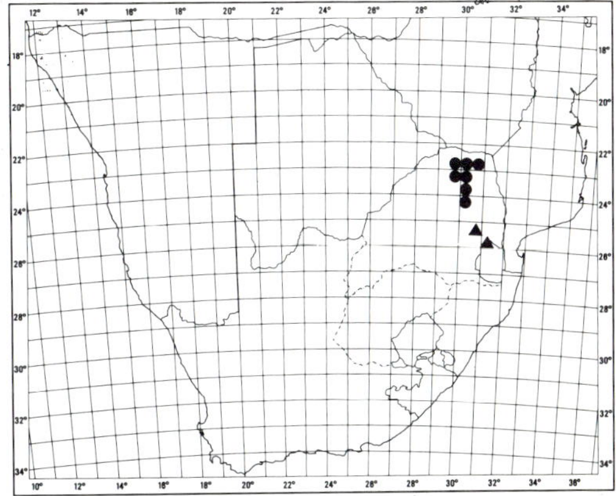

MAP 118. — ▲ **Hemizygia persimilis**
● **H. petiolata**

24. **Hemizygia petiolata** *Ashby* in J. Bot., Lond. 73: 355 (1935); Codd in Bothalia 12: 17 (1976). Type: Transvaal, Soutpansberg, Tshakoma, *Obermeyer* sub TRV 31571 (PRE, holo.!).

Soft shrub up to 1 m tall, branching usually from the base; stems few to many, glandular-pilose. *Leaves* petiolate; blade ovate to ovate-lanceolate, 20—55 × 6—30 mm, upper surface dark brown and shortly glandular-pubescent, under-surface canescent, apex acute to obtuse, base cuneate to obtuse, margin regularly serrate-dentate in the upper two-thirds; petiole 6—14 mm long. *Inflorescence* usually paniculate, lax, 100—300 mm long; verticils 4—6-flowered; apical bracts sometimes persisting as a purple coma, usually rather small, 5—10 × 3—5 mm, more often the apex of the raceme is broken off, lower bracts caducous. *Calyx* 8—9 mm long at maturity, glandular-tomentose. *Corolla* pale mauve to lilac, 17—20 mm long; tube 13—16 mm long, expanding to 3—4 mm wide at the mouth; lower lip 4 mm long. *Stamens* well exserted beyond the lower lip; upper pair attached about 3 mm from base of tube, filaments puberulous near the base; lower pair united for more than half their length. *Stigma* swollen, emarginate.

Recorded from the north-eastern Transvaal from the Soutpansberg to The Downs, at altitudes of 1 000 to 1 600 m; on wooded hillsides and at forest margins. Map 118.

Vouchers: *Codd* 8331; 9423; *Scheepers* 387.

A remarkably aromatic plant, smelling of mint and coconut. Allied to the next species, *H. canescens,* but has more ovate leaves, longer petioles and longer corolla.

25. Hemizygia canescens *(Gürke) Ashby* in J. Bot., Lond. 73: 354 (1935); Ross, Fl. Natal 306 (1972); Codd in Bothalia 12: 17 (1976); Compton, Fl. Swaziland 509 (1976). Lectotype: Transvaal, Wonderboompoort, *Rehmann* 4507 (Z, lecto.; K).

Orthosiphon canescens Gürke in Bull. Herb. Boissier 6: 557 (1898); N.E. Br. in F.C. 5,1: 259 (1910).

O. affinis N.E. Br., l.c. 257 (1910). Syntypes: Transvaal, Woodbush Mts, *Schlechter* 4737 (K; PRE!); near Potgietersrus, *Bolus* 11146 (BOL!).

Herb, probably a weak perennial, 0,3—0,6 m tall, woody below and often branched; stems spreading to ascending, shortly greyish-tomentose, hairs often crisped or occasionally sparse but not villous. *Leaves* subsessile or shortly petiolate; blade linear or linear-lanceolate to lanceolate or rarely ovate-lanceolate, 25—55 × 3—15 mm, densely grey-tomentose on both surfaces to sparingly crisped tomentulose, apex acute, base cuneate to attenuate, margin finely to fairly coarsely toothed in the upper half; petiole up to 5 mm long. *Inflorescence* simple to paniculate, lax, 70—250 mm long; verticils 4—6-flowered; bracts early deciduous, small, c. 2 × 1 mm. *Calyx* 7—8 mm long at maturity, glandular-tomentulose to hispidulous. *Corolla* white to pale mauve or purplish, 14—17 mm long; tube 10—13 mm long, expanding to 3—4 mm wide at the mouth; lower lip 3—4 mm long. *Stamens* well exserted beyond the lower lip; upper pair attached about 4 mm from the base of the tube, filaments puberulous near the base; lower pair united for most of their length. *Stigma* somewhat clavate. Fig. 36.

Distributed in a broad band from the Mafikeng region of Bophuthatswana across south-western and central Transvaal to eastern Transvaal, avoiding the high mountains, extending to Swaziland and northern KwaZulu; among rocks in open arid to moist woodland

and marginal grassland at altitudes of 300 to 1 700 m. Map 119.

Vouchers: *Codd* 9840; *Galpin* M286; *Mogg* 16475; *Schlechter* 4070.

A good deal of variation is included in *H. canescens,* from linear leaves (3—4 mm broad) in the dry western extremity of its range in the Mafikeng region to lanceolate and ovate-lanceolate in more mesophytic areas. It is diagnosed by the short often crisped tomentum of stems and leaves. See also notes after the following species, *H. linearis* and *H. petrensis* (no. 27). Superficially *H. canescens* and *H. petrensis* are very similar, but the latter has long villous hairs on the stems, though the pubescence of the leaves is often similar. *H. petrensis* is a more western species, entering the northern and eastern Transvaal lowveld. However, two specimens from the Waterberg in S.W.A./Namibia, *Boss* sub TRV 35003 and *De Winter* 2799, have pubescence resembling *H. canescens,* and this area should be investigated further.

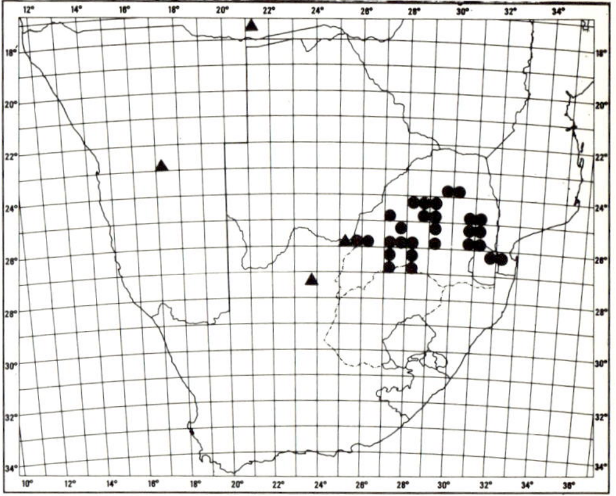

MAP 119. — ● **Hemizygia canescens**
▲ **H. linearis**

26. Hemizygia linearis *(Benth.) Briq.* in Bull. Herb. Boissier sér. 2, 3: 997 (1903); Ashby in J. Bot., Lond. 73: 354 (1935); Codd in Bothalia 12: 18 (1976). Type: Zimbabwe, Matabeleland, *Oates* s.n. (K, holo.).

Orthosiphon linearis Benth. in Hooker's Icon. Pl. t.1274 (1878); Rolfe in Oates, Matabeleland edn 2: 407 (1889); Bak. in F.T.A. 5: 374 (1900).

Herb, probably a weak perennial, 0,3—0,5 m tall, somewhat woody and branching near the base; stems subglabrous

---

FIG. 36. — 1, **Hemizygia canescens,** flowering stem, × 1; a, leaf, × 1; b, mature calyx, × 4; c, section through corolla, × 3 (*Leistner* 3553, Pretoria District).

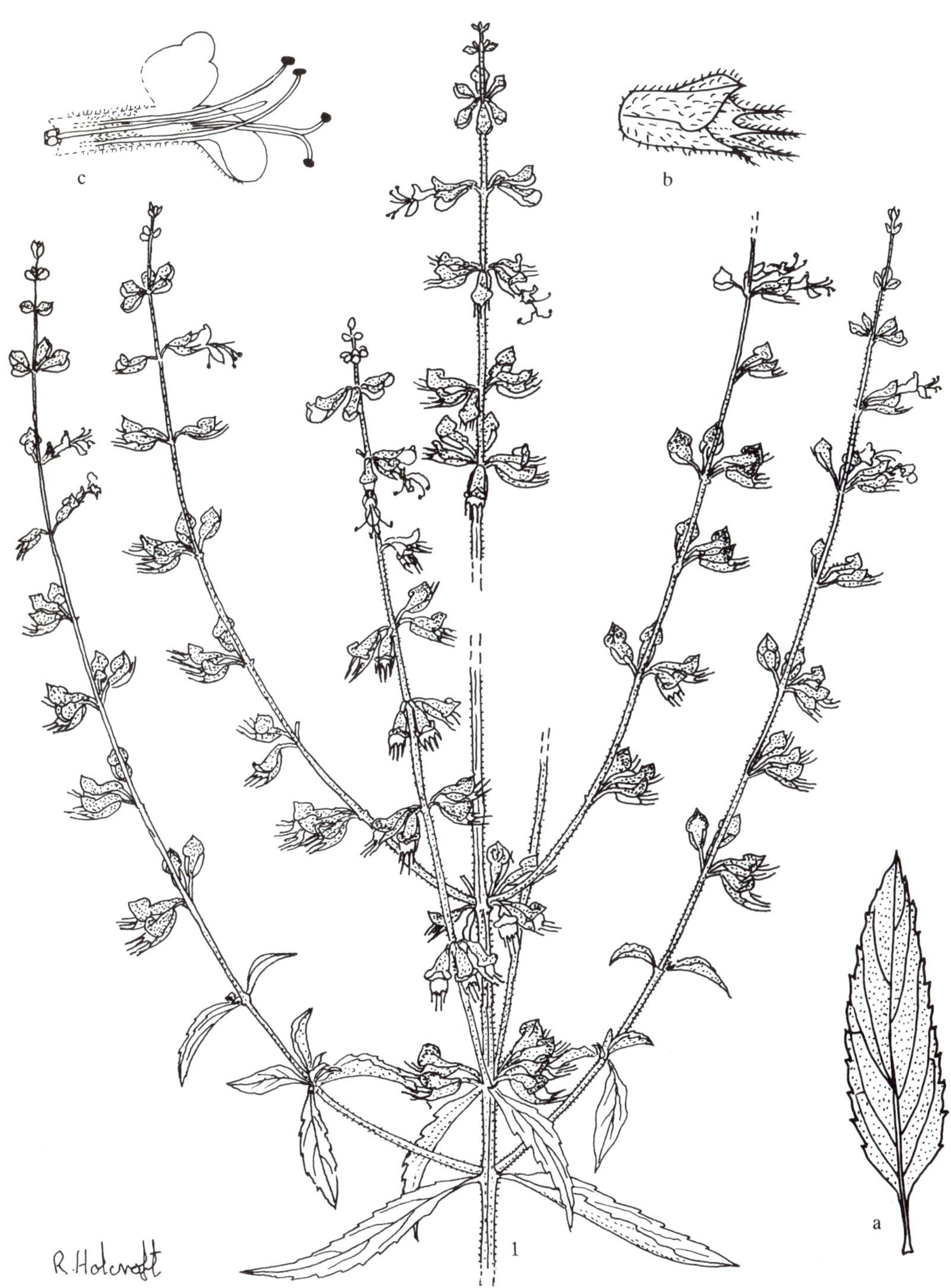

R. Holcroft

to sparingly villous, usually with a a somewhat varnished appearance. *Leaves* sessile or subsessile; blade linear, 20−30 × 2−4 (−5) mm, puberulous to sparingly hispid, often folded along the midrib or with margins inrolled, apex acute, base attenuate, margin finely and distantly toothed. *Inflorescence* simple or branched near the base, lax, 120−200 mm long; verticils 4−6-flowered; bracts early deciduous, small, c. 2 × 1 mm. *Calyx* 7−8 mm long at maturity, hispidulous. *Corolla* mauve, often with violet stripes, 12−13 mm long; tube 9−10 mm long, expanding to 3 mm wide at the mouth; lower lip 3 mm long. *Stamens* exserted beyond the lower lip; upper pair attached about 3 mm from the base of the tube, filaments puberulous near the base; lower pair united nearly to the apex. *Stigma* somewhat clavate.

Found in open places in dry woodland in S.W.A./Namibia, Botswana and northern Cape Province; also in Angola and Zimbabwe. Map 119.

Vouchers: *Burtt Davy* 13961; *De Winter & Marais* 4789; *Strey* 2571.

Diagnostic features are the linear, subglabrous leaves and the subglabrous to sparingly villous stems which have a somewhat varnished appearance. See also notes after *H. canescens* (above) and *H. petrensis* (below).

27. Hemizygia petrensis *(Hiern) Ashby* in J. Bot., Lond. 73: 353 (1935); Launert & Schreiber in F.S.W.A. 123: 13 (1969); Codd in Bothalia 8: 159 (1964); ibid. 12: 18 (1976). Type: Angola, *Welwitsch* 5494 (BM, holo.).

Orthosiphon petrensis Hiern, Cat. Afr. Pl. Welw. 1: 859 (1900); Bak. in F.T.A. 5: 524 (1900).

H. dinteri Briq. in Bull. Herb. Boissier sér. 2, 3: 995 (1903). Type: S.W.A./Namibia, 10 km E. of Orumbe, *Dinter* 1320.

O. varians N.E. Br. in F.C. 5,1: 256 (1910); Ashby, l.c. 357 (1935). Type: Transvaal, Komatipoort, *Schlechter* 11746 (BOL, holo.!).

O. holubii N.E. Br., l.c. 258 (1910). Type: Cape, Molopo River, *Holub* s.n. (K, holo.).

O. engleri Perkins in Bot. Jb. 54: 34 (1917). Type: S.W.A./Namibia, Okahandja, *Engler* 6475.

O. mossianus Good in J. Bot., Lond. 63: 175 (1925). *H. mossiana* (Good) Ashby, l.c. 356 (1935). Type: Transvaal, Messina, *Moss & Rogers* 193 (BM, holo.; PRE!).

Annual or weak perennial herb 0,2−0,6 m tall, somewhat woody and branching near the base; stems sparingly to densely villous with long spreading hairs, or rarely almost glabrous. *Leaves* subsessile or shortly petiolate; blade linear-lanceolate to oblong-lanceolate or ovate-lanceolate, 20−50 × 5−15 mm, sparingly to densely pilose or canescent, often with long and short hairs intermingled, apex acute, base cuneate to attenuate, margin distinctly to obscurely and somewhat distantly toothed. *Inflorescence* simple or with a pair of branches near the base, lax, 80−200 mm long; verticils 4−6-flowered; bracts early deciduous, ovate, 3 × 2 mm. *Calyx* 6−8 mm long at maturity, glandular-hispid to villous. *Corolla* pinkish to lilac or violet, 13−15 mm long; tube 9−12 mm long, expanding to 2,5−3 mm wide at the mouth; lower lip 3 mm long. *Stamens* exserted beyond the lower lip; upper pair attached 2−3 mm from the base of the tube, filaments puberulous near the base; upper pair united for the greater part of their length. *Stigma* somewhat clavate.

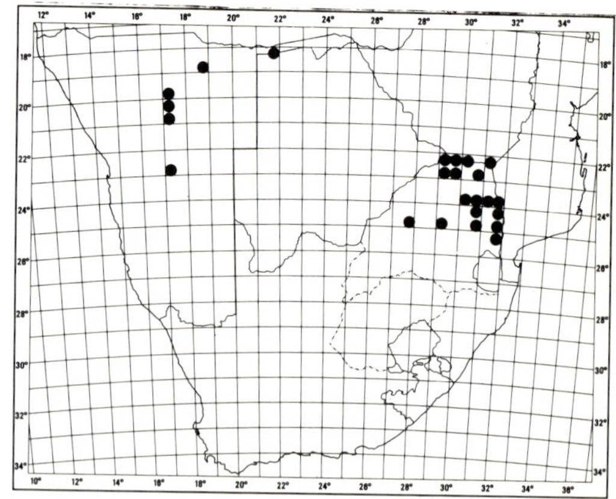

MAP 120. — **Hemizygia petrensis**

Recorded from northern S.W.A./Namibia and northern and eastern Transvaal lowveld; among rocks and in open places and water-courses in semi-arid woodland. Also in Angola and Zimbabwe. Map 120.

Vouchers: *Codd & Dyer* 3832; *De Winter* 2799; *Giess* 11675.

Together with *H. canescens* (no. 25) and *H. linearis* (above), the three species form a closely related group with almost identical floral characters and small inconspicuous bracts. *H. canescens* may be distinguished on the basis of the dense, short and often crisped pubescence on stems and leaves and is distributed more on the high plateau formed by the northern Cape and south-western and central Transvaal, but extending to eastern Transvaal (where the two may overlap), Swaziland and Natal.

In *H. linearis* (no. 26) the leaves are linear to filiform and the leaves and stems are glabrous or with a few scattered long hairs. It overlaps with *H. petrensis* in Angola, Zimbabwe, S.W.A./Namibia and Botswana and occasional intermediates may be found.

28. **Hemizygia bracteosa** *(Benth.) Briq.*

in Annu. Conserv. Jard. bot. Genève 2: 248 (1898); Ashby in J. Bot., Lond. 73: 352 (1935); Morton in F.W.T.A. edn 2, 2: 455 (1963); Launert & Schreiber in F.S.W.A. 123: 12 (1969); Codd in Bothalia 12: 19 (1976). Type: Senegal, *Le Prieur & Perrottet* s.n. (G, holo.).

Ocimum bracteosum Benth., Lab. 14 (1832); in Hooker's Icon. Pl. t.455 (1842); in DC., Prodr. 12: 41 (1848). *Orthosiphon bracteosus* (Benth.) Bak. in F.T.A. 5: 375 (1900); N.E. Br. in F.C. 5,1: 248 (1910).

Orthosiphon schinzianus Briq. in Bot. Jb. 19: 173 (1894). Type: S.W.A./Namibia, Amboland, *Schinz 45* (Z, holo.).

H. junodii Briq. in Annu. Conserv. Jard. bot. Genève 2: 249 (1898). Syntypes: Mozambique, Delagoa Bay, *Junod 61; 235.*

— var. *quintasii* Briq., l.c. 249 (1898). Type: Mozambique, Delagoa Bay, *Quintas* s.n.

H. hoepfneri Briq. in Bull. Herb. Boissier sér. 2, 3: 994 (1903). Type: S.W.A./Namibia, Hereroland, *Höpfner 85.*

H. serrata Briq., l.c. 996 (1903). Syntypes: S.W.A./Namibia, Amboland, *Rautanen* s.n.; *Wulfhorst* 1.

Orthosiphon rhodesianus S. Moore in J. Bot., Lond. 43: 50 (1905). Type: Zimbabwe, Wankie, *Eyles 132* (BM, holo.).

Bouetia ocimoides A. Chev. in Mém. Soc. bot. Fr. 8: 200 (1912). Type: from Dahomey.

Herb, probably annual, sometimes woody at the base, 0,25—0,7 m tall; stems sparingly to densely pilose with long weak multicellular hairs. *Leaves* sessile; blade narrowly lanceolate to oblong-lanceolate, 40—90 × 8—24 mm, upper surface hispidulous, under-surface sparingly to densely canescent, apex acute, base attenuate,

margin usually distinctly toothed. *Inflorescence* simple or paniculate, lax, 120—300 mm long; verticils 4—6-flowered; bracts large, broadly ovate, 5—10 × 4—8 mm, persisting as an apical coma. *Calyx* 7 mm long when mature, glandular-villous. *Corolla* white or tinged with mauve, rarely violet, 10—11 mm long; tube 7—8 mm long, expanding gradually to 2,5—3 mm wide at the mouth; lower lip 3 mm long. *Stamens* exserted shortly beyond the lower lip; upper pair attached below the middle of the tube, finely puberulous for half or more of their length; lower pair united to near the apex. *Stigma* somewhat clavate.

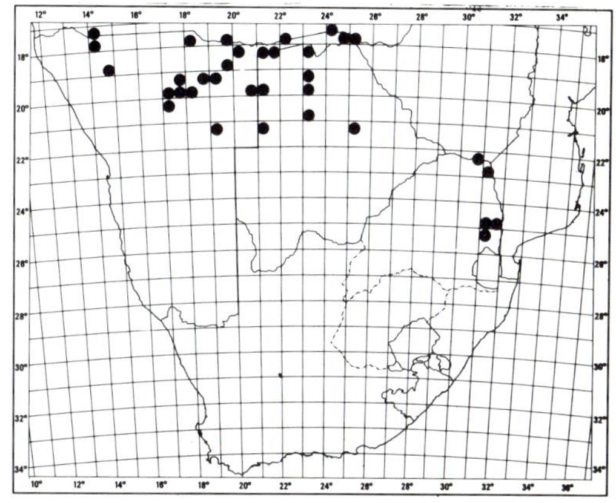

MAP 121. — **Hemizygia bracteosa**

Widespread from Senegal and Tanzania southwards to northern S.W.A./Namibia, northern Botswana, the eastern Transvaal lowveld and Mozambique; among rocks, in watercourses and in open sandy places in dry tropical woodland. Map 121.

Vouchers: *Acocks* 16668; *Codd* 4260; 5198; *De Winter* 4390; *Giess* 9819; 12522.

In habit, ecology and distribution within our area it resembles *H. petrensis* (above), but may be distinguished by the conspicuous coma of large whitish to rose-purple bracts and by the usually whitish corolla which is slightly shorter than the mauve to violet corolla of *H. petrensis*.

7362   **31. ACROCEPHALUS**

Acrocephalus *Benth.* in Edwards's Bot. Reg. sub. t.1282 (1829); Lab. 23 (1832); in DC., Prodr. 12: 47 (1848); Benth. & Hook. f., Gen. Pl. 2,2: 1173 (1876); Briq. in Natürl. PflFam. 4,3a: 365 (1897); Bak. in F.T.A. 5: 354 (1900); Robyns & Lebrun in Annls Soc. scient. Brux. sér. B, 48: 169 (1928); Robyns in Bot. Notiser 119: 185 (1966); Launert & Schreiber in F.S.W.A. 123: 4 (1969); R.A. Dyer, Gen. 534 (1975). Type species: *A. scariosus* Benth.

Haumaniastrum Duvign. & Plancke in Biol. Jaarb. 27: 222 (1959); Morton in J. Linn. Soc., Bot. 58: 239 (1962); in F.W.T.A. edn 2,2: 455 (1963); Gilli in Annln naturh. Mus. Wien 77: 33 (1973); Agnew, Upland Kenya Wild Flow. 643 (1974). Type species: *H. polyneurum* (S. Moore) Duvign. & Plancke.

Perennial herbs or shrubs. *Leaves* opposite or whorled, usually narrow, sometimes in basal rosettes. *Inflorescence* capitate, usually corymbose, subtended by leafy bracts which are often coloured; floral bracts small. *Calyx* bilabiate, compressed; tube subcylindric, often arcuate; upper lip shortly 3-toothed or entire; lower lip shortly 2-toothed or entire. *Corolla* bilabiate, slightly longer than the calyx; tube short; upper lip shortly 4-lobed; lower lip entire, flat. *Stamens* 4, didynamous, declinate, scarcely exserted; the lower pair attached near the throat, the upper pair about the middle of the corolla; anthers 1-thecous. *Style* shortly bilobed. *Nutlets* ovoid or oblong, smooth.

About 70 or more species, mainly in tropical Africa and a few in Asia; 1 species in Southern Africa. Duvigneaud & Plancke, l.c., working with the Congo species, considered the African species to be worthy of separate generic status. However, only a few of the species names have been transferred to *Haumaniastrum* and, until a thorough revision of the whole group is undertaken, it is preferred to retain the name *Acrocephalus* for our solitary species.

Acrocephalus sericeus *Briq.* in Bot. Jb. 19: 170 (1894); Hiern, Cat. Afr. Pl. Welw. 1,4: 857 (1900); Bak. in F.T.A. 5: 362 (1900); Launert & Schreiber in F.S.W.A. 123: 5 (1969). Type: Angola, Huilla, *Welwitsch* 5603 (PRE, iso.!).

Stems 1—several from a perennial base, erect, virgate, semi-woody, sparingly branched, 4-angled, sericeous, 0,3—1,2 m tall. *Leaves* opposite, subsessile; blade linear to linear-lanceolate, 30—70 × 3—6 mm, appressed sericeous, gland-dotted below, tapering gradually to apex and base, margin entire to obscurely toothed. *Inflorescence* corymbose; flower-heads subglobose, 8—10 mm in diameter, densely pubescent; subtending bracts lanceolate, 10—15 mm long, not coloured; floral bracts imbricate, broadly ovate, apiculate, 6 × 5 mm, densely villous; flowers in 3-flowered subsessile cymes. *Calyx* densely villous, 1,5—2 mm long at flowering, enlarging to 5 mm long; upper lip minutely 3-toothed; lower lip shortly 2-toothed. *Corolla* mauve, densely villous, 5—6 mm long. *Stamens* exserted by up to 2 mm. *Style* exserted by 2—3 mm. *Nutlets* oblong, 1 mm long.

Recorded from north-eastern S.W.A./Namibia and the Caprivi Strip, in moist areas in open woodland. Also in Angola, Zambia, Zimbabwe and Malawi. Map 122.

Vouchers: *Dinter* 7211; *Merxmüller & Giess* 2148.

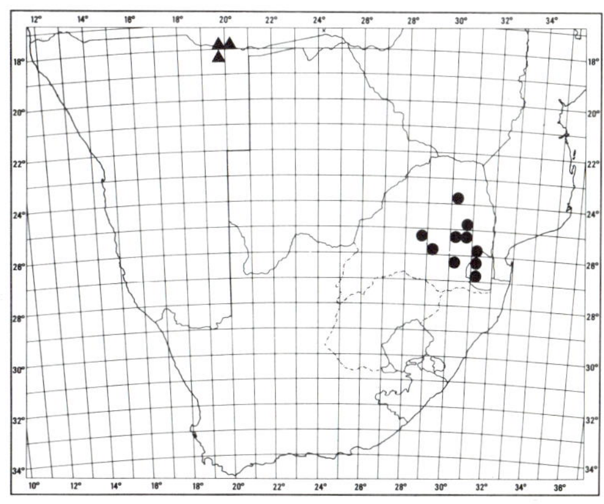

MAP 122. — ▲ **Acrocephalus sericeus**
● **Geniosporum angolense**

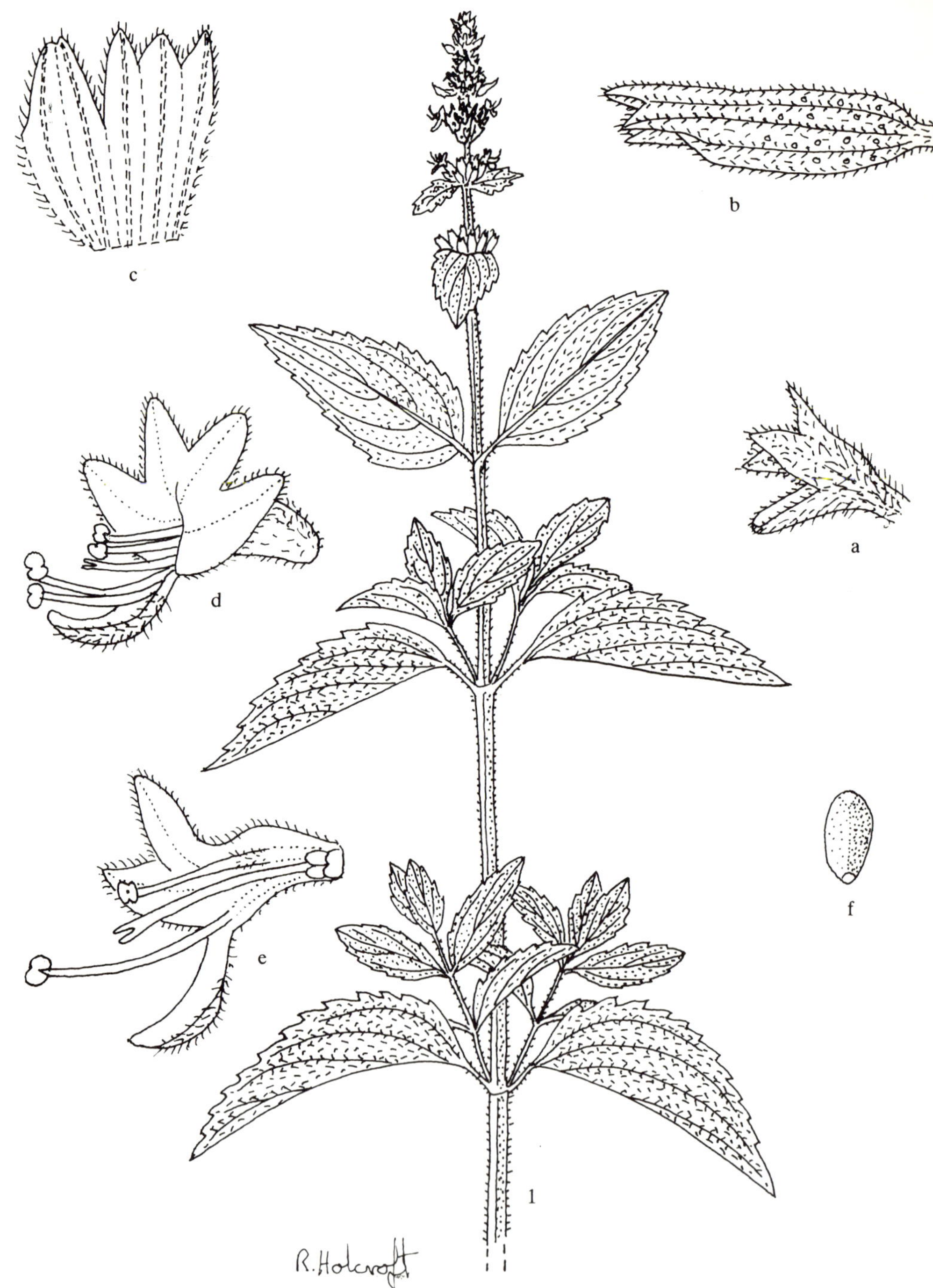

FIG. 37. — 1, **Geniosporum angolense,** flowering stem, × 1; a, flowering calyx, × 9; b, mature calyx, × 9; c, mature calyx, opened longitudinally, × 9; d, corolla, × 9; e, section through corolla, × 9; f, nutlet, × 9 (*Burtt Davy* 8099).

7363

32. GENIOSPORUM

Geniosporum *Wall. ex Benth.* in Bot. Reg. sub t.1300 (1830); Benth., Lab. 19 (1832); in DC., Prodr. 12: 44 (1848); Briq. in Natürl. PflFam. 4,3a: 367 (1897); Bak. in F.T.A. 5: 351 (1900); N.E. Br. in F.C. 5,1: 293 (1910); Morton in F.W.T.A. edn 2,2: 453 (1963); R.A. Dyer, Gen. 534 (1975). Lectotype: *G. coloratum* (D. Don) Kuntze (= *G. strobiliferum* Benth., nom. illeg.)

Perennial herbs. *Leaves* opposite or ternate. *Inflorescence* terminal, spike-like; flowers in dense, opposite, many-flowered, cymose clusters subtended by relatively large bracts, the lower bracts leaf-like, often blotched with white or mauve. *Calyx* sub-bilabiate; tube at first campanulate, elongating and becoming tubular; upper lip of 3 subequal teeth; lower lip smaller, emarginate. *Corolla* small, bilabiate; tube campanulate; upper lip short, broad, subequally 4-lobed; lower lip narrow, oblong, concave. *Stamens* 4, didynamous, declinate, exserted; filaments pubescent in the lower half, the upper pair inserted below the middle of the tube, the lower pair inserted near the throat; anthers 1-thecous. *Disc* saucer-shaped, slightly produced in front. *Style* filiform, exserted, 2-lobed. *Nutlets* ellipsoid, compressed, brown.

About 20 species, in Asia, Africa and Malagasy Republic; 1 species in Southern Africa.

Geniosporum angolense *Briq.* in Bot. Jb. 19: 164 (1894); Bak. in F.T.A. 5: 351 (1900); Hiern, Cat. Afr. Pl. Welw. 1,4: 852 (1900); Compton, Fl. Swaziland 508 (1976). Type: Angola, Huilla, *Welwitsch* 5491 (PRE, iso.!).

Stems 1—several from the base, erect, 0,5—1,2 m tall, sparingly branched, retrorse-pubescent. *Leaves* petiolate; blade lanceolate to ovate-lanceolate, 50—70 × 15—25 mm, sparingly appressed-pubescent especially on the nerves, copiously gland-dotted on both surfaces, apex acute, base obtuse to cuneate, margin serrate. *Inflorescence* dense, spike-like, 50—100 mm long; rhachis densely retrorse-pubescent; bracts broadly ovate, acute, the lowermost pair 10—25 × 8—15 mm, becoming progressively smaller towards the apex. *Calyx* densely pubescent, 2 mm long at flowering, increasing to 4—5 mm at fruiting stage. *Corolla* white or mauve, 5—6 mm long; upper lip 2,5 mm long, 3 mm broad; lower lip 2,5 × 1 mm. *Stamens* exserted by 2,5 mm. *Style* exserted by 2 mm. *Nutlets* 1 mm long. Fig. 37.

Found in the Transvaal, from Magoebaskloof along the eastern escarpment to Barberton and westward to Witbank and Pretoria districts, extending into Swaziland as far south as Mankaiana, growing with sedges and other moisture-loving plants on stream banks and marshy places. Also in Angola, Zimbabwe and Malawi. Map 122.

Vouchers: *Compton* 26780; *Galpin* 1317; *Schlechter* 4118 (also erroneously distributed as 2118).

Apparently the plant is not noticeably aromatic.

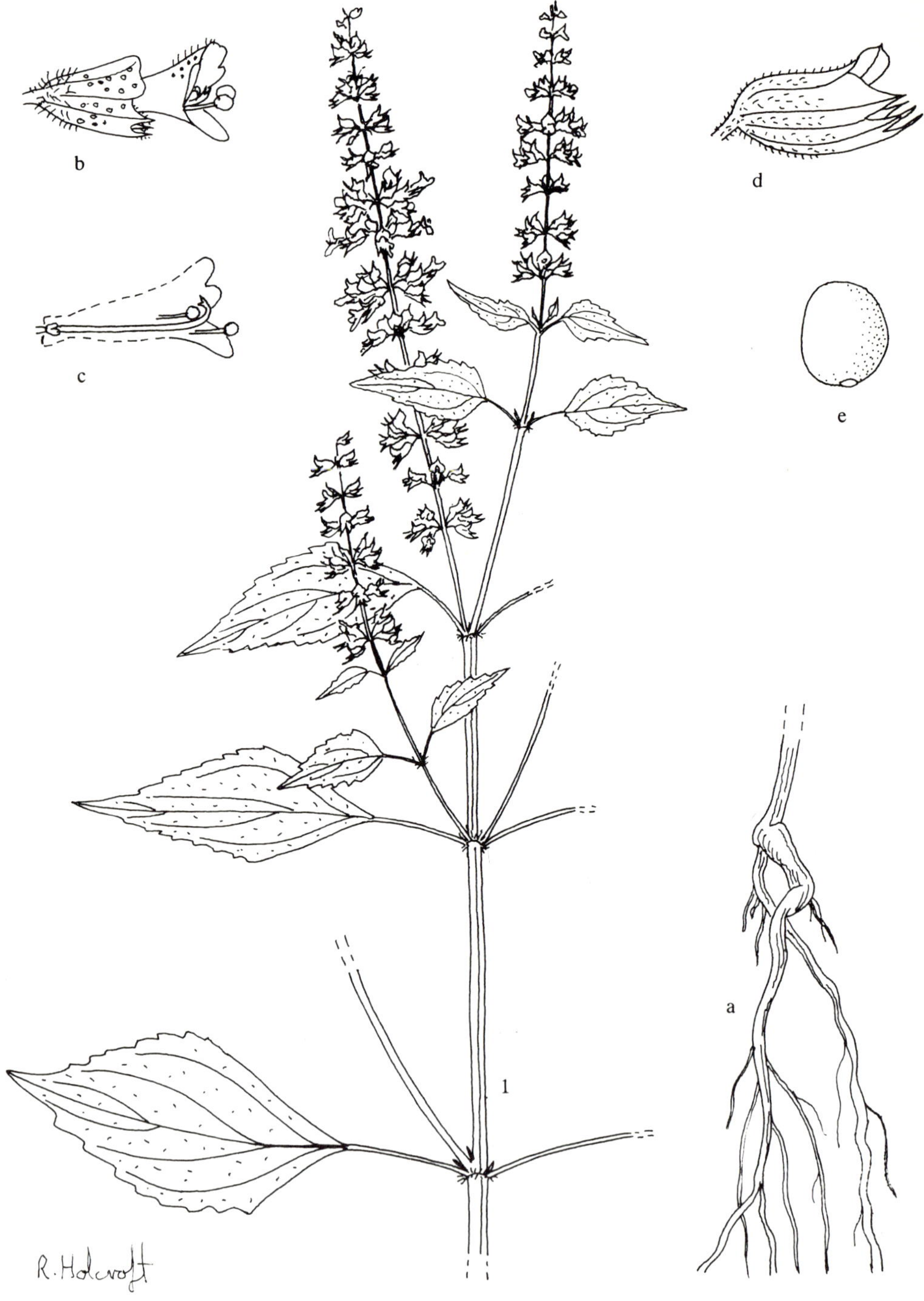

FIG. 38. — 1, **Basilicum polystachyon,** flowering stem, × 1; a, base of plant, × 1; b, flower, × 6; c, section through corolla, × 6; d, mature calyx, × 6; e, nutlet, × 9 (*Culverwell* 1145).

7364   **33. BASILICUM**

Basilicum *Moench,* Suppl. Meth. Pl. 143 (1802); Kuntze, Rev. Gen. Pl. 2: 512 (1891); Andrews, Flow. Pl. Sudan 3: 205 (1956); Morton in J. Linn. Soc., Bot. 58: 238 (1962); in F.W.T.A. edn 2, 2: 454 (1963); R. A. Dyer, Gen. 535 (1975). Type species: *B. polystachyon* (L.) Moench (= *Ocimum polystachyon* L.).

Moschosma Reichb., Consp. 171 (1828); Benth., Lab. 24 (1832); in DC., Prodr. 12: 48 (1848); Benth. & Hook. f., Gen. Pl. 2, 2: 1173 (1876); Hook. f., Fl. Brit. Ind. 4: 612 (1885); Briq. in Natürl. PflFam. 4, 3a: 368 (1897); Bak. in F.T.A. 5: 352 (1900). Type species: *M. polystachyon* (L.) Benth. (= *Ocimum polystachyon* L.).

Annual, erect, aromatic herbs. *Leaves* membranous, petiolate. *Inflorescence* a slender, many-flowered spike-like raceme, terminal and on side branches, rarely branched; flowers shortly pedicellate, in usually 6-flowered verticils; bracts much smaller than the leaves, persistent. *Calyx* bilabiate, somewhat declinate, accrescent, 5-toothed; tube campanulate; upper tooth the largest, ovate, slightly decurrent, two lateral teeth deltoid, two lower teeth lanceolate, subulate. *Corolla* small, obscurely bilabiate; tube short; upper lip short and broad, 4-lobed; lower lip oblong, nearly flat, entire. *Stamens* 4, didynamous, declinate; filaments not kneed or crested near the base; anthers 1-thecous. *Disc* saucer-shaped. *Ovary* glabrous; style shortly exserted; stigma bifid. *Nutlets* ovoid, somewhat compressed, smooth, pale brown.

2 or 3 species of the Old World tropics; 1 species extends into Southern Africa.

Related to *Ocimum* (no. 34) but the corolla is less markedly bilabiate and the filaments lack a knee, crest or teeth near the base.

A proposal to conserve *Moschosma* Reichb. was turned down by the Nomenclatural Committee (Taxon 19: 481, 1970; 21: 534, 1972).

Basilicum polystachyon *(L.) Moench,* Suppl. Meth. Pl. 143 (1802); Kuntze, Rev. Gen. Pl. 2: 512 (1891); Andrews, Flow. Pl. Sudan 3: 205 (1956); Morton in J. Linn. Soc., Bot. 58: 238 (1962); in F.W.T.A. edn 2, 2: 454 (1963); Ross, Fl. Natal 306 (1972); Keng in Fl. Males. 8: 366 (1978); Cramer in Fl. Ceylon 3: 122 (1981). Type: from India.

Ocimum polystachyon L., Mant. Alt. 567 (1771). *Moschosma polystachyon* (L.) Benth. in Wall., Pl. As. Rar. 2: 13 (1830; Lab. 24 (1832); in DC., Prodr. 12: 48 (1848); Hook. f., Fl. Brit. Ind. 4: 612 (1885); Bak. in F.T.A. 5: 352 (1900); Mansfeld in Bot. Jb. 62: 380 (1929); Dalziel, Useful Pl. W. Trop. Afr. 462 (1955). Type: as above.

Freely branched, glabrous herb, 0,3—0,5 m tall; stems 4-angled. *Leaves* petiolate; blade ovate, 20—50 × 10—35 mm, under-surface freely gland-dotted, apex acuminate, base cuneate to obtuse, entire, upper margin crenate-dentate; petiole slender, 20—30 mm long. *Inflorescence* lax to fairly dense, 50—100 × 8 mm; bracts minute, ovate; pedicels 1—2 mm long. *Calyx* glandular-puberulous, 1,25 mm long at flowering, enlarging to 3 mm long. *Corolla* white or mauve, 1,5—2 mm long. *Stamens* scarcely exserted; lower pair attached at the corolla throat, the upper pair about the middle of the tube. Fig. 38.

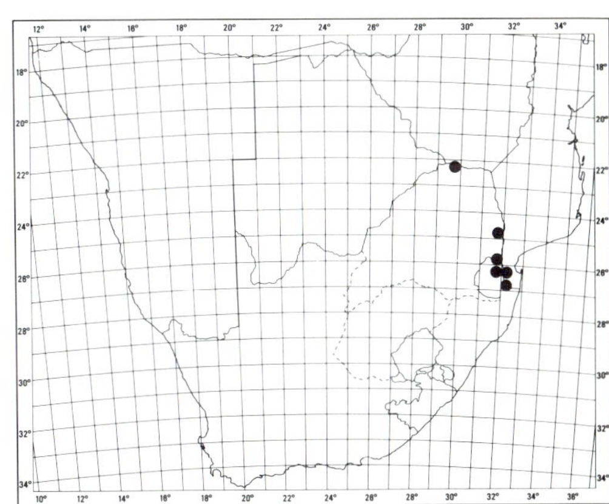

MAP 123. — **Basilicum polystachyon**

Recorded from the eastern Transvaal lowveld, Swaziland and northern KwaZulu; in damp situations, often in disturbed places; widespread in tropical Africa, tropical Asia and Malesia. Map 123.

Vouchers: *Culverwell* 1145; *Van der Schijff* 634; 3474; *Ward* 3699.

7366

34. OCIMUM

Ocimum *L.*, Sp. Pl. 597 (1753); Gen. Pl. edn 5: 259 (1754); Benth., Lab. 1 (1832); in DC., Prodr. 12: 31 (1848); Benth. & Hook. f., Gen. Pl. 2, 2: 1171 (1876); Briq. in Natürl. PflFam. 4, 3a: 369 (1897); Bak. in F.T.A. 5: 334 (1900); N.E. Br. in F.C. 5,1: 233 (1910); Morton in F.W.T.A. edn 2, 2: 451 (1963); Launert & Schreiber in F.S.W.A. 123: 20 (1969); R. A. Dyer, Gen. 536 (1975). Sometimes spelt *Ocymum* in earlier literature. Type species: *O. basilicum* L.

Herbs or soft shrublets. *Leaves* opposite, simple. *Inflorescence* a terminal, spike-like raceme; flowers shortly pedicellate in usually 6-flowered verticils; bracts much smaller than the leaves; pedicels ascending. *Calyx* bilabiate, 5-toothed; tube short, campanulate; upper tooth much larger than the rest, broadly ovate to subrotund, decurrent on the tube; two lateral teeth small, subulate; two lower teeth fused, forming an oblong, emarginate or 2-toothed lip. *Corolla* small, bilabiate, mauve to whitish; tube campanulate, slightly gibbous at the base; upper lip 4-lobed, with lobes more or less equal; lower lip spreading, concave. *Stamens* 4, didynamous, shortly exserted; upper pair attached near the base of the corolla tube, kneed and with a tuft of hairs near the base of the filaments; lower pair inserted in the corolla mouth; anthers 1-thecous. *Disc* saucer-shaped, 4-lobed. *Ovary* glabrous; style exserted; stigma shortly 2-lobed. *Nutlets* globose, mucilaginous when wetted.

About 6 species, used medicinally and as culinary herbs; 2 species indigenous in Southern Africa. In addition, *O. basilicum* L. (Sweet Basil) is grown as a pot-herb for its aromatic foliage and a purple-leaved cultivar is grown as an ornamental garden plant.

In the genera *Ocimum* and *Becium* (no. 35), the two upper filaments are attached near the base of the corolla tube and are kneed, crested or toothed not far from the base. In *Becium* the upper and lower lips of the calyx are separated by a wide sinus and the stamens and corolla are more markedly exserted than in *Ocimum*.

1 Calyx tube glabrous inside; lower lip of calyx emarginate or shortly toothed, eventually closing the mouth of the calyx:

   2 Leaves dentate, (40−) 50−120 × 25−65 mm; inflorescence (60−) 70−150 mm long
    .... 1(a). *O. urticifolium* subsp. *urticifolium*

   2 Leaves obscurely toothed in the upper half, 25−45 × 12−30 mm; inflorescence 50−70 mm long
    .... 1(b). *O. urticifolium* subsp. *caryophyllatum*

1 Calyx tube hairy inside; lower lip of calyx deeply 2-toothed, spreading .... 2. *O. canum*

1. **Ocimum urticifolium** *Roth*, Catalecta Bot. 2: 52 (1800). Type: from India.

Erect perennial herb or soft shrub, 0,4−2 m tall, with few to several stems from the base or branching mainly in the upper half; stems sparingly to densely pilose. *Leaves* petiolate, soft; blade ovate to ovate-lanceolate or elliptic, 25−120 × 12−65 mm (see subspecies), subglabrous or sparingly to densely pubescent on both surfaces, copiously gland-dotted below, apex acute to acuminate, base cuneate, margin dentate or obscurely dentate only in the upper half; petiole 10−40 mm long. *Inflorescence* simple or sparingly branched, 50−150 mm long, verticils 4−6 mm apart; bracts persistent, broadly ovate, 4−8 mm long, abruptly tapering to base and apex; pedicels 3 mm long. *Calyx* densely pubescent on the outside, glabrous within, at maturity 5−6 mm long with broadly ovate to subrotund, somewhat concave upper tooth, minute lateral teeth and oblong, shortly toothed lower lip which eventually bends upwards closing the mouth of the calyx. *Corolla* usually white, 4−5 mm long. *Stamens* exserted by 4 mm.

Found in the warmer parts of Southern Africa, in northern S.W.A./Namibia, northern Botswana, northern, central and eastern Transvaal, low-lying parts of Swaziland and coastal to midland parts of Natal as far south as Durban and Pietermaritzburg; widespread in tropical Africa and in southern Asia.

For key to subspecies, see key to species.

(a) subsp. **urticifolium.**
Codd in Bothalia 14: 219 (1983).

O. urticifolium Roth, Catalecta Bot. 2: 52 (1800); Launert & Schreiber in F.S.W.A. 123: 21 (1969); Ross,

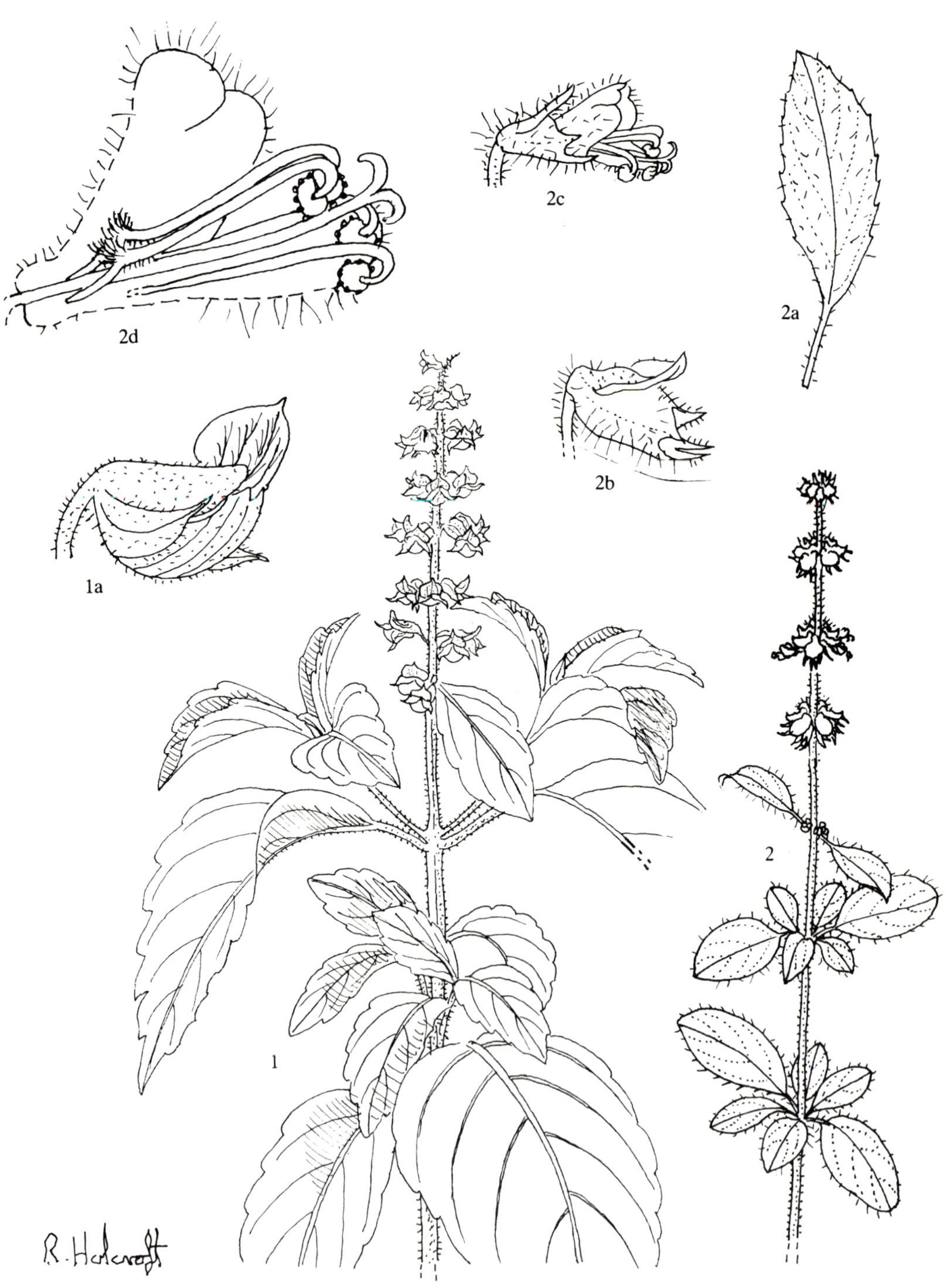

Fl. Natal 306 (1972); Compton, Fl. Swaziland 511 (1976). Type: from India.

O. suave Willd., Enum. Pl. Hort. Berol. 629 (1809); Benth., Lab. 7 (1832); in DC., Prodr. 12: 35 (1848); Bak. in F.T.A. 5: 338 (1900); Wood, Natal Pl. 4: t. 325 (1903); N.E. Br. in F.C. 5,1: 234 (1910); Morton in F.W.T.A. edn 2, 2: 451 (1963). *O. gratissimum* L. var. *suave* (Willd.) Hook. f., Fl. Brit. India 4: 609 (1885). Type: a cultivated plant.

— var. *distantidens* Briq. in Bull. Herb. Boissier sér. 2, 3: 980 (1903). Syntypes: S.W.A./Namibia, Olukonda, *Schinz* 57 (Z); Angola, Omupanda, *Wulfhorst* s.n. (Z).

O. micranthum Dinter ex Launert in Launert & Schreiber, F.S.W.A. 123: 21 (1969), nom. nud. in syn., non Willd.

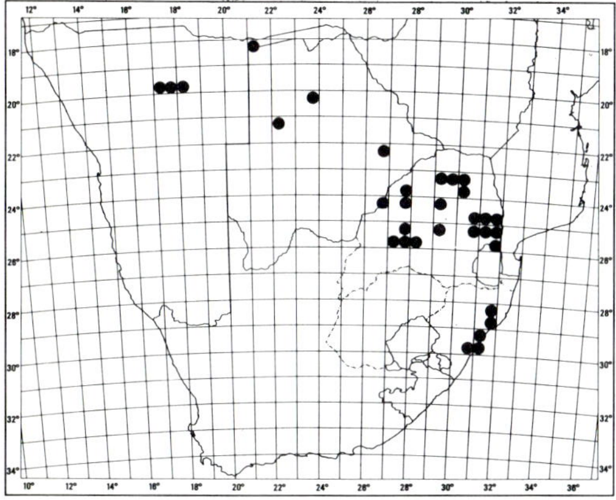

MAP 124. — **Ocimum urticifolium** subsp. **urticifolium**

Erect perennial herb or soft shrub 0,4−1,5 m tall, usually branching near the base; leaf blade (40−) 50−100 (−130) × 25−65 mm, margin distinctly dentate for almost the whole length; inflorescence fairly compact or elongate, (60−) 70−140 mm long.

Distribution as for the species. Map 124.

Vouchers: *Acocks* 12439; *Giess* 11299; *Scheepers* 182; *Wild & Drummond* 7201.

The leaves are usually described as lemon-scented, though occasional specimens are recorded as having clove-scented leaves. In nothern S.W.A./Namibia a medicinal tea is made from the dried leaves.

(b) subsp. **caryophyllatum** *Codd* in Bothalia 14: 219 (1983). Type: Natal, Mapelana Forest, south of St Lucia Estuary, *Cooper* 119 (PRE, holo.!).

Erect soft shrub 1−2 m tall, branching mainly in the upper part; leaf blade ovate-elliptic to elliptic, 25−45 × 12−30 mm, margin subentire or obscurely toothed in the upper half; inflorescence fairly compact, 50−70 mm long. Fig. 39: 1.

Found at the margins of dune forest in northern Natal. Map 125.

Vouchers: *Strey* 6450; *Venter* 4088.

The leaves are described as having the scent of cloves or nutmeg.

2. **Ocimum canum** *Sims* in Curtis's bot. Mag. t.2452 (1823); Benth., Lab. 3 (1832); in E. Mey., Comm. 226 (1837); in DC., Prodr. 12: 32 (1848); Hook. f., Fl. Brit. India 4: 607 (1885); Bak. in F.T.A. 5: 337 (1900); Morton in J. Linn. Soc., Bot. 58: 232, 234 (1962); in F.W.T.A. edn 2, 2: 451 (1963); Launert & Schreiber in F.S.W.A. 123: 21 (1969); Ross, Fl. Natal 306 (1972); Compton, Fl. Swaziland 511 (1976). Type: from China, a cultivated plant.

O. stamineum Sims in Curtis's bot. Mag. sub t.2452 (1823), sphalm.

O. fruticulosum Burch., Trav. 2: 264 (1824); Benth. in DC., Prodr. 12: 34 (1848); N.E. Br. in F.C. 5, 1: 236 (1910). Type: Cape, Griqualand West, near Klipfontein, *Burchell* 2160.

O. serpyllifolium sensu Benth. in E. Mey., Comm. 226 (1837).

O. serpyllifolium Forssk. var. *glabrior* Benth. in E. Mey., Comm. 226 (1837), partly. Type: Cape, Griqualand West, near Klipfontein, *Burchell* 2160.

O. canum var. *integrifolium* Engl. in Bot. Jb. 10: 267 (1888). Syntypes: Cape, Griqualand West, near Kimberley, *Marloth* 763; S.W.A./Namibia, near Otiimbingwe, *Marloth* 1288 (PRE!).

O. dinteri Briq. in Bull. Herb. Boissier sér 2, 3: 980 (1903); N.E. Br. in F.C. 5, 1: 236 (1910). Type: S.W.A./Namibia, Great Namaqualand, *Dinter* 1549.

O. simile N.E. Br. in F.C. 5, 1: 234 (1910). Syntypes: Transvaal, Madjadjes Mountains, *Burtt Davy* 2714 (PRE!); 5288.

O. americanum sensu N.E. Br. in F.C. 5, 1: 235 (1910); sensu Hutch. & Dalz., F.W.T.A. 2, 1: 285 (1931).

FIG. 39. — 1, **Ocimum urticifolium** subsp. **caryophyllatum**, flowering stem, × 1; 1a, mature calyx, × 4 (*Stirton* 8793, Mtunzini, BRI garden No. 26607). 2, **O. canum**, flowering stem, × 1; 2a, leaf from near base, × 1; 2b, mature calyx, × 4; 2c, flower, × 4; 2d, corolla opened longitudinally, × 10 (2a from *Van Vuuren* 570; remainder cult. *Mrs. E. Jenkins*).

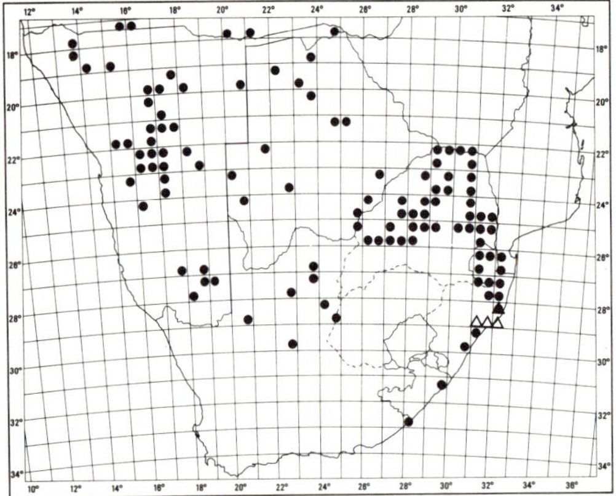

MAP 125. — △ **Ocimum urticifolium** subsp.
caryophyllatum
● **O. canum**

Perennial herb or soft shrublet, often woody below, 0,15—0,5 (—0,8) m tall, freely branched; stems subglabrous or sparingly pubescent to villous, particularly at the nodes. *Leaves* petiolate; blade very variable in size, linear-lanceolate to ovate-lanceolate or elliptic, (8—) 15—50 (—70) × (3—) 5—15 (—25) mm, subglabrous to pilose, copiously gland-dotted below, apex acute, base cuneate, margin entire to obscurely few-toothed. *Inflorescence* simple or sparingly branched below, 60—200 mm long, verticils 8—20 mm apart; bracts elliptic to ovate, persistent, 3—6 mm long, tapering at each end; pedicels 3 mm long, ascending. *Calyx* pilose on the outside, densely hispid within, 5—7 mm long at maturity, upper lip subrotund, somewhat concave, lateral teeth broad-based, subulate, 2 mm long, lower lip spreading, longer than the upper, deeply 2-toothed. *Corolla* lilac to mauve or white, 4—6 mm long. *Stamens* exserted by 4—6 mm. Fig. 39: 2.

Common in the warmer parts of Southern Africa, for example in S.W.A./Namibia, Botswana, northern Cape Province, Transvaal (except on the Highveld), at low and medium altitudes in Swaziland, Natal coast and midlands, and recorded from the Port St Johns and Komga districts in Transkei. Map 125.

Vouchers: *Codd & De Winter* 4974; *De Winter* 2521; 4080; *Rodin* 2779; *Schlechter* 4235.

In northern S.W.A./Namibia the leaves are cooked with meat and fish and are also used as a tea. In northern Transvaal the dried leaves are smoked in a pipe for chest complaints. The plants may be strongly to slightly aromatic and the description of the scent varies from that of thyme or mint to aniseed, liquorice or eucalyptus oil. Plants tend to spread on overgrazed or disturbed areas and, when such plants are collected, they may be described as annuals.

There is considerable variation in leaf size according to growing conditions, and specimens with larger leaves begin to resemble the cultivated Basil, *O. basilicum* L., which has the same floral structure as *O. canum* but in which the calyx, corolla, bracts and leaves are larger. According to Morton in J. Linn. Soc., Bot. 58: 234 (1962), the type of *O. americanum* L. is an immature specimen of *O. basilicum* L.

7366a   **35. BECIUM**

Becium *Lindl.* in Bot. Reg. Misc. 28: 42 (1842); N.E. Br. in F.C. 5, 1: 230 (1910); Andrews, Flow. Pl. Sudan 3: 206 (1956); Morton in F.W.T.A. edn 2, 2: 453 (1963); Cufodontis in Bull. Jard. bot. État. Brux. 33 (Suppl.): 849 (1963); Launert & Schreiber in F.S.W.A. 123: 9 (1969); R. A. Dyer, Gen. 1: 536 (1975); Codd in Taxon 32: 490 (1983); nom. cons. prop. Type species: *B. bicolor* Lindl.

Ocimum sensu Benth., Lab. 1 (1832), partly.

Ocimum Sect. *Hiantia* Benth. in DC., Prodr. 12: 35 (1848); Briq. in Natürl. PflFam. 4, 3a: 369 (1897); Bak. in F.T.A. 5: 334 (1900).

Perennial herbs or shrublets. *Leaves* opposite or whorled, often very small. *Inflorescence* a terminal spike-like raceme, sometimes subcapitate; verticils spaced or crowded; bracts present as an apical coma in the bud stage, early deciduous often leaving a conspicuous circular gland-like scar; verticils usually 6-flowered. *Calyx* bilabiate; tube campanulate; upper lip broadly ovate, decurrent on the tube; lateral pair of teeth obsolete but replaced by a wide shoulder-like sinus often with a fimbriate margin; lowest pair of teeth subulate or bristle-like. *Corolla* bilabiate; tube exceeding the calyx, expanding towards the mouth; upper lip erect, 4-lobed; lower lip spreading, entire, concave. *Stamens* 4, subequal, well exserted, declinate; filaments free, upper pair attached near base of corolla tube with a hairy knee-bend near the base, lower pair attached near the throat. *Disc* cup-shaped. *Ovary* glabrous; style well exserted, deeply 2-lobed. *Nutlets* ellipsoid or oblong, somewhat compressed.

A mainly African genus of 10 or more species, extending into the southern Arabian Peninsula and India; 4 species in Southern Africa. Lindley states that *Becium* is derived from the Greek Bekion, "one of the ancient names for sage", but it is apparent that the name was applied in classical times to various plants used in the treatment of chest complaints.

1 Inflorescence with the verticils crowded near the apex and usually only 1—3 of the lower ones separate; stems several, arising annually from a woody rootstock, sparingly branched above:

   2 Leaves subglabrous to sparingly pubescent .... 1(a). *B. obovatum* var. *obovatum*

   2 Leaves villous .... 1(b). *B. obovatum* var. *galpinii*

1 Inflorescence elongate with 5—20 verticils more or less evenly spaced 5—15 mm apart; stems usually solitary at the base, rarely several from ground level, branching freely above:

   3 Leaves more than 5 mm broad; stamens exserted by 10 mm or more .... 2. *B. knyanum*

   3 Leaves less than 5 mm broad; stamens exserted by less than 10 mm:

     4 Leaves linear to linear-lanceolate, thinly pubescent, usually more than 10 mm long; calyx tube campanulate .... 3. *B. angustifolium*

     4 Leaves subspathulate to oblanceolate, densely white-puberulous, usually less than 12 mm long; calyx tube tubular .... 4. *B. burchellianum*

1. **Becium obovatum** *(E. Mey. ex Benth.) N.E. Br.* in F.C. 5, 1: 230 (1910); Ross, Fl. Natal 306 (1972); Compton, Fl. Swaziland 512 (1976). Type: Natal, near Umzimkulu, *Drège* (K, holo.).

Ocimum obovatum E. Mey. ex Benth. in E. Mey., Comm. 226 (1838); in DC., Prodr. 12: 35 (1848).

Perennial with several stems arising annually from a woody rootstock; stems erect or ascending, rarely spreading, slender, simple or sparingly branched, puberulous to villous, 0,1—0,25 (—0,3) m tall.

Leaves subsessile or shortly petiolate; blade very variable in shape from linear-elliptic to lanceolate, lanceolate-oblong, ovate, subrotund or obovate, (10—) 15—40 (—60) × (3—) 5—20 (—30) mm, subglabrous to villous, gland-dotted, apex acute to rounded, base cuneate to obtuse, margin entire or with few shallow teeth; petiole 0—5 mm long. *Inflorescence* often subcapitate or with 1—3 spaced verticils below the crowded apex; pedicels c. 1 mm long. *Calyx* 4—5 mm long at flowering, enlarging to 7—10 mm long, reticulate-veined, pubescent; tube campanulate. *Corolla* white to pale mauve, (8—) 10—17 mm long with longitudinal violet lines on the upper lip. *Stamens* exserted by 14—20 mm.

Found in dense grassland on the higher parts of the Transvaal, extending into Swaziland, Natal, extreme eastern Orange Free State and eastern Cape as far south as East London. Also in Zimbabwe to east tropical Africa and possibly also the higher parts of Angola.

B. obovatum is a typical pyrophyte, adapted to grassland which is periodically burnt and, even if the grass is not burnt for several years, the stems die in winter and regenerate annually from the woody subterranean rootstock. The extent to which the plants behave in this way in tropical Africa, where several closely related species have been described, is not clear. Some of these "species" may prove to be local forms of *B. obovatum* but the tendency to take a very broad view of the species, e.g. by Morton in F.W.T.A. edn 2, 2: 453 (1963), Cufodontis in Bull. Jard. bot. État. Brux. 33: 849 (1963) and Launert & Schreiber in F.S.W.A. 123: 9 (1969), seems scarcely justified. See also note under *B. knyanum* (no. 2).

For key to varieties see key to species.

(a) var. **obovatum.**

Ocimum obovatum E. Mey. ex Benth. in E. Mey., Comm. 226 (1838); in DC., Prodr. 12: 35 (1848); Wood, Natal Pl. 3: t.257 (1902); Handb. Fl. Natal 105 (1907). *Becium obovatum* (E. Mey. ex Benth.) N.E. Br. in F.C. 5, 1: 230 (1910); Bews, Fl. Natal 177 (1921); Martineau & Phear, Rhod. Wild Flow. 69, t.30 (1930); Letty, Wild Flow. Transv. 288, t.143 (1962); Batten & Bokelmann, Wild Flow. E. Cape 125, t.100 (1966); Lucas & Pike, Wild Flow. Witwatersrand 75 (1971); Ross, Fl. Natal 306 (1972); Compton, Fl. Swaziland 512 (1976); Tredgold & Biegel, Rhod. Wild Flow. 46, t.30 (1979).

O. serpyllifolium Forssk. var. *glabrius* Benth. in E. Mey., Comm. 226 (1838) (as var. *glabrior*), partly, excl. syn. *O. fruticulosum* Burch. *O. hians* Benth. in DC., Prodr. 12: 36 (1848); S. Moore in J. Bot., Lond. 41: 405 (1903). *B. obovatum* var. *hians* (Benth.) N.E. Br. in F.C. 5, 1: 231 (1901). *B. obovatum* var. *glabrius* (Benth.) Cufod. in Bull. Jard. bot. État. Brux. 33: 850 (1963). Type: Cape, between Gekau (Butterworth) and Bashee River, *Drège* (K, holo.).

O. striatum Hochst. in Flora 28: 66 (1845). Syntypes: Natal, Port Natal, *Krauss* 390a; 390b (in K, *Krauss* 390, 2 sheets).

Leaves very variable in shape and size as in the description of the species, but rarely exceeding 40 mm in length, subglabrous to sparingly pubescent. Fig. 40: 1.

Distribution as for the species. Map 126.

Vouchers: *Flanagan* 2806; *Mauve* 4943; *C. A. Smith* 866; *Tyson* 471.

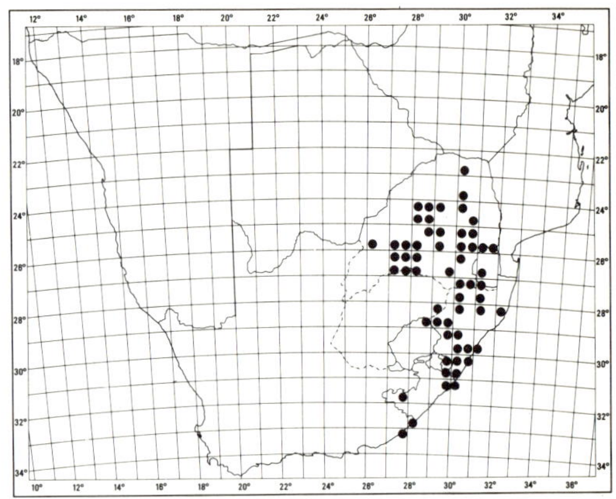

MAP 126. — **Becium obovatum** var. **obovatum**

(b) var. **galpinii** *(Gürke) N.E. Br.* in F.C. 5, 1: 231 (1910); Bews, Fl. Natal 177 (1921); Compton, Check List Fl. Swaziland 67 (1966). Type: Transvaal, Barberton, Saddleback Range, *Galpin* 413 (K, PRE).

Ocimum galpinii Gürke in Bot. Jb. 26: 78 (1888).

Leaves relatively large, lanceolate or oblanceolate to broadly ovate or obovate, 25—60 × 10—30 mm, villous, especially on the rather conspicuous veins on the underside of the leaf; margin often distinctly toothed.

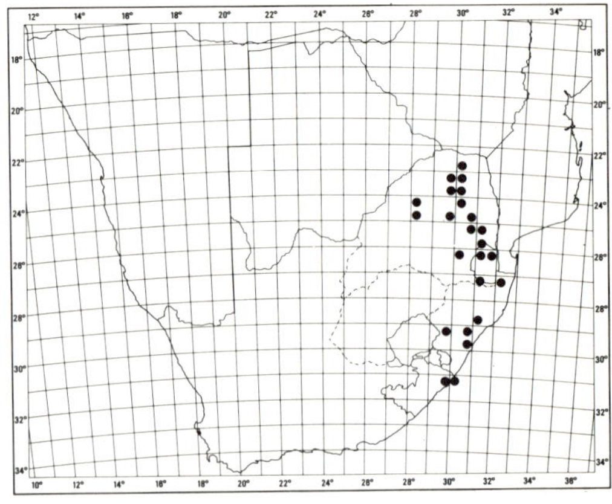

MAP 127. — **Becium obovatum** var. **galpinii**

FIG. 40. — 1, **Becium obovatum** var. **obovatum**, flowering stem, × 1; 1a, base of plant, × 1; 1b, mature calyx, × 3; 1c, section through corolla, × 4 (plant growing naturally in BRI garden). 2, **B. angustifolium,** flowering branch, × 1; 2a, flower, × 4 (plant growing naturally in BRI garden).

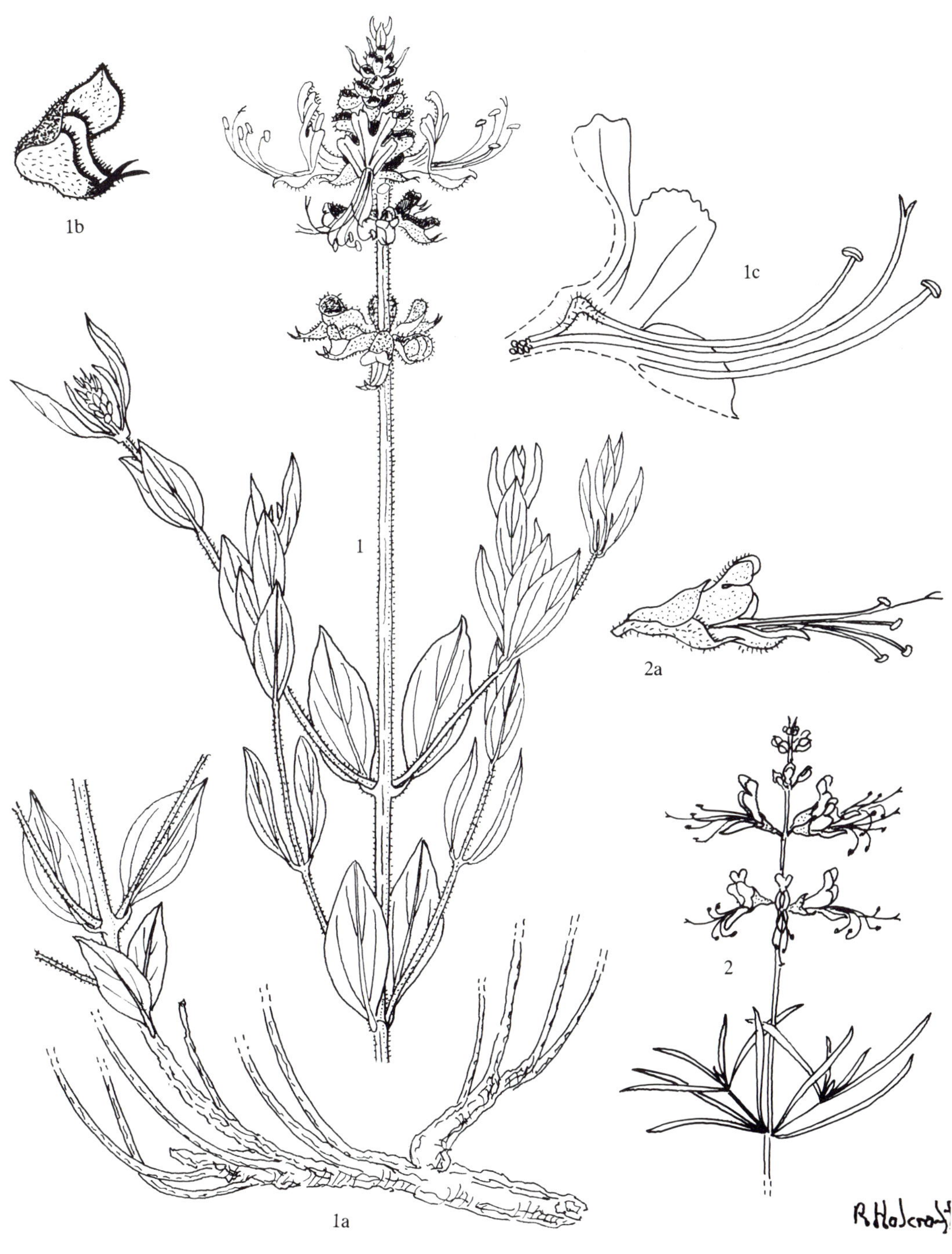

Found together with var. *obovatum* on the Waterberg and along the eastern escarpment of the Transvaal to Swaziland, Natal and coastal Transkei; also on the eastern mountains of Zimbabwe. Map 127.

Vouchers: *Acocks* 13156; *Codd* 4205; 4729; *Galpin* 10283; 12072.

2. **Becium knyanum** *(Vatke) N.E. Br. ex Broun & Massey,* Fl. Sudan. 357 (1929); Letty, Wild Flow. Transv. 288 (1962); Compton, Fl. Swaziland 512 (1976). Type: Ethiopia, *Schimper* 387 (K; PRE, photo.).

Ocimum knyanum Vatke in Linnaea 37: 315 (1871); Bak. in F.T.A. 5: 346 (1900); Hiern, Cat. Afr. Pl. Welw. 1, 4: 850 (1900). *B. obovatum* var. *knyanum* (Vatke) Cufod. in Bull. Jard. bot. État Brux. 32 (Suppl.) 850 (1963).

O. stenoglossum Briq. in Bull. Herb. Boissier sér. 2, 3: 981 (1903). Type: S.W.A./Namibia, Windhoek, *Dinter* 344.

O. rautanenii Briq., l.c. 982 (1903). Type: S.W.A./Namibia, Outjo, *Rautanen.*

O. fissilabrum Briq., l.c. 984 (1903). Type: S.W.A./Namibia, Okahandja, *Höpfner* 90.

B. obovatum sensu Launert & Schreiber in F.S.W.A. 123: 9 (1969).

Perennial, erect, soft shrub 0,3–0,8 (–1) m tall and of nearly equal diameter, usually single-stemmed and branching above, occasionally with several stems from a basal rootstock; stems relatively stout, puberulous. *Leaves* shortly petiolate; blade lanceolate to lanceolate-elliptic, ovate-elliptic or obovate, (12–) 20–40 (–50) × (4–) 7–18 (–22) mm, under-surface glabrous, gland-dotted, upper surface usually puberulous on the nerves, occasionally puberulous on both surfaces, apex acute to obtuse, base cuneate to obtuse, margin shallowly toothed to subentire; petiole 2–7 mm long. *Inflorescence* slender, 80–200 mm long, of 6–20 verticils regularly spaced along the rhachis 10–20 mm apart; pedicels 2–3 mm long. *Calyx* 3–4 mm long at flowering, enlarging to 6–7 mm long, reticulate-veined, puberulous; tube campanulate. *Corolla* white to mauve, 8–10 (–12) mm long. *Stamens* exserted by 12–17 mm.

Found in northern S.W.A./Namibia, Botswana, northern and eastern Transvaal, Swaziland and coastal Natal as far south as Empangeni; in open woodland or thorn scrub on sandy or stony places, often gregarious under thorn trees. Also in Angola and east tropical Africa to Ethiopia. Map 128.

Vouchers: *Codd & Dyer* 4597; *De Winter* 2743; *Giess* 12601; *Moll* 4134.

Ecologically, this species replaces *B. obovatum* (above) in the warmer and usually drier parts of the country, but taxonomically the two appear distinct and no difficulty was experienced in separating the two in the herbarium. *B. knyanum* is a stouter plant which shows little tendency to regenerate annually from the base and the inflorescences tend to be more elongate with regularly spaced verticils. The leaves show considerably less variation in shape, size and hairiness than in *B. obovatum*, while the corolla tends to be smaller.

3. **Becium angustifolium** *(Benth.) N.E. Br.* in F.C. 5, 1: 231 (1910); Letty, Wild Flow. Transv. 288 (1962). Type: Transvaal, Magaliesberg, *Burke* (K, holo.).

Ocimum angustifolium Benth. in DC., Prodr. 37 (1848).

O. filiforme Gürke in Bull. Herb. Boissier 6: 556 (1898). Syntypes: Transvaal, Pretoria, near Apies River, *Rehmann* 4272 (Z); Kuduspoort, *Rehmann* 4614 (Z).

O. polycladum Briq. in Bull. Herb. Boissier sér. 2, 3: 982 (1903). Type: Transvaal, Klippan, *Rehmann* 5309 (Z, holo.).

Erect perennial herb or soft shrublet 0,2–0,6 (–0,8) m tall, single-stemmed (rarely several-stemmed from the base), branching freely above; stems slender, puberulous. *Leaves* subsessile to shortly petiolate; blade linear to linear-lanceolate, 12–25 × 1,5–4 mm, grey-green, often folded longitudinally, puberulous and densely gland-dotted on both surfaces, apex subacute, base attenuate, margin entire. *Inflorescence* slender, 40–120 mm long with 4–12 verticils regularly spaced 8–15 mm apart; pedicels 1–2 mm long, eventually

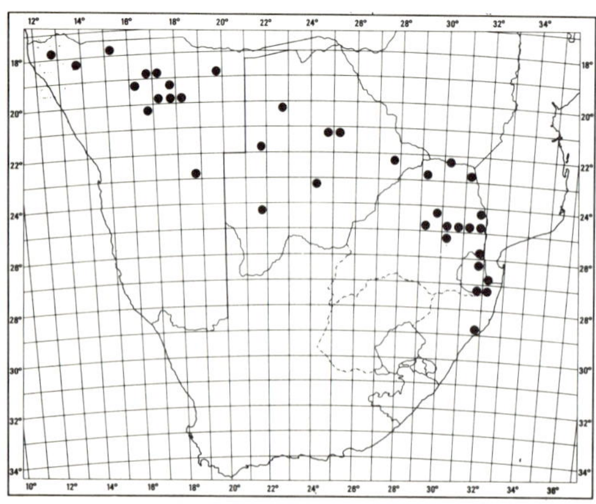

MAP 128. — **Becium knyanum**

deflexed. *Calyx* 2,5 mm long at flowering, enlarging to 5 mm long, puberulous, gland-dotted and often with white globular glands; tube campanulate. *Corolla* white, 4—4,5 mm long. *Stamens* exserted by 5—8 mm, eventually coiled. Fig. 40: 2.

Found at medium altitudes in south-western, central and northern Transvaal, and eastern and north-eastern Botswana; usually in open woodland among rocks, but sometimes on stream banks. Also in Zimbabwe, Zambia, Malawi and Tanzania. Map 129.

Vouchers: *Acocks* 12429; *Galpin* 11649; *Hutchinson* 2584; *Schlechter* 3644.

In the tropics this species frequently grows in moist, grassy places, almost invariably with many stems from a swollen woody rootstock, but this is not usually the case in Southern Africa, except for the specimen from the Chobe National Park in northern Botswana which shows this character. A specimen from Angola, which appears to be this species, has exceptionally long leaves, up to 40 mm long. The leaves are said to be strongly mint-scented.

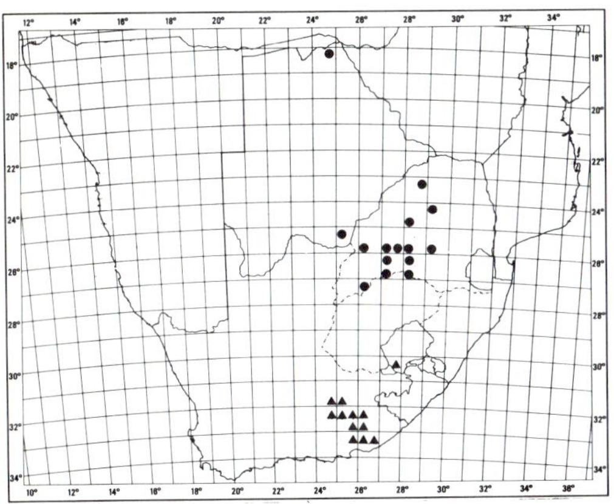

MAP 129. — ● **Becium angustifolium**
▲ **B. burchellianum**

4. **Becium burchellianum** *(Benth.)* *N.E. Br.* in F.C. 5, 1: 232 (1910). Lectotype: Cape, Middelburg district, *Burchell* 2812 (K, lecto.; PRE!).

Ocimum burchellianum Benth., Lab. 8 (1832); in DC., Prodr. 12: 36 (1848).

O. serpyllifolium sensu Benth., Lab. 707 (1835); in E. Mey., Comm. 226 (1838).

O. helianthemifolium Hochst. in Flora 28: 67 (1845); Benth. in DC., Prodr. 12: 36 (1848). Type: Cape, Uitenhage, *Krauss* 1121.

Twiggy soft shrub 0,5—1 m tall, single-stemmed at the base and woody below, freely branched above; stems white-puberulous. *Leaves* subfasciculate at the nodes, subsessile; blade subspathulate to oblanceolate, 5—14 × 2—5 mm, rather coriaceous, grey-green, gland-dotted and densely white-puberulous especially on the under-surface, often folded longitudinally, apex rounded, base attenuate, margin entire. *Inflorescence* slender, 40—120 mm long, of 4—8 spaced verticils; pedicels 2—2,5 mm long. *Calyx* 3 mm long at flowering, enlarging to 7 mm long, fairly densely white-tomentulose; tube tubular; upper lip curving upwards, median sinus rounded with a densely ciliate margin. *Corolla* white to mauve or dull purple, 8—9 mm long. *Stamens* exserted by 7—9 mm.

Found in a restricted area in the eastern Cape Province from Middelburg and Graaff-Reinet southwards to Uitenhage and Grahamstown, where it is a frequent constituent of karroid scrub. Map 129.

Vouchers: *Acocks* 12001; *Galpin* 10589; *Rogers* 3589.

N. E. Brown, l.c., included *O. helianthemifolium* in synonymy and this treatment is followed here.

7367
36. ORTHOSIPHON

Orthosiphon *Benth.* in Bot. Reg. sub t.1300 (1830); Lab. 25 (1832); in DC., Prodr. 12: 49 (1848); Benth. & Hook. f., Gen. Pl. 2, 2: 1174 (1876); Briq. in Natürl. PflFam. 4, 3a: 372 (1897); Bak. in F.T.A. 5: 365 (1900); N.E. Br. in F.C. 5, 1: 237 (1910); Ashby in J. Bot., Lond. 76: 1 (1938); Morton in F.W.T.A. edn 2, 2: 454 (1963); Codd in Bothalia 8: 149 (1964); R. A. Dyer, Gen. 536 (1975). Type species: not designated.

Nautochilus Brem. in Ann. Transv. Mus. 15: 253 (1933). Type species: *Nautochilus labiatus* (N.E. Br.) Brem.

Herbs or undershrubs, sometimes with tuberous roots. *Leaves* opposite or rarely ternate. *Inflorescence* terminal, racemose or paniculate; bracts small, persistent; flowers 1−6 in the axils of each bract, forming 2−12-flowered verticils 10−20 mm apart. *Calyx* 5-toothed, 2-lipped; tube cylindric to campanulate; upper tooth broadly ovate-orbicular with the margin more or less decurrent on the tube; lower 4 teeth subequal, ovate-deltoid, acuminate to subulate, the 2 lowest usually longer than the 2 laterals. *Corolla* bilabiate; tube narrowly to broadly cylindric, straight or curved; upper lip erect, 3−4-lobed; lower lip horizontal to recurved, concave to boat-shaped. *Stamens* 4, free, didynamous, declinate, exserted; lower pair inserted near the mouth; upper pair inserted further back in the corolla tube; filaments glabrous or pilose at the base; anthers 1-thecous. *Disc* saucer-shaped, often with a ventral lobe. *Style* filiform, more or less capitate, lying together with and subequal to the stamens. *Nutlets* suborbicular to oblong, glabrous.

In some non-Southern African species the leaves form a large basal rosette.

Species about 50 of which 9 occur in Southern Africa.

Related to *Ocimum* L. (no. 34) but corolla tube longer, and style more or less capitate.

1 Upper pair of stamens attached about 2 mm from the throat of the corolla tube and exserted by about 2 mm; floral bracts up to 4 mm long (subgen. *Orthosiphon*):

  2 Rhachis pubescent but lacking stipitate glands; leaves glabrous to pubescent, but under-surface not conspicuously dotted with red sessile glands nor appressed canescent:

    3 Petioles 4−20 mm long; leaf blade ovate to broadly ovate, rarely exceeding 40 mm in length, base truncate to obtuse .... 1. *O. suffrutescens*

    3 Petioles 0−3 mm long; leaves ovate-elliptical to ovate-lanceolate, 35−80 mm long, base cuneate to obtuse .... 2. *O. rubicundus*

  2 Rhachis with numerous short stipitate glands; under-surface of leaves either dotted with red sessile glands or appressed canescent:

    4 Leaf blade 20−30 mm long, scabrid, under-surface with numerous red sessile glands; stems simple, usually less than 0,5 m long .... 3. *O. vernalis*

    4 Leaf blade 8−12 mm long, appressed canescent especially on the lower surface; stems much branched, usually exceeding 0,5 m .... 4. *O. fruticosus*

1 Upper pair of stamens attached near the base of the corolla tube and exserted by 4−12 mm; bracts usually exceeding 4 mm long (occasionally less in *O. pseudoserratus* and *O. amabilis*):

  5 Leaf blade lanceolate or elliptic to obovate or if ovate then margin distinctly serrate; lower lip of corolla less than 8 mm long:

    6 Corolla tube exceeding 20 mm in length; leaf blade usually less than 20 × 10 mm .... 5. *O. tubiformis*

    6 Corolla tube 5−16 mm long; leaf blade usually exceeding 20 × 10 mm:

      7 Leaf blade usually exceeding 40 × 20 mm; stamens exserted from the throat of the corolla by 4−6 mm .... 6. *O. serratus*

      7 Leaf blade usually less than 40 × 20 mm; stamens exserted from the throat of the corolla by 7−8 mm .... 7. *O. pseudoserratus*

  5 Leaf blade broadly ovate to subrotund, margin crenate; lower lip of corolla 8−12 mm long:

    8 Leaf blade less than 25 × 20 mm .... 8. *O. amabilis*

    8 Leaf blade exceeding 25 × 20 mm .... 9. *O. labiatus*

1a
1b
1
2a
2b
2c
2
R. Holcroft

1. Orthosiphon suffrutescens *(Thonn.)*
J. K. Morton in J. Linn. Soc., Bot. 58: 238,
266 (1962); in F.W.T.A. edn 2, 2: 454
(1963); Ross, Fl. Natal 307 (1972); Agnew,
Upland Kenya Wild Flow. 648 (1974);
Compton, Fl. Swaziland 514 (1976). Type:
Ghana, *Thonning* 288 (C, fide Junghans in
Bot. Tidsskr. 57, 340, 1961).

Ocimum thonningii Thonn. in Schumach., Beskr.
Guin. Pl. 269 (1827) (as "thoningii"), non *O.
thonningii* Schumach. & Thonn., l.c. 265 (1827). *O.
suffrutescens* Thonn. in K. danske Vidensk. Selsk. Skr.
4: 330 (1829).

Orthosiphon glabratus Benth. var. *africanus* Benth.
in DC., Prodr. 12: 51 (1848). Type: Transvaal,
Magaliesberg, Crocodile River, *Burke* 162 (K, holo.;
PRE!).

O. australis Vatke in Linnaea 40: 179 (1876); ibid.
43: 86 (1881−82); Bak. in F.T.A. 5: 373 (1900); Ashby
in J. Bot., Lond. 76: 40 (1938); Andrews, Flower. Pl.
Sudan 3: 221 (1956); Codd in Bothalia 8: 150 (1964).
Type: Mozambique, Rios de Sena, *Peters.*

O. wilmsii Gürke in Bot. Jb. 26: 81 (1898); N.E. Br.
in F.C. 5, 1: 255 (1910); Ashby, l.c. 44 (1938). Type:
Transvaal, near Lydenburg, *Wilms* (K; BM).

O. neglectus Briq. in Bull. Herb. Boissier sér 2, 3:
988 (1903). Type: Pretoria, Wonderboompoort, *Rehmann* 4510 (Z, holo.).

O. inconcinnus Briq., l.c. 991 (1903); N.E. Br., l.c.
256 (1910). Type: Natal, Camperdown, *Wood* 4963 (K;
NH).

Plectranthus bolusii T. Cooke in Kew Bull. 1909: 377
(1909), partly, as to syntype, Potgietersrus, *Bolus*
11011 (BOL!).

O. wilmsii var. *komghensis* N.E. Br., l.c. 256 (1910).
Type: Transkei, Komga, *Flanagan* 477 (K, holo.;
PRE!).

Herb, branching at or near the base,
0,15−0,6 m tall; stems semi-woody pubescent, with short simple and multicellular
hairs, often glabrescent. *Leaves* petiolate;
blade thin-textured, ovate to broadly ovate,
15−40 (−50) × 10−30 mm, subglabrous to
pubescent, under-surface with pale brownish gland-dots, apex acute to obtuse, base
truncate to abruptly cuneate, margin subentire to sparingly serrate-dentate; petiole
5−25 mm long. *Inflorescence* simple,
40−150 mm long; verticils 2−6-flowered,
10−20 mm apart; bracts subrotund, abruptly acuminate, 2,5−3 mm long. *Calyx* 7−9

mm long at fruiting stage, pubescent.
Corolla white to mauve; tube 6−7 mm long,
straight; upper lip erect, 3−3,5 mm long;
lower lip concave, 4−5,5 mm long. *Stamens*
exserted from the throat by 2 mm; upper 2
filaments attached 1,5−2 mm from the
throat, glabrous. *Stigma* thickened, minutely bifid. Fig. 41: 1.

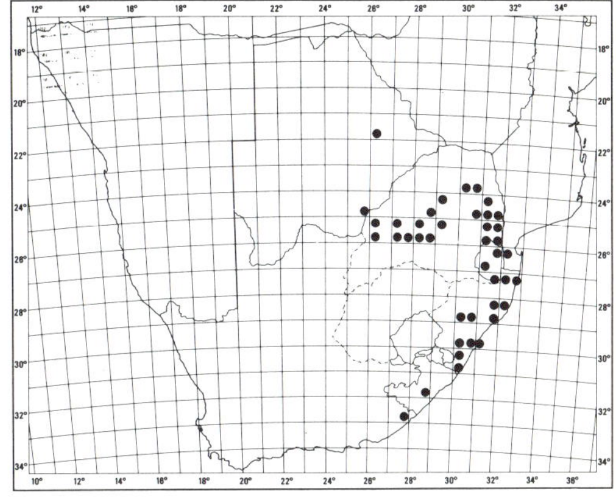

MAP 130. — **Orthosiphon suffrutescens**

Found in dry, wooded country in Botswana,
Transvaal, Swaziland, Natal and Transkei; extends
through east tropical Africa to Ethiopia and Sudan.
Map 130.

Vouchers: *Codd* 8604; 9370; *Medley Wood* 11972;
Ward 2276.

O. wilmsii is the form found in the central
Transvaal where plants are subjected to cooler
conditions and periodic burning; the plants branch
more freely from the base, with shorter stems and
smaller, thicker-textured and almost entire leaves with
the gland-dots more densely placed in the somewhat
wrinkled under-surface; there is, however, a complete
gradation linking it with the more typical *O.
suffrutescens* of the warmer lowveld.

See Ashby, l.c., for tropical African synonyms.
Morton, l.c., says the species is very closely allied to
and probably only racially distinct from the Indian *O.
glabratus* Benth.

2. Orthosiphon rubicundus *(D. Don)*
Benth. in Wall., Pl. As. Rar. 2: 14 (1831);
Lab. 26 (1832); in DC., Prodr. 12: 51

FIG. 41. — 1, **Orthosiphon suffrutescens,** flowering stem, × 1; 1a, mature calyx, × 2; 1b, section through
corolla, × 2 (*Codd* 5059). 2, **O. labiatus,** leaf, × 1; 2a, section through corolla, × 2; 2b, flowering calyx, × 2; 2c,
mature calyx, × 2 (living plant, BRI garden).

(1848); Hook. f., Fl. Brit. India 4: 614 (1885); Ashby in J. Bot., Lond. 76: 41 (1938); Morton in J. Linn. Soc., Bot. 58: 239 (1962); in F.W.T.A. edn 2, 2: 454 (1963); Codd in Bothalia 8: 152 (1962); Agnew, Upland Kenya Wild Flow. 648 (1974). Type: India, Nepal, *Wallich* (BM).

Plectranthus rubicundus D. Don, Prodr. Fl. Nepal. 116 (1825). *Lumnitzera rubicunda* (D. Don) Spreng., Syst. 4, cur. post.: 223 (1827).

Perennial herb with 1−several stems 0,25−0,6 m long arising from a woody or tuberous rootstock; stems simple or sparingly branched, glabrous to pilose. *Leaves* sessile to shortly petiolate; blade ovate to ovate-lanceolate, 35−80 × 18−40 mm, glabrous to sparingly pubescent, lower surface with scattered yellowish gland-dots, apex acute to obtuse, base cuneate to obtuse, margin coarsely crenate-dentate mainly in the upper two-thirds; petiole 0−3 mm long. *Inflorescence* usually simple, 50−200 mm long; verticils 4−6-flowered, 5−15 mm apart; bracts broadly ovate, acuminate, 2,5−3 mm long. *Calyx* 8−9 mm long at fruiting stage, pubescent. *Corolla* white to mauve; tube 6 mm long, straight; upper lip erect, 3,5 mm long; lower lip concave, 4 mm long. *Stamens* exserted from the throat by 2 mm; upper 2 filaments attached 1−1,5 mm from the throat, glabrous. *Stigma* thickened, minutely bifid.

Widespread from China and India, throughout tropical Africa, and reaching its southernmost limit in northern Transvaal; in grassy places in fairly dense woodland. Map 131.

Vouchers: *Dryfhout* 833; *Obermeyer* sub TRV 29238.

Several tropical African synonyms are listed by Ashby, l.c.

3. Orthosiphon vernalis *Codd* in Bothalia 8: 152 (1964); Compton, Fl. Swaziland 514 (1976). Type: Swaziland, Manzini district, Malkerns, *I'Ons* 60/43 (PRE, holo.!).

Perennial herb 0,2−0,3 m tall; stems 1−3 arising annually from a woody rootstock, erect, simple or sparingly branched, pubescent with simple hairs, long multicellular hairs and red gland-dots. *Leaves* sessile to shortly petiolate; blade ovate to ovate-lanceolate or elliptic, 20−30 × 10−15 mm, scabrid pubescent, under-

surface with numerous red gland-dots, apex acute to obtuse, base rounded to truncate, margin slightly thickened, obscurely crenate-dentate to subentire; petiole 0−4 mm long. *Inflorescence* simple, 80−150 mm long; verticils 2-flowered, 3−10 mm apart; rhachis glandular-pubescent with dense stipitate glands and long multicellular hairs; bracts lanceolate, acuminate, 3−4 mm long. *Calyx* 9−11 mm long at fruiting stage, glandular-hispid. *Corolla* purple, gland-dotted; tube 7−8 mm long, straight; upper lip erect, 5−6 mm long; lower lip concave, 6−7 mm long. *Stamens* exserted from the throat by 2 mm; upper 2 filaments attached 1,5 mm from the throat, glabrous. *Stigma* minutely bilobed.

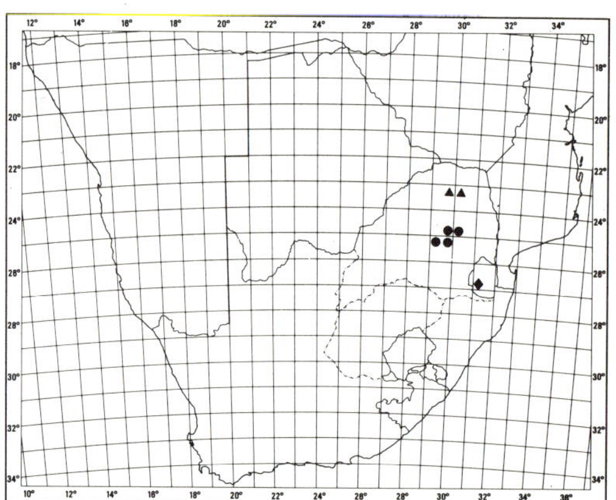

MAP 131. — ▲ **Orthosiphon rubicundus**
♦ **O. vernalis**
● **O. fruticosus**

Recorded only from the Manzini and Mankaiana districts of Swaziland; on grassy slopes where it is apparently subjected to regular burning. Map 131.

Vouchers: *Compton* 29167; 31098.

Related to *O. rubicundus* (above) but has smaller leaves which are freely red gland-dotted beneath, densely stipitate-glandular rhachis and 2-flowered verticils.

4. Orthosiphon fruticosus *Codd* in Bothalia 8: 153 (1964). Type: Transvaal, near Steelpoort Station, *Codd* 9777 (PRE, holo.!).

Twiggy shrub 0,5−1,2 m tall; young stems tomentulose, glabrescent; bark on old stems often splitting off in thin strips.

Leaves shortly petiolate, often fasciculate on short shoots; blade coriaceous, lanceolate to oblanceolate, 8−18 × 2,5−6 mm, canescent, lower surface reticulate, gland-dotted, apex acute, base cuneate, margin entire; petiole 1−2 mm long. *Inflorescence* simple, 40−90 mm long; verticils 2−4-flowered, 10−15 mm apart; rhachis glandular-puberulous; bracts ovate to lanceolate, 2−4 mm long. *Calyx* 7−9 mm long at fruiting stage. *Corolla* purple, gland-dotted; tube 7−8 mm long, straight; upper lip erect, 5−6 mm long; lower lip concave, 6 mm long. *Stamens* exserted from the throat by 2 mm; upper 2 filaments attached 1,5 mm from the throat, glabrous. *Stigma* minutely bilobed.

Grows in dry bushveld on stony slopes, from Loskop Dam to Steelpoort Valley. Map 131.

Vouchers: *Acocks* 20952; *Codd* 8797.

The small leathery leaves and twiggy habit are reminiscent of *O. tubiformis* (below), which has a very much longer corolla tube and in which the stamens are attached near the base of the corolla tube.

5. **Orthosiphon tubiformis** *R. Good* in J. Bot., Lond. 63: 173 (1925); Ashby in J. Bot., Lond. 76: 10 (1938); Letty, Wild Flow. Transv. 285, t.142 (1962); Codd in Bothalia 8: 154 (1964); in Flower. Pl. Afr. 43: t.1697 (1974). Type: Transvaal, Pilgrims Rest, Vaalhoek, *Rogers* 25104 (BM, holo.; PRE!).

Virgate shrub 0,3−0,9 m tall; stems ascending, branched, subglabrous to hispid. *Leaves* opposite or ternate, often fasciculate on short shoots, shortly petiolate; blade subcoriaceous, lanceolate-elliptic to ovate or obovate, 14−20 × 7−10 mm, subglabrous to pubescent, lower surface reticulate, gland-dots not obvious, apex acute to obtuse, base cuneate to obtuse, margin finely serrate to subentire; petiole 2−4 mm long. *Inflorescence* simple, 50−170 mm long; verticils 3−6 (−8)-flowered, 10−15 mm apart; rhachis finely glandular-puberulous; bracts ovate, acuminate, 5−8 mm long. *Calyx* 10−16 mm long at fruiting stage, puberulous. *Corolla* whitish to pale or deep mauve; tube narrowly cylindrical, straight, 20−35 mm long; upper lip erect, 5−7 mm long; lower lip concave, 6 mm long. *Stamens* exserted from the throat by 5−6 mm; upper 2 filaments attached near

the base of the tube, pubescent below. *Stigma* shortly bilobed.

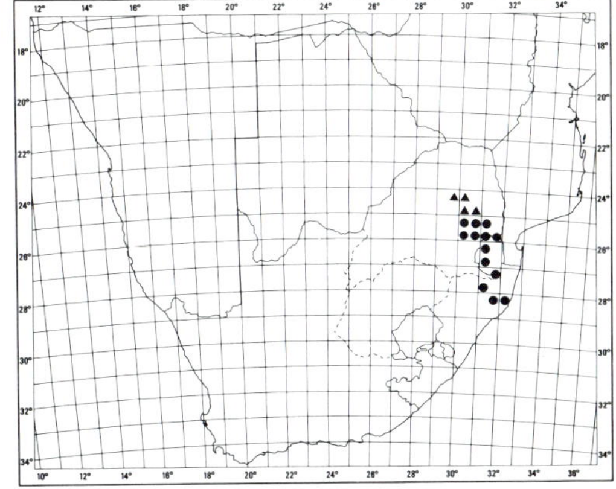

Map 132. — ▲ **Orthosiphon tubiformis**
● **O. serratus**

Grows on wooded, stony slopes in relatively dry parts at medium altitudes in the Lydenburg, Pilgrims Rest and Letaba districts of Transvaal. Map 132.

Vouchers: *Codd & Dyer* 7714; 7696.

6. **Orthosiphon serratus** *Schltr.* in J. Bot., Lond. 35: 431 (1897); N.E. Br. in F.C. 5,1: 260 (1910); Ashby in J. Bot., Lond. 76: 9 (1938); Codd in Bothalia 8: 155 (1964); Ross, Fl. Natal 307 (1972); Compton, Fl. Swaziland 513 (1976). Type: Transvaal, Barberton, *Galpin* 499 (K, holo.; PRE!).

Shrub 0,3−0,9 m tall, often with several stems arising from a woody rootstock; stems erect, sparingly branched, densely hispid. *Leaves* opposite or ternate, shortly petiolate; blade broadly ovate, ovate-oblong or obovate-elliptic, 40−90 × 20−35 mm, densely pubescent to subglabrous, apex acute to obtuse, base cuneate to obtuse, margin distinctly and regularly serrate; petiole 3−8 mm long. *Inflorescence* simple, 80−320 mm long; verticils 4−12-flowered, 10−30 mm apart; rhachis glandular-hispid; bracts ovate, acuminate, 6−10 (−16) mm long. *Calyx* up to 15 mm long at fruiting stage, glandular-hispid. *Corolla* mauve to purple; tube straight, cylindrical, (6−) 9−16 mm long; enlarging slightly towards the throat; upper lip erect,

6−7 mm long; lower lip concave, 5−6 mm long. *Stamens* exserted from the throat by 4−5 mm; upper 2 filaments attached near the base of the tube, pubescent below. *Stigma* bilobed, lobes spreading, 0,5 mm long. Fig. 42.

Recorded from eastern Transvaal, Swaziland and northern KwaZulu, in dense grass on stony hillsides at medium altitudes where it is usually subjected to periodic burning. Map 132.

Vouchers: *Codd* 4727; 9786; *Rogers* 14304; 18341; *Schlechter* 3866.

With its numerous fairly large purple flowers, it is a striking species when in flower and worth trying in cultivation.

7. Orthosiphon pseudoserratus *Ashby*

in J. Bot., Lond. 76: 8 (1938); Codd in Bothalia 8: 156 (1964); in Flower. Pl. Afr. 42: t.1657 (1973). Type: Transvaal, Potgietersrus district, Moorddrift, *Leendertz* 2243 (BM, holo.; PRE!).

Shrublet 0,3 m or more tall; stems ascending, sparingly branched, glandular-hispid. *Leaves* subsessile to petiolate; blade ovate to broadly elliptic, 20−40 × 12−20

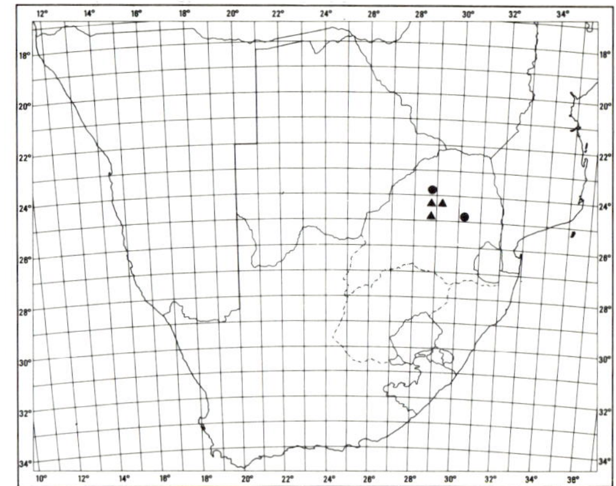

MAP 133. — ▲ **Orthosiphon pseudoserratus**
● **O. amabilis**

mm, glandular-pubescent on both surfaces, lower surface with yellowish gland-dots, apex obtuse, base rounded, margin finely serrate; petiole 2−5 mm long. *Inflorescence* simple, 30−150 mm long; verticils 2−6-flowered, 7−20 mm apart; rhachis

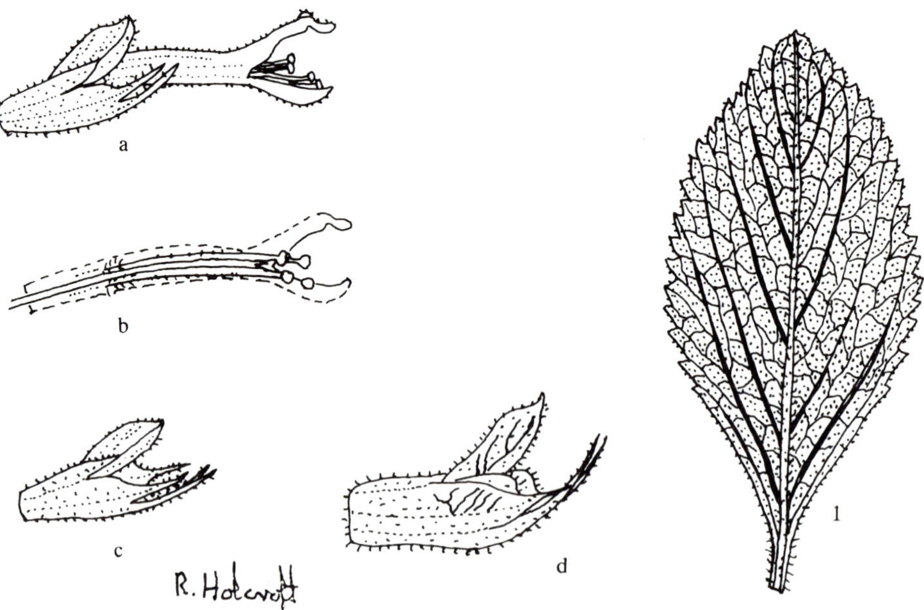

FIG. 42. — 1, **Orthosiphon serratus,** leaf, × 1; a, flower, × 2; b, section through corolla, × 2; c, flowering calyx, × 2; d, fruiting calyx, × 2; *(Onderstall s.n.).*

glandular-hispidulous; bracts ovate-lanceolate, acuminate, 3—6 mm long. *Calyx* 9—11 mm long at fruiting stage. *Corolla* whitish to mauve or pink; tube somewhat decurved, widening towards the throat, 5—7 mm long; upper lip erect, 3—4 mm long; lower lip boat-shaped, 5—6 mm long. *Stamens* exserted from the throat by 7—8 mm; upper 2 filaments attached near the base of the tube, slightly thickened and pubescent below. *Stigma* not thickened, minutely bilobed.

Apparently restricted to the Potgietersrus district in Transvaal; on rocky, wooded slopes at medium altitudes. Map 133.

Vouchers: *Galpin* 9065; 9154; 13455.

8. **Orthosiphon amabilis** (*Brem.*) *Codd* in Bothalia 8: 157 (1964). Lectotype: Transvaal, Potgietersrus district, Swerwerskraal, *Bremekamp* sub PRU 1220 (PRE, lecto.!).

Nautochilus amabilis Brem. in Ann. Transv. Mus. 15: 254 (1933).

Twiggy shrub 0,6—0,9 m tall; branches ascending, pubescent with numerous multicellular hairs. *Leaves* petiolate; lamina broadly ovate to subrotund, 10—15 (—20) × 10—14 (—18) mm, pubescent on both surfaces, under-surface greyish with long interwoven multicellular hairs and yellowish gland-dots, apex rounded, base truncate, margin finely crenate; petiole 5—10 mm long. *Inflorescence* simple, 70—140 mm long; verticils 2—6-flowered, 10—15 mm apart; rhachis glandular-hispidulous; bracts ovate, acuminate, 4—6 mm long. *Calyx* up to 9 mm long at fruiting stage. *Corolla* mauve or pink; tube decurved, widening towards the throat, 8—9 mm long; upper lip erect to recurved, 5 mm long; lower lip boat-shaped, 8—9 mm long. *Stamens* exserted from the throat by 9 mm; upper 2 filaments attached near the base of the tube, slightly thickened and pubescent below. *Stigma* not thickened, entire.

Recorded from the Potgietersrus and Lydenburg districts of Transvaal; on dry, wooded slopes. Map 133.

Vouchers: *Barnard* 339; 421.

Closely related to *O. labiatus* (below) but has smaller, more tomentose leaves and smaller flowers. From *O. pseudoserratus* (above) it may be distinguished by the more rotund leaves with crenate margins and longer petioles, and the longer corolla lobes. *Pole*

Evans 3094, which was cited by Bremekamp as a syntype, is *O. pseudoserratus*.

9. **Orthosiphon labiatus** *N.E. Br.* in F.C. 5,1: 245 (1910); Codd in Bothalia 8: 157 (1964); Ross, Fl. Natal 307 (1972); Compton, Fl. Swaziland 513 (1976). Type: Transvaal, Woodbush, *Schlechter* 4434 (K, holo.; PRE!).

Plectranthus bolusii sensu T. Cooke in Kew Bull. 1909: 377 (1909); in F.C. 5,1: 282 (1910), partly, as to *Rehmann* 6167 and *Wood* 4488.

Nautochilus labiatus (N.E. Br.) Brem. in Ann. Transv. Mus. 15: 253 (1933); Verdoorn in Flower. Pl. S. Afr. 23: t.901 (1943); Letty, Wild Flow. Transv. 288, t.143,1 (1962).

N. breyeri Brem., l.c. 254 (1933). Type: Transvaal, Louis Trichardt, *Breyer* sub TRV 19400 (PRE, holo.!).

N. urticaefolia Brem., l.c. 254 (1933). Type: Transvaal, Blouberg, Leipzig Mission, *Bremekamp & Schweickerdt* 131 (PRE, holo.!).

Soft shrub 0,6—1,8 m tall, branching from the base; stems ascending, freely branched, sparingly pubescent, denser at the nodes. *Leaves* petiolate; soft in texture, broadly ovate to subrotund, 30—80 × 20—60 mm, upper surface sparingly pubescent, lower surface with long multicellular hairs and yellow gland-dots, apex acute to rounded, base truncate to abruptly and shortly cuneate, margin regularly and coarsely crenate; petiole slender, 5—30 mm long. *Inflorescence* simple or occasionally with a pair of branches at the base, 50—180 mm long; verticils 2—6 (—8)-flowered, 10—20 mm apart; rhachis glandular-pubescent; bracts ovate, acuminate, 8—10

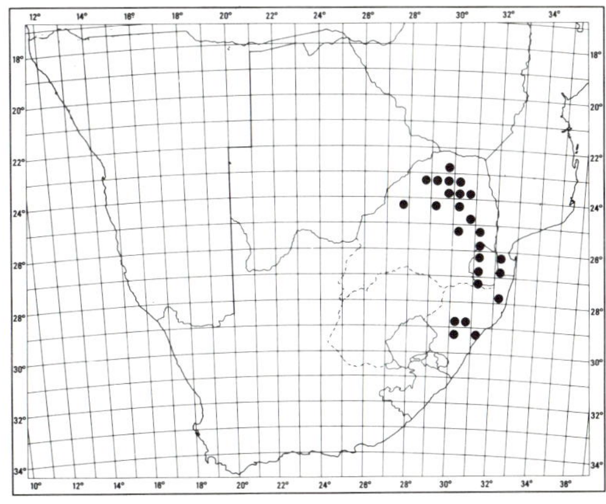

MAP 134. — **Orthosiphon labiatus**

mm long. *Calyx* 10—15 mm long at fruiting stage, glandular-puberulous. *Corolla* pale mauve to pink; tube decurved, widening towards the throat, 10—12 mm long; upper lip erect to recurved, 7—8 mm long; lower lip boat-shaped, later deflexed, 8—12 mm long. *Stamens* exserted from the throat by 9—12 mm; upper 2 filaments attached near the base of the tube, slightly thickened and pubescent below. *Stigma* minutely bifid. Fig. 41:2.

Recorded from Transvaal, Swaziland and northern Natal and extends into Zimbabwe; on dry, rocky, wooded hillsides and wooded watercourses at medium altitudes. Map 134.

Vouchers: *Acocks* 10151; 16700; *Codd & Dyer* 7736; 9138; *Medley Wood* 4488.

A distinctive species with its bushy habit, large leaves on slender petioles and its large, declinate corolla with well exserted stamens; popular as a garden plant.

37. THORNCROFTIA

Thorncroftia *N.E. Br.* in Kew Bull. 1912: 281 (1912); Codd in Bothalia 7: 429 (1961); R.A. Dyer, Gen. 537 (1975). Type species: *T. longiflora* N.E. Br.

Perennial herbs or soft shrubs, semisucculent. *Leaves* opposite, often crowded on short shoots. *Inflorescence* paniculate or racemose; flowers solitary in the axils of persistent bracts; bracts semisucculent, not sharply differentiated from the leaves, becoming progressively smaller towards the apex of the inflorescence. *Calyx* bilabiate, 5-toothed; tube campanulate; upper tooth larger than the lower 4, ovate-lanceolate to broadly ovate; lower 4 teeth subequal, narrowly deltoid, acuminate. *Corolla* bilabiate, 4-lobed; tube campanulate to long-cylindric; upper lip erect, oblong, emarginate; lower lip concave, spreading to reflexed; lateral lobes strap-shaped, spreading on each side of the lower lip. *Stamens* 4, didynamous, declinate, inserted at the throat of the corolla tube; filaments free to the base; anthers 1-thecous. *Disc* small, swollen into a gland in front. *Style* bifid. *Nutlets* ellipsoid.

3 species, found in the northern and eastern Transvaal, with 1 species extending to Swaziland. Allied to *Plectranthus* (no. 23) and *Orthosiphon* (no. 36), differing from both in the bracts being leaf-like below and becoming progressively smaller towards the apex of the inflorescence, and in the 4-lobed corolla limb; from *Plectranthus* it differs in the flowers being borne solitarily in the axils of the bracts, and from *Orthosiphon* in the upper pair of stamens being attached at the throat of the corolla tube.

All 3 species are parasitized by a weevil, *Apion rectangulum* Wagn. which causes thickened swellings in the stems. Such swellings have also been seen in the stems of *Plectranthus cylindraceus* which have a somewhat similar texture to those of *Thorncroftia* spp.

1 Corolla tube less than 10 mm long; plants up to 0,25 m tall .... 1. *T. thorncroftii*

1 Corolla tube 15−38 mm long; plants 0,3−1,2 m tall:

  2 Corolla tube 30−38 mm long; pubescence consisting of simple or multicellular, not dendroid hairs
    .... 2. *T. longiflora*

  2 Corolla tube 15−20 mm long; dendroid hairs present, mixed with simple and multicellular straight
    hairs .... 3. *T. succulenta*

1. **Thorncroftia thorncroftii** *(S. Moore) Codd* in Bothalia 7: 430 (1961). Type: Transvaal, Barberton, *Thorncroft* sub Rogers 16987 (BM, holo.; PRE! On the PRE specimen the number has been altered to 14987).

Plectranthus thorncroftii S. Moore in J. Bot., Lond. 56: 39 (1918).

Semisucculent herb 0,1−0,25 m tall, sparingly branched at the base; stems ascending, about 8 mm in diameter at the base, glandular-pubescent. *Leaves* shortly petiolate; blade fleshy, drying subcoriaceous, obovate to oblong-obovate, 15−20 × 6−8 mm, pilose and gland-dotted especially on the under-surface, apex obtuse, base cuneate, margin sparingly toothed in the upper half. *Inflorescence* often simple, 50−80 mm long; lower bracts leaf-like, smaller and about 3−5 mm long near the apex. *Calyx* 7 mm long at fruiting stage; upper tooth ovate, acuminate, more or less decurrent on the tube. *Corolla* whitish with purple spots; tube campanulate, 4,5−5 mm long, enlarging abruptly at the base; upper lip erect, 6 mm long; 2 lateral lobes oblong, 3−4 mm long, deflexed; lower lip boat-shaped, horizontal, 6−7 mm long. *Stamens* up to 6−7 mm long.

Known so far only from the mountain massif between Barberton in the Transvaal and Piggs Peak in Swaziland; among rocks in mountain grassland.

Vouchers: *Clarke* 41; *Compton* 30002; *Werdermann* 2197.

2. **Thorncroftia longiflora** *N.E. Br.* in Kew Bull. 1912: 281 (1912); Prain in Curtis's bot. Mag. t.8824 (1919); Codd in Bothalia 7: 430 (1961); in Flower. Pl. Afr. 36: t.1425 (1964). Type: Transvaal, near Barberton, *Thorncroft* 795 (K, holo.; PRE!).

Semisucculent herb or soft shrublet 0,3−0,6 m tall with several stems arising from a thickened rootstock about 40 mm in

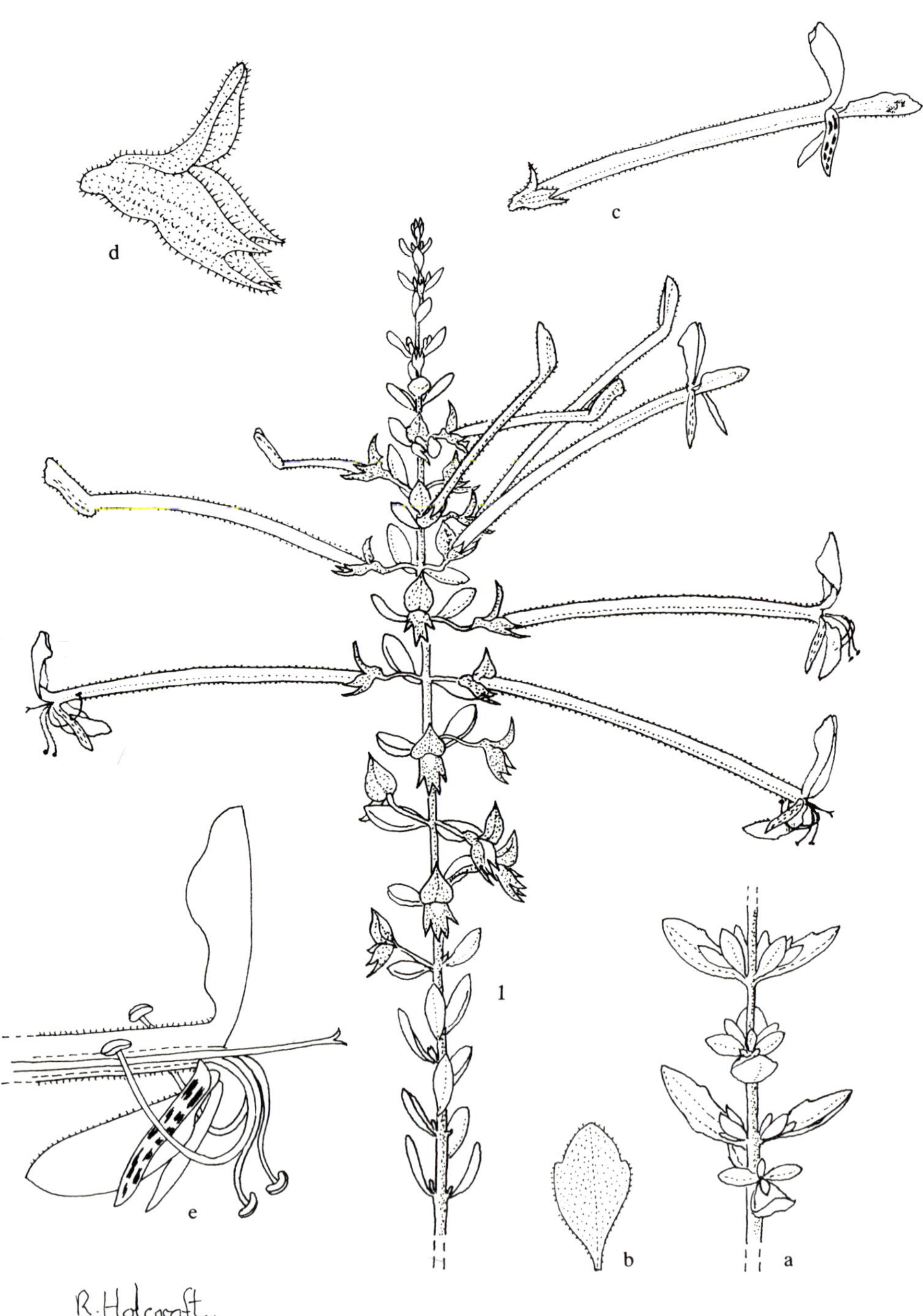

R. Holcroft.

diameter; stems ascending, 10 mm in diameter at the base, sparingly branched, densely grey tomentose, hairs simple. *Leaves* shortly petiolate; blade fleshy, drying subcoriaceous, elliptic to obovate, 10−20 × 4−10 mm, finely grey tomentose, gland-dotted, apex rounded, base cuneate, margin subentire or obscurely toothed in the upper half. *Inflorescence* lax or dense, up to 90 mm long; lower bracts leaf-like, progressively smaller and about 3 mm long near the apex. *Calyx* 7 mm long at fruiting stage; upper tooth ovate, acuminate, more or less decurrent on the tube. *Corolla* pink to mauve-pink with deeper flecks on the lateral lobes; tube narrowly cylindrical, 30−38 mm long, not expanding at the base nor towards the throat; upper lobe erect, 7−8 mm long; lateral lobes oblong, deflexed, 5−6,5 mm long; lower lip at first horizontal and boat-shaped, soon reflexed, 6−8 mm long. *Stamens* up to 8 mm long. Fig. 43.

Evidently a rare species, known only from the rocky hillside above Joe's Luck siding, south-eastern Transvaal, at about 1 200 m altitude; in pockets of humus on rock slabs.

Voucher: *Thorncroft & Clarke* s.n.

3. **Thorncroftia succulenta** *(Dyer & Bruce) Codd* in Bothalia 7: 431 (1961). Type: Transvaal, Entabeni, *Loock* in PRE 27461 (PRE, holo.!).

Plectranthus succulentus Dyer & Bruce in Flower. Pl. Afr. 27: t.1073 (1949).

Semisucculent herb or soft shrub 0,6−1,2 m tall with several stems arising from a thickened rootstock; stems ascending, 7−15 mm in diameter at the base, sparingly branched, densely grey tomentose, hairs dendroid (branched). *Leaves* shortly petiolate; blade fleshy, drying subcoriaceous, ovate-elliptic to obovate, 16−30 × 15−20 mm, thinly to densely tomentose and gland-dotted on both surfaces, hairs dendroid and simple, apex rounded, base cuneate to obtuse, margin crenate in the upper two-thirds. *Inflorescence* congested, 80−140 mm long; bracts somewhat leaf-like, ovate, 12 mm long near the base, progressively smaller towards the apex. *Calyx* about 7 mm long at fruiting stage; upper tooth ovate-lanceolate, acuminate, not decurrent on the tube. *Corolla* bluish mauve with darker spots on the upper lip and lateral lobes; tube narrowly cylindrical, 15−20 mm long, not expanding at the base; upper lip erect, 6−8 mm long; lateral lobes 5−6 mm long; lower lip at first horizontal and boat-shaped, later reflexed, 5−6 mm long. *Stamens* up to 5 mm long.

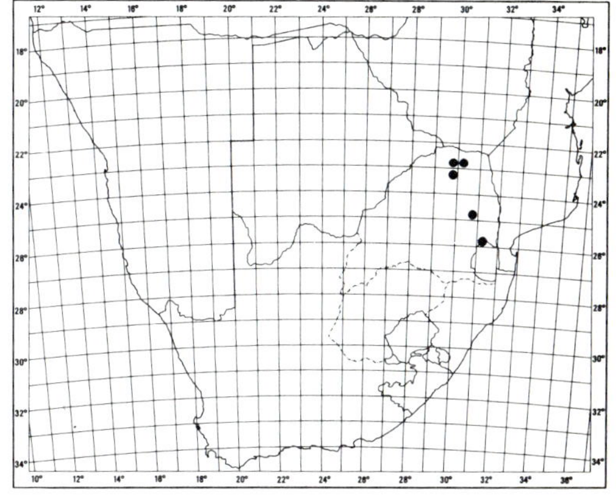

Map 135. — **Thorncroftia succulenta**

Found in mountains of northern and eastern Transvaal; in humus-filled crevices of bare rock outcrops. Map 135.

Vouchers: *Clarke* 213; *Codd* 4194; 7904.

Habit and ecology similar to *T. longiflora* (above) but leaves larger and more crenate with dendroid (branched) hairs, and corolla bluish mauve with a shorter tube.

Fig. 43. — 1, **Thorncroftia longiflora**, flowering stem, × 1; a, lower part of stem, × 1; b, leaf, × 1; c, flower, × 1; d, mature calyx, × 4; e, section through apex of corolla, × 3 (*Thorncroft & Clarke* s.n., cult. BRI garden).

INDEX

---

* An asterisk signifies exotic species which are not naturalised